CRUISING IN CATAMARANS

32 years of experience
cruising
delivering
sailing
surveying
over 1000 different sailboats

Charles E. Kanter, AMS

*SAIL*co Press
Key Largo, Florida
United States of America

Cover photo by author
Little Harbor, Abaco, Bahamas
in front of Pete's Pub.

SAILco **Press**
Key Largo, Florida

Publishers Cataloging-In-Publication Data

Kanter, Charles E.
ISBN 0-9618406-6-8
Library of Congress Catalog: 2002091539
$29.95 Softcover

Cover Design: Corinne C. Kanter
Edited by: Robert J. and Lynn Kanter, Lupe Eyde Tucker, www.sailmiami.com,
www.AtlanticCoastBoating.com
Printed: in the United States of America
Illustrations and photos by the author unless otherwise noted.

Questions and inquiries regarding the content of this book should be addressed to:
*SAIL*co Press
P.O. Box 2099
Key Largo, FL. 33037
U.S.A.
(305)-743-0626 Tel & Fax
email: publisher@sailcopress.com
website: http://www.sailcopress.com

Cruising in Catamarans

PREFACE

Focus of this book
This book is focused on cruising and the characteristics of cruising boats, no matter how many hulls they have. Its fundamental premise is that the highest and best usage of catamaran technology is for cruising. It presumes that there are five different classifications of recreational sailing vessels (no matter how many hulls they have) the differentiation being the intent of the designer/builder/marketer and the design compromises required to achieve that intent. Book One contains a full definition of these classifications.

Permission to copy forms
Permission is here by given to all individual purchasers of this book to reproduce the forms printed herein for their own use. Anyone wishing to make reproductions for commercial purposes must get permission from the author in writing.

New or special terms used in this book

Aka	Frugalphile
Ama	Speedophile
Camberspar®	Trampoline
Crossbeam	Underdeck slamming
Fat-head sails	

Use the extensive glossary
All terms are covered in the glossary

Disclaimer
Measurements, weights, clearances are all supplied by the builders of the described vessels. Many are owner modified. Classifications of usage such as blue water or inshore are the province of the designers and builders. The author bears no responsibility for the accuracy of those figures or statements of usage.

Other books by *SAIL*co Press

The Galley K.I.S.S. Cookbook
The 13th Trip and Other Sea Stories
Cruising on More Than One Hull
The Cruising K.I.S.S. Cookbook
Tales From the Decks of Winddancer
Cruising is Contagious

Cruising in Catamarans

Acknowledgments

Some of the material is published elsewhere in various periodicals. The following is a list of them and I gratefully acknowledge the opportunities they afforded me.

Boating Industry Magazine
Cruising World Magazine
Exchange for Marine Professionals
Good Old Boat Magazine
Latitudes and Attitudes Magazine
Living Aboard Magazine
MULTIHULLS Magazine
Professional Boat Builder
Soundings
Soundings Trade Only
Southwinds
Tidal Times
Waterfront News

I must give special note to Charles Chiodi, Editor Publisher of MULTIHULLS Magazine for whom I have been writing articles since 1976. And to Doran Cushing of Southwinds for whom I have been writing steadily since 1994.

Technical support:
Steve Lambert, of OTSI, Robert J. Kanter, Lupe Eyde-Tucker, www.sailmiami.com

Special noteworthy:
The Society of Accredited Marine Surveyors (SAMS) is a major factor in my outlook and attitude towards the vessels I review in this book. I can sum up that focus by quoting old Joe Friday of Dragnet fame: *"Just the facts ma'am, just the facts."*

Special circumstances:
Moral support, (prodding) typing, research, fabulous cuisine and most of all, infinite patience: my wife, Corinne C. Kanter

Cruising in Catamarans
Table of Contents

Book Two: How To Sail a Cruising Catamaran

Book Three: Catamarans

Book Four: Trimarans

Appendices

Cruising in Catamarans

Introduction

Cruising under sail is an exhilarating, satisfying and adventuresome occupation. Tales of adventurers exploring the outermost reaches of the world are legion. I believe that the most satisfying part of adventuring under sail is the sense of complete freedom. Freed from the mores of shore bound civilization, your only restraint and constant companion is the indifference of nature.

Cruisers faces a number of considerations in addition to an indifferent nature. Contrary currents, shallow waters, drawbridges, stuffy officials, are among them. However, in the search for cruising adventure and experience, there is one item of great importance over which you have control and decision: your choice of vessel. The vessel you choose can help or hinder your cruising experience. Your vessel can make the difference between an adventure and an ordeal.

Based upon thirty-two years of experience dealing with cruising catamarans, I believe that they have the most virtues and the fewest vices for the cruising life. This book focuses upon the practical and the pragmatic. Its driving force is to provide information in a format understandable by all readers. There is important information not available in any other reference book. Safety, security, ease of operation, practical application of equipment are the bases of the book. There are no sacred cows, I explain in detail just why one technology or another is satisfactory or unsatisfactory. This volume is divided into four books.

Book One, *Cruising,* describes the various aspects of cruising and the characteristics most valuable in a cruising boat.

Book Two, *How To Sail a Cruising Catamaran,* details the nuances of sail trim and aspects of safety and performance as they apply to cruisers.

Book Three, *Catamarans,* details 151 cruising catamarans with which I have considerable experience. In this book there are many short articles and vignettes of actual experiences sailing the vessels.

Book Four, *My Relationships With Trimarans,* describes considerable mileage sailed giving me the opportunity to make valid comparisons between technologies.

Appendices include a thorough exposition on anchoring technique; a briefing on surveying; a very extensive bibliography designed to give additional information not ordinarily found in a bibliography and the most extensive glossary in the industry.

Cruising In Catamarans

Defining Cruising

Before we can begin any meaningful discussion and understand why there is such raging debate about whether or not a boat is a cruising boat, we first must define the word cruising.

Through the ages, poets, scholars and sages have tried to define the word love. No one has ever come up with a fully satisfactory definition. Thus, it is with cruising.

In the opinion and experience of this author, cruising is more a social experience than it is a sailing experience. Cruising means looking, seeking, exploring and adventuring. Webster further defines cruising as "To sail or drive about from place to place as for pleasure or in search of something." None of these necessarily relate directly to a specific type of vessel, sailing or otherwise.

Cruising under sail, as defined by the popular mythology, is a romantic, idealistic, intellectual exercise. Ah, if only that were true. The people who choose sailing vessels over powerboats do it for two basic reasons: The economy of sail (though that economy is probably false) and the romance of sail, which includes all the mythology that goes with it.

A cruising boat is a means to an end. A fishing boat is a means to an end. When a person buys a cruising boat, he intends to go cruising, just as when he buys a fishing boat he intends to go fishing. The same principles apply to your car. You buy a car for its intended use. Of course, there are those who are willing to make sacrifices in comfort and convenience for high performance, both in their cars and their boats. However, just because a person uses an item out of its intended design function, does not reclassify that item to match the use. If you use a knife to shave with, it does not automatically become a razor. If you use a sport boat or a racing boat to cruise with, it does not automatically become a cruiser. The picture below is of a family cruising the Intracoastal Waterway on a motorized Hobie 16. I doubt anyone would classify it a cruising boat, yet they were cruising aboard it and having a good time.

For the purpose of this book, I will accept Webster's definition of cruising "To sail or drive about from place to place as for pleasure or in search of something." Therefore, if you want to argue about the definition, argue with Webster!

The Requirements of A Good Cruising Boat (no matter how many hulls, power or sail)

By Priority

1. Good visibility from the helm. A comfortable, protected seat from which you can see all four corners of the vessel, see underneath the sails, have no blind spots and be protected from the sun and inclement weather is required. Good visibility of instruments, easy access to controls and proper management of lines leading to helm area are a necessity.. (ABYC H-1 has specific standards for visibility) The American Boat and Yacht Council (ABYC) designs the industry standards.

2. A sturdy, well placed rub rail. The rub rail should have a protective sacrificial strip which is considered a maintenance item. When you are inextricably blown against a rough concrete piling while entering a fuel dock, the rub rail should be properly positioned and adequate to prevent damage to the hull proper. The sacrificial metal strip protects the rub rail proper.

3. Strong, well placed life lines and/or bulwarks, adequate toe rail. Lifelines should be strong enough to haul yourself out of the water by them. That means reasonable numbers of stanchions. ABYC recommends stanchions spacing not to exceed 84 inches and height not less than 24 inches for a recreational vessel.

4. An adequate engine with these attributes

a. Serviceability. Easy to reach all parts, easy to change maintenance items, easy to check oil and water daily.

b. Parts availability. Popular brand with established service dealers.

c. Preferably not in living space. Engines under bunks are often neglected because of the difficulty of reaching them and the need to disturb bedding.

d. If an outboard, it must be placed so it will not cavitate.

e. Easy to operate.

f. Good visibility and protection of instruments and controls.

5. **Accessible, uncluttered decks,** full walking access to the entire boat.

6. **Strong, well placed cleats,** preferably without the need for chocks.

7. **Appropriate anchor handling facilities,** from the bow, without gimmicks. Nothing hidden, no openings in the forward underside of the bridgedeck. They cannot be dedicated to a single operating system.

8. **Easily handled sails and rigging.**

9. **A galley with sufficient counter and storage space,** good ventilation, comfortable seating, visibility and proximity to social activity.

10. **Good protection for underwater parts,** especially from crab/lobster trap lines. (Nothing should break when you bump the bottom)

11. **Low Maintenance design,** easy to care for.

12. **Good handling characteristics,** easy to dock, easy to kedge off bottom.

13. **Sea-kindly ride,** dry, stable, takes powerboat wakes well.

14. **Reasonable, easy boarding access** from dinghy, swimming or dockside.

Some questions about cruising sailboat performance

- Does going 20 knots mean high performance to you?
- Does going to windward at 15 knots create visions of euphoria? Or terror?
- What are the parameters of performance in the real world of cruising?
- What is cruising compared to racing performance?
- What might be a reasonable workable definition(s) of cruising boat performance?

Have you ever gone 20 knots on the water? Have you gone that fast in a powerboat? Calm day? Flat seas? How many conceptualize that to get a sailboat to achieve 20 knots, you need at least 20 knots of wind. If it is blowing 20 knots it most likely is quite rough. Unless you have a really long sailboat, absolute minimum length of forty-five feet, you will be bounced around. Even a large boat can get uncomfortably rough at that speed, at least for cruising. At that speed, even in ideal conditions, in a powerboat forty feet or less, it still is hardly possible to maintain any semblance of cruising normality and certainly not hold a relaxed conversation.

When we talk about cruising boats, what describes the people that sail them and their motives, desires and perhaps fantasies? Obviously, the younger person that belongs to the work-a-day world but takes an extended vacation to sail from California to Hawaii for the sheer pleasure of sailing is in a quite different category than the person who sells the family home and moves permanently aboard the boat to cruise off into the sunset.

Just by definition the person sailing to Hawaii is willing to sacrifice amenities for performance that the permanent liveaboard is not. There is no question about it, sailing fast for long passages is a thrill and experience that is equal to none. Personally, I am willing to sacrifice a certain amount of luxury for increased performance. However, I am realistic enough to fully understand that the vast majority of both cruisers and potential cruisers are not willing to make sacrifices to the level that I have made. The scores of people that come aboard my boat and the hundreds of people I meet at boat shows tell me that.

One very important personal factor in evaluating cruising sailboat performance is the level of commitment of you and your crew. A large percentage of cruisers are couples who have gone to adventure under sail and one spouse is either non committal or possibly even hostile to the new life style, avocation or recreation. Perhaps one or both simply do not have the physical capability of handling a sailboat at high speed, of grinding powerful winches or hauling in on ratchet block main sheets.

Based upon the reality of cruising under sail for the vast majority of present and potential cruisers I talk with, I have developed a personal checklist to help you understand the parameters of cruising sailboat performance. Thus we come to the inevitable conclusion: the choice of vessel is a result of the reasons for choosing it. The reason for the vessel's existence is to take you comfortably and safely to your desired destinations, be they near or far, modest or determined. Look at choosing a cruising boat from the perspective of your objectives, then a certain list of priorities falls neatly into place.

Define Cruising Sailboats According to This Theme

No piece of equipment should be beyond the physical limits of the smallest, weakest, oldest, crew member.

This is especially applicable to sheet winches, sails and sail handling, dinghy handling and anchor tackle. Nothing, but nothing will destroy the budding romance of cruising under sail faster than asking a person to grind a winch or haul an anchor that is beyond their physical capability.

Use the following criteria as a guide to your personal need for boat size

- Where will it be used?
- When will it be used?
- How will it be used?
- Why will it be used?
- What number of people will be sleeping aboard?
- How many people will be entertained?
- Will these people be relatives or friends?
- Is your need a big, lavish vessel?
- Is your need a small, frugal vessel?

Are you an upwardly mobile professional using the vessel to entertain friends, family or business relations? Are you a grandparent looking forward to entertaining your grandchildren and teaching them the ways of the sea? Are you an early retired couple planning on seeing the world on your own terms? In each case the concept of performance is a different concept and I will attempt in this book to clarify the design attributes that best meet the needs of those individual users.

Each part of the boat should be safely accessible from behind sturdy lifelines. There should be reasonable deck area around the complete perimeter of the boat. Safe, easy access for docking, mooring, rafting and emergency measures are the preeminent virtues of a cruising boat's design. Boats with trampoline deck areas must have absolutely strong, stable nets, fully capable of supporting twice the weight of the entire crew. They should be constructed of material that you can walk upon barefoot. These features clearly separate the various classes of boats.

Contemporary literature is replete with articles of sardonic complaints from people that have sailed aboard the so-called performance cruising boats that are really water borne rocket ships. Many of these complaints involve trying to tie to docks without any proper handholds, inability to raft because of see-saw syndrome and needing to wear foul weather gear on even the most benign of sailing days.

Definitions

Simply by asking: "what is cruising catamaran or trimaran performance?" it becomes obvious there is no single answer. Based on my observations, there are almost as many definitions for cruising as there are people defining it.

Therefore, I have cataloged a one through five scale upon which you can plot where you fit in the universe of cruising. This will help define which sailboat attributes are most or least desirable for your purposes and which aspects of performance define performance for you.

The aspects of performance as viewed by a cruiser

- docility and safety at anchor, subsumes minimal anchor-sailing, does not roll excessively to wakes
- downwind stability, no tendency to broach, no excessive lee or weather helm, does not bury bows
- seakindliness, takes seas easily, little or no underdeck slamming, things stay in place
- average passage speed potential, consistently sails above average passage times
- dryness/wetness, no green water on deck, little if any spray
- windward ability easily maintains 35 - 40 degrees to apparent wind
- maximum speed potential makes occasional brilliant passages
- light air speed potential keeps going well without need to motor sail

Place yourself in the following description

1. Little or no organized activity. Probably not a club member or serious club participant. Has little consuming interest in sailing, when other activities actively compete with sailing as a pastime or leisure activity. Sail when and where the fancy strikes, little concern for competition, real or imagined.

2. Mostly organized activity. Stays close to home. Some racing, some group cruising, some weekends with family, friends. The boat is often berthed at a marina which has its own social milieu. There is play at informal racing.

3. Avid club member. Participates in all racing and cruising activities, looks forward to longer cruises, rendezvous, socializing, competition, status, and recognition preeminent.

4. Extended vacations, long weekends. Preparing for extended cruising or adventuring under sail. Cruising under sail is a passion. Often charters in foreign waters.

5. Adventuring under sail, full time liveaboard, long cruises. Actually out there, cruising.

Let me comment that my own experience is overwhelmingly class five. I have had considerable experience as an average wage earner with a weekend sailing schedule and done plenty of racing.

Some numbers

If we look at the base cruising milieu from my perspective, you see over 5000 Seven Seas Cruising Association (SSCA) members, with thousands spread around the world. You see approximately 10,000 cruisers migrating up and down the Intracoastal waterway on the East Coast of the U.S. each Spring and Fall, according to a survey taken by U.S. *News & World Report*. I find at least the same amount of cruisers sprinkled throughout the various tropical paradises I manage to visit. The U.S. West Coast also has its migrants as does Europe and the rest of the world. Perhaps more than the U.S. East coast, though, to my knowledge, no one has attempted a count. My guestimate of a number would be somewhere in the 150,000 range, give or take, worldwide. I base this on actual numbers of cruising sailboats manufactured and sold, magazine subscriptions, tax revenues and a gut feeling. This then is my base. All else is extrapolated from that data.

A large share of these cruising sailboats are between thirty-five and forty-five feet and sailed by cruising couples. About 10 percent of the cruisers have children on board. A percentage have others on board at one time or another, either visiting friends and relatives or crew for long passages. Most are mature people.

Early retirement couples head the list. Therefore, I extrapolate my median cruising boat to be forty feet long and crewed by a couple in their fifties.

Even if you cut that median back to a thirty-five foot boat crewed by a couple in their forties with a large contingent of adventurous types in their thirties, it would not alter the formula.

In the real world of cruising, those with monohulls motorsail almost exclusively except during portions of their transoceanic passages and when possible, in short local stints. (Trans-oceanic passages are often the smallest part of a cruising itinerary.)

The reasons are distressingly simple. Most passages are of short duration and are timed to take advantage of weather, daylight, tides, currents and consider schedules, appointments or other extenuating circumstances. Most cruisers only sail in weather conditions in which they feel comfortable. Those conditions most often are light winds. If you have light winds, you can't sail fast. If you can't sail fast, you must run your engine to make up the difference in speed.

When you are really living aboard, it is difficult or tiring to sail all day heeled over being bashed around. You cannot comfortably cook, bake, use the head or perform any of the other routine chores of daily living. Few want to arrive at their destination needing to first return their boats (homes) to a semblance of order before starting cooking, etc. Pushing too hard, from my observation, is the single most important cause of unfulfilled cruising dreams and the breakup of various cruising couples. Motor-sailing syndrome is intensified by the fact that most sailboats used as cruisers do not sail very well anyway, especially in light air.

Therefore, cruising performance reality for my median cruiser is an average passage speed of five knots. My personal experience and the hundreds of cruisers with whom I personally associate, confirm that average. The highest percentage of cruisers appear in my defined class five, a lesser percentage in class four and the smallest percentage in classes two and three.

Obviously, even a small increase in that average speed would be of an enormous advantage! The reality is that any of the presently available production cruising catamarans can increase that average five knot monohull speed easily to six knots and possibly to seven without sacrificing one iota of comfort or convenience. Quite the contrary. Because the catamaran does not heel and is more sea kindly, you can go faster in rougher seas and be far more comfortable than your fellow cruisers on their monohulls.

Proof

Corinne and I proved this to countless people time after time. Throughout the subtropical winters, we sailed as much as 50 percent more miles than others in our cruising social circle. We achieve this extra mileage simply because we perceive conditions to be better, based on the sea kindly characteristics of our catamaran and our higher speed. In addition, we use our engine about 75 percent less because our cat sails better than it motors. This as opposed to the average thirty-five foot monohulls which apparently motor better than they sail, under the same conditions.

Countless times we have sailed right by monohulls motor sailing toward the same destination. Our average speed for the same period and the same weather conditions for a ten year period of cruising was seven (7) knots. That is forty percent (40%) faster than our monohull fellow cruisers' speed. Is that a significant difference? They seem to think so. That speed is so significantly better than their own performance they tend to disbelieve it even though they have witnessed it with their own eyes!

I know only a tiny handful of multihull cruisers that glory at rushing along at top speed. Most others either get their speed kicks at racing or buy powerboats. Therefore, I propose the following verity to ponder:

Measure cruising sailboat performance by the average speed attainable while sailing in a cruiser friendly mode of operation.

This definition of cruising sailboat performance leaves leeway for the desires and fantasies of the user. It is understood that the boat that has the higher potential top speed will most likely, but not always, maintain a higher average speed. It also considers the fact that bigger is better. Some *speedophile*'s who keep talking about surfing at twenty knots conveniently neglect to mention that their boats are much longer than yours or mine.

However, tempering the measure of actual cruising performance is *cruiser friendly mode*. Thus, immediately eliminating classes one and class two racing boats, and many class three boats as well.

In a popular book on multihulls, the author claims his 54-foot long by 32 feet wide open wing deck sailing trimaran is a cruising boat, and further explains:

"Sailing is the reason I go cruising in the first place, and cruising in a boat that sails well is important to me."

I believe this quotation represents the crux of the problem defining the performance of multihulls as cruising boats. This sailor, just like so many speedophile's, has not yet come to grips with the fact that the vast majority of cruisers will gladly forgo the top end speed for a bit more amenity. On the other hand, he is a person with a commitment to a certain lifestyle and is willing to make the required sacrifices and compromises to achieve it, just as I have. For the rest of the cruisers, note that the book's cover picture of his boat shows spray shields along the cockpit sides and that though he is a well known designer, there are no other similar vessels. I assume that most likely because others are not willing to make such sacrifices for the additional speed under sail.

Most people buy cruising sailboats to cruise! Cruising is what they are doing. Sailing performance is a secondary consideration.

The utility of the vessel as a secure cruising envelope is the primary vessel function. They buy sailboats because they believe that there is economy in sail and for the romance of sail as a believable notion. Otherwise, they would buy motor vessels. Most cruisers *do* buy motor vessels. The *speedophile* view of multihull sailboats tends to discourage the average person not only with multihulls, but with sailing itself! Only a tiny fraction of persons wanting to cruise under sail are willing to make the enormous sacrifices demanded of them by the *speedophiles.*

(In certain circles this attitude is called "the hair-shirt mentality." Further defined by the fable that sailing is throwing hundred dollar bills into a hole in the water whilst taking a cold shower)

The constant emphasis on maximum performance and maximum windward ability in combination with the increasing sophistication and concomitant expense necessary to achieve that performance, in modern sailboats, has affected the entire industry.

As stated elsewhere in this book: If you need a station wagon or a van to meet your personal requirements and the salesperson looks down his nose at you because all he wants to sell you are expensive high performance sports cars, you are going to get discouraged and go elsewhere. That has been precisely the contemporary multihull dilemma. That is precisely the entire sailing industry dilemma!

Elsewhere in this book, I have attempted to give you an impression of the sailing characteristics of various cruising boats. Each has its own unique attributes, strong and weak points. Interestingly, when you talk with people out there, doing it, the boats that are the most criticized by the *speedophile* fraternity seem to be the ones that people most enjoy cruising. They also have the overwhelming majority of offshore passages to their credit. Those owners and sailors who will sacrifice everything for the potential of making an occasional, brilliant, high speed passage, are a tiny minority with a voice far out of proportion to their numbers.

I used to be a speedophile racing sailor myself. Weekends were for racing. My family acted as crew and cruising consisted of the overnight stays at various destination points or between harbors of necessity. All compromises and basic expenses were for making the boat go faster. To be sure, we had lots of fun and I can unhesitatingly recommend that as a fulfilling lifestyle. I love to sail a fast boat and I love to sail fast. However, the needs of the go fast sailor and the true cruiser are at cross purposes.

Now, after living aboard and cruising for fifteen years, it is clearly in focus. When we were just sailing, at the end of the sail, be it a daysail or a weekends sail, we went back to a comfortable home, stretched out on a recliner and contemplated the good times experienced. Perhaps even quaffing a tad of attitude enhancement elixir, making the good times seem even better. This was especially true on a cold winter night when reminiscences of summer sailing were better than the actual experience.

As a liveaboard cruiser, there is no getting off the boat to go somewhere else for a hot shower and a little relaxation. My boat is prepared to make me comfortable. It is also far more difficult to suffer some of the inherent problems of really fast boats when cruising. Finishing the day soaking wet from spray and not having a place to properly care for wet gear or not having enough room for minimum comfort are other examples.

Nevertheless, I still like a good turn of speed. That speed, however, cannot be at the sacrifice of our basic cruising equilibrium which means no streaking around down waves at 15 plus knots, no heroic sails, and few sacrifices for top speed. It means reefing early and avoiding deliberate exposure to bad weather. After all, we are cruising, not transporting ourselves somewhere. As transportation, a sailboat, any sailboat, is a little irrelevant.

What it does mean is that on a catamaran I routinely arrive at my destination 40 percent faster than my fellow cruisers. That translates into more time to enjoy the experience of cruising.

There is no question about it. I do not in any way mean to diminish the value of superior performance in a cruising boat. I mean to define what that performance is, what is reasonable to expect and what is achievable under cruising circumstances.

Using Racing Tactics

Now and then you will read articles that try to convey the idea that you can use racing tactics in your ordinary cruising. The one most promulgated tactic is tacking downwind. To be sure, there are some who will do it. I respectfully suggest, however, that tacking downwind has all the same problems as tacking upwind! In the real world of cruising, which subsumes going someplace, it has little or only occasional value. If your cruising requires following a marked channel, tacking is not an option.

Actual Passages

The listed set of conditions defines actual cruising passages and circumstances. These are limitations and challenges, usually enjoyed in the ability to subdue them. The magazine *Latitudes and Attitudes* calls this the difference between an ordeal and an adventure. This means navigation and piloting, knowledge and application of the Rules of The Road and courtesy and understanding toward your fellow cruisers. Quite unlike a racing situation, where you are sailing toward a predetermined point with all deliberate speed and interfering with fellow sailors is part of the sport.

A Few Difficulties With Tacking (up wind or down)

- Marked channels.
- Possible (or impossible) courses.
- Water borne obstructions/crab pot floats , fish nets, anchored fishing boats, etc.
- Navigation problems.
- Shoals.
- Heavy boat traffic.
- Prohibited areas.

Therefore, when a salesperson discusses performance and only performance, ask him to explain how his definition of performance meshes with your definition of cruising? The way he (or she) answers that question will give you the clues you need to evaluate the boat he is trying to sell, as a cruising boat.

> *The numbers of speedophiles fit the count of your fingers and toes. The number of ultra purists, fit the count of your fingers. Yet these gurus affect our industry and concomitantly, our ability to realistically address the vessels offered as cruising boats.*
>
> *Circa 2002, it is virtually impossible to cruise in an engineless boat larger than a day sailor. The difficulties are impossibilities for most mere mortals, simply because the ports and harbors have been laid out by engineers who have given no consideration at all to the needs of sailboats. Docks crosswise to currents, bridges across narrow channels creating swift currents, huge building's shielding the wind from need to sail critical areas, inlet jetties laid out without even the slightest consideration for the prevailing winds compose some difficulties. Only if you are content to never approach either modern facilities or populated areas can you continue to sail an engineless boat (or, if you have unlimited time and a penchant to ask for tows).*

Why People Buy Boats

Contrary to certain assumptions, people buy boats, especially sailboats, for many more reasons, and often conflicting reasons, than simply for comfort and convenience. Based on my assessment of people for whom I work as a surveyor, for whom I deliver boats, and from people I speak with at boat shows, people buy boats for the following reasons and roughly in the arranged priorities:

1. Social significance.
2. Recreation and enjoyment.
3. Racing.
4. Cruising.
5. Adventuring under sail.
6. Living aboard.

Living aboard, and adventuring under sail are the smallest and lowest priority. Social significance and its satellite psychological portraits are far more important, my estimate is about five to one. This is extremely important and those of us who are enthusiastic about catamarans for cruising must keep this in perspective.

To grasp a perspective of this phenomenon for yourself, I suggest the following: Personally observe how many lavish boats you see tied to docks in your sailing area are actually being used?

After spending fifteen sailing seasons racing, cruising and teaching in the Annapolis area, it is all too obvious that the ratio of boats used to boats not used is about six or maybe even, ten to one. Annapolis is a world premier sailing location. In a way, it's a good thing. If all those boats went out simultaneously, we would have the first water borne gridlock.

If we accept that premise as correct, then it becomes obvious why there is still a market for teak plated museum replicas. (Uh-oh, my bias is showing) It translates to, for every person who really likes to sail, there are five who find alternate methods of enjoying their craft.

Recreation and Enjoyment

The recreation and enjoyment group is about the same size group as the social significance group. For the most part, they have smaller, less expensive and less ostentatious vessels. This group represents good candidates for some of the catamarans and trimarans I classify as class two or class three (racing, racing/cruising).

Racing

Racing boats make up the next category. True, many of these boats occasionally cruise on weekends when they are not racing (rare). Frequently, a big party at the end of the race, placates the family who happens to be the crew, therefore, calling the race a cruise. I am sure by now many of you are laughing, thinking about how often you have been a party to that ploy.

This group is a mixed market for multihulls. The reason being, as I have witnessed time after time myself, the racing handicapping system stands in the way. There are comparatively few multihull boats racing. Many multihull owners give up in disgust and either quit sailing or buy monohulls, mostly one-design. I personally have been at races where there were upwards of 13 different multihull sailboats entered. Talk about handicapper's nightmare! I remember once on the beach at Montauk,

Long Island, New York. We almost came to blows over handicapping. Even today, more than thirty years later, animosities generated that day are still around. That is surely not conducive to having fun under sail.

**Fortunately, this is changing rapidly! The arrival of the new, sport cruiser trimarans, has created an entirely new group of performance multihull sailboats. These boats are at once, practical, handsome, seaworthy, high performance and best of all: Affordable!*

The race committees try as hard as possible. What winds up happening all too often is that the slowest boats win all the races, so the faster boats give up in disgust! Or, the faster boats win all the time, so the slower ones don't show up. You can see this happening presently. Research the results of organizations and racing associations past, Examine the histories of **The Long Island Multihull Association* or the **Ocean Racing Catamaran Association* and several others for corroboration. This same phenomenon happens to a certain extent in monohull circles. However, there are so many more monohull boats that usually there are enough to have level racing within a club or fleet. Most clubs will take five boats and make them a separate class, some will even do it with three. It has not been the usual case that you get into a cruising multihull race and find three boats alike. This has changed considerably with the coming of the Gemini Catamaran, at least on the U.S. East Coast, and the Farrier trimarans worldwide. (*presently inactive)

Since this is a book about cruising, I will not detail the multihull ocean racing greyhounds that now dominate the worlds' great ocean races. That is a subject for another time and, perhaps, another author. The exploits of *Playstation* and *Club Med* are in current headlines.

Cruising

Now we come to the real cruising catamaran market segment, those who really want to cruise, gunkhole, live aboard or adventure under sail. It is for this group that the cruising catamaran or trimaran has all, and I do mean all the advantages. Those advantages are clear, definable and, I think, indisputable. Beyond comfort and convenience are shallow draft, seakindliness, ease of handling, safety and performance.

Years ago, there was a TV series called *Route 66*. This series featured two young men driving across the United States in a Chevy Corvette®. (Sports car) They did it, had their various adventures, managed to find women and whatever. But, what stands out in my mind is the price they paid for driving the "Vette."

All their luggage was strapped to a luggage carrier on the trunk lid and to stretch out for a snooze was impossible. In the excess heat or cold, they were squeezed into that tiny space, freezing or roasting and getting leg cramps. Now, my question is this: Would they have had fewer adventures, found fewer women or whatever if they were comfortably ensconced in a station wagon? Would they have made less road time? Who cares? That is not why they were out there. They were out there for the adventure, the women what do you do when you meet women and you only have a two seated sports car? The point of this silliness is, don't drive a sports car when what you really need is a station wagon. Why sail some deep draft, unstable racing boat or its laughingly called cruising clone. Why suffer all the indignities of that choice when you can be secure and comfortable in a proper cruising catamaran?

La Forza transits the infamous Rock Pile on the ICW. Corinne is steering

What indignities?

As I write this chapter, I am cruising through the Dismal Swamp Canal (part of the U.S. Intracoastal Waterway). Seven sailboats came through the locks into the canal with us. The main reason they came this way was to avoid the powerboat wakes on the Virginia Cut route. Since the Dismal Swamp Canal has a strict six mile per hour speed limit for its twenty two mile length, the powerboats shun it like the plague, thus leaving it a sailboat paradise, except for the very shallow draft! (six-foot controlling draft). The monohullers make a choice: white-knuckle it through the swamp, or get tossed and thrown by the powerboats on the Virginia Cut route. We, in our 32-foot catamaran with two-foot draft laugh at both choices. To Corinne and me, neither the controlling depth nor powerboat wakes are a concern. That is comfort! We believe that maintaining a constant vigil that borders upon terror is an indignity and an affront to the enjoyment of cruising. I think my premise will be undisputed by my peers, that premise being, the greatest indignity suffered by cruisers is rolling and heeling, both power and sail. Now, if you can eliminate that indignity by your choice of vessel?, I rest my case.

The Compromises of Size

Until this point, we have only discussed compromises of performance versus amenities. However, perhaps a more important area of compromise is total size. How large a boat do you need for your requirements? How large can you go or how small should you go for your chosen cruising lifestyle? Let us divide size into its four components and discuss them both separately and as the whole.

1. Length
2. Beam
3. Draft
4. Cubic Volume
5. Gestalt, the entirety

Length

First thought about length is that boats are assessed fees based upon their length. Some examples of such fees are haul outs, transient and permanent dockage, taxation and registration, insurance, importation, Customs, both foreign and domestic, port , mooring, and survey fees,etc. In other words, from a monetary point of view, longer is absolutely more expensive and probably always will be.

On the other hand, the single most important number in a vessels design is waterline length. There may be controversy associated with that remark and talking about a single number is always an oversimplification. Yet for the aim of this book, I believe that number is the single most important performance arbiter. Waterline length is the determinate of speed, a major factor in seakindliness, a factor in acceleration, pitching moment, etc. With waterline length, longer is better. No one will question or deny that fact. Longer is better.

Overall length as measured on deck is but one factor of total length. The ratio of waterline length to overall length is an important factor in pitching moment. Those who increase their overall length by adding bowsprits, boomkins, dinghies in davits, etc., increase the ratio of overall length to waterline length, therefore, increase their pitching moment of inertia. Increasingly about the world, they also increase the monetary length. Many places include the overhanging appendages as part of the measurable length. Therefore, they assess the appendages as part of overall length and charge accordingly.

Beam

Beam is a major factor, but not the only factor, in measuring stability. I think, when a catamaran exceeds the length/beam ratio of: BMAX=LWL/2, you move into the realm of performance, class one, two, & three catamarans. As the beam gets wider, the ability to carry tall masts with the concomitant sail area increases dramatically. However, there are serious and potentially dangerous aspects to this trend. First, the ratio of fore and aft stability changes and the possibility of burying the lee bow increases. Second, the pitching moment increases. Third, the requirement for bridgedeck clearance increases considerably. Forth, increased bridgedeck height will increase windage. Fifth, more structure and more sophisticated structures are needed to keep the boat together.

Those five characteristics are perhaps technical in nature. From the user point of view, excessive wide beam has other disadvantages.

a. Turning radius

Turning radius is determined by scribing a line from the center of lateral resistance to the outermost point on the boat. This distance is one half the diagonal of the rectangle the boat encompasses. It is greater than one-half the boat's length. Wide beam significantly increases that distance.

b. Haulouts

The days of marine railways have ended. Almost to a railroad, they have been replaced with travel lifts. There are precious few lifts that can take anything wider than 20 foot beam and many of those that can are found in commercial yards. To be assured of not having problems finding haulout facilities or dockage, consider 16 feet as maximum beam. Between 16 and 20 feet your options rapidly diminish. Over 20 feet wide, you most likely will require a commercial shipyard or crane.

c. Dockage

I must assume that many of you will be using your boats in highly populated areas. I think that is a valid assumption since that is how those places got crowded in the first place! In the more densely cruised locations, any wider beam than 16 feet will virtually preclude your getting a marina slip. Most modern marinas consider the Hatteras 42 as their "scratch boat." Therefore, the slips will begrudgingly accommodate a 16-foot wide boat. Older marinas will have narrower slips. In the good old days, when transient marinas were not dominated by the Harvard Business School approach to maximizing cash flow, there were lots of end slips (face docks) and vacant fuel docks available. There has been a change in that whole scene since many have eliminated the face docks with pilings for additional slips and fuel docks by keeping them open twenty four hours a day.

Additionally many marinas, reserve the face docks for the transient mega-yachts; those 60 feet or better that buys fuel by the thousands of gallons. There are very few face docks left at all. Those that remain, a distressingly large percentage have little interest in tying up their prime space with sailboats. Mom and Pop who love boats and boaters are no longer running the marinas. Increasingly, remote corporations and condominium associations are running them and the ambience and feeling has changed.

Fuel docks that used to close promptly at 5:00 PM and not reopen until 7:00 a.m. are now open longer hours. The reasons for this are as profound and complicated. Many marinas have stopped selling fuel. The million dollar pollution insurance requirement and the new stringent EPA regulations have caused retrenchment. Those that are still selling fuel, must sell a great deal more to justify the dramatic increases in the cost of providing this fuel. Therefore, longer hours and harder sell tactics. This means that so many transient docks we could use during the closed hours are now nonexistent, just at a time when wide boats are really proliferating.

Hull fineness ratio

Hull fineness ratio,(HB) is another aspect of beam. That ratio is the measure of fatness that your hulls have. Skinny hulls are faster. Fat hulls are slower but are able to carry more weight and are more stable. It is also this ratio that creates the theoretical hull speed limitation factor.

Catamarans and trimarans also have a theoretical hull speed limitation factor. However, catamaran hull speed limitation factor is sufficiently high that consideration as a speed limitation is of little consequence. On the other hand, as hulls get "fatter," their bow waves become more pronounced. Narrow beam catamarans with fat hulls are apt to have the potential of hull wave interference as the waves collide between the hulls. This phenomenon can definitely limit their speed. In most boats, that critical convergence point is beyond the point of average cruising speed. However, overloading and dirty bottoms have a serious negative effect on wave interference and tend to increase it, therefore, becoming a factor in limiting speed.

In the tables that are in the boat description sections, you will find the hull fineness ratio of the vessels for which I could obtain information. Keep in mind in your comparisons, hull fineness ratio is but one measurement and only a partial one at that. The purpose of even mentioning it at all is to provide you with a rough yardstick.

Rule of thumb, a set of gross approximations concerning fineness ratio:

16:1 = Race Boats
10:1 = Racer/cruiser (could go either way, depending on other #s)
8:1 = Cruiser
4:1 = Monohull

There are several other sets of numbers that arguably are equal in importance. I have avoided them in the attempt to keep this book simple and to the point of cruising. Those who wish to delve further into this fascinating subject can obtain a suggested reading list in the bibliography. The referenced numbers refer to waterplane shape, the fullness or fineness of bow entry and stern exit, and the shape of the underwater parts including rocker. There are various formulas and coefficients that go into figuring the volume displacement and thus the carrying capacity.

Draft

Shallow draft is a most profound and important part of a good cruising boat. First, shallow draft allows one to find room at crowded anchorages, often to windward of the anchored fleet. Being to windward at anchor, as at racing, has every advantage. In anchoring, these advantages are more profound. Protection from those who drag, less noise, no smoke or odors from generators, barbecues, cigars, etc. are all yours with your shallow draft.

I remember once at Block Island, Rhode Island, on our way to Newport. It was a major race weekend and there were no vacant moorings. Every corner of the harbor in which anchoring was permitted was filled with revelers. Many of these boats, I suspected, were anchored with uncertain technique. Corinne and I had no problem at all. In one corner of the harbor there was a little creek. At the mouth of this creek were some rental skiffs on moorings. We simply sailed right past the skiffs and anchored in the creek in about three feet of water, as safe as the proverbial bug in a rug!

The Chesapeake Bay and its tributaries are unquestionably some of the finest cruising grounds in the world. Forty percent (40%) of this vast area is four feet deep or less. The Outer Banks of North Carolina, another of the worlds most intriguing cruising grounds are also a shallow water area. In the Fabulous Florida Keys, four feet (4') is considered deep water. If you closely examine just about all the worlds exotic cruising grounds, you will find there are places you cannot reach without shallow draft.

Safety at sea is another important consideration. Without a deep keel to trip over, the safety considerations inherent in your vessel rise dramatically. (See Beating a Dead Horse for additional comment)

Weight carrying ability, fact and fiction

The two similar sized sailboats diagramed below have roughly the same ability to carry weight based upon their displacement. The difference in *opportunity* to carry weight is blatant in the catamaran. Just the potential for a dinghy in davits and major machinery alone mark the onset of a potential problem

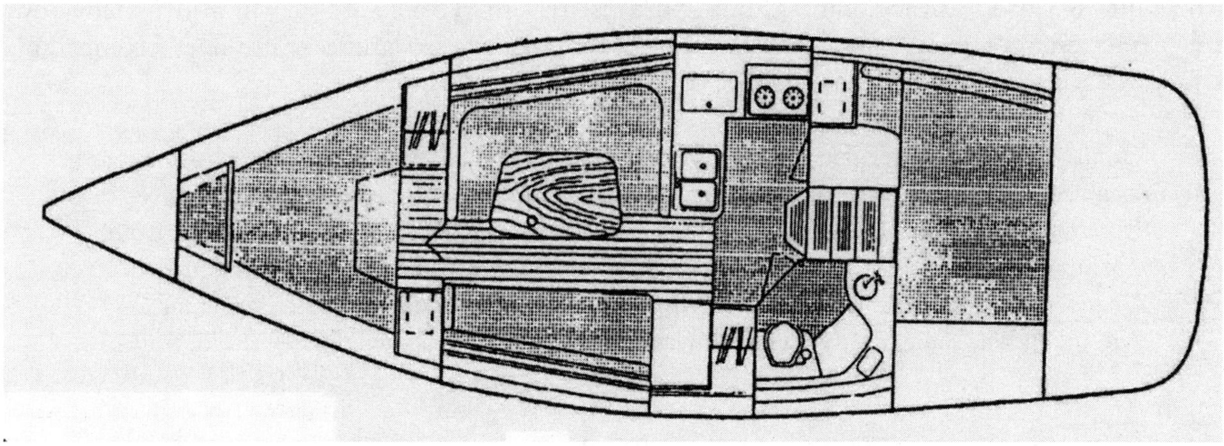

Popular 34' Catamaran		Popular 34' Monohull	
LOA:	33'6"	LOA:	34'6"
BMAX	14' 0"	BMAX	11' 9"
SA:	510^{ft2}	SA:	528^{ft2}
DISP:	7,300	DISP	12,550
Hull/beam (HB) ratio: 10:1		Hull/beam (HB) ratio: 3:1	

Functionally related characteristics are weight carrying capacity and cubic volume. A quick rule of thumb states that a proper cruising vessel ought to be capable of carrying 1000 pounds per bunk.

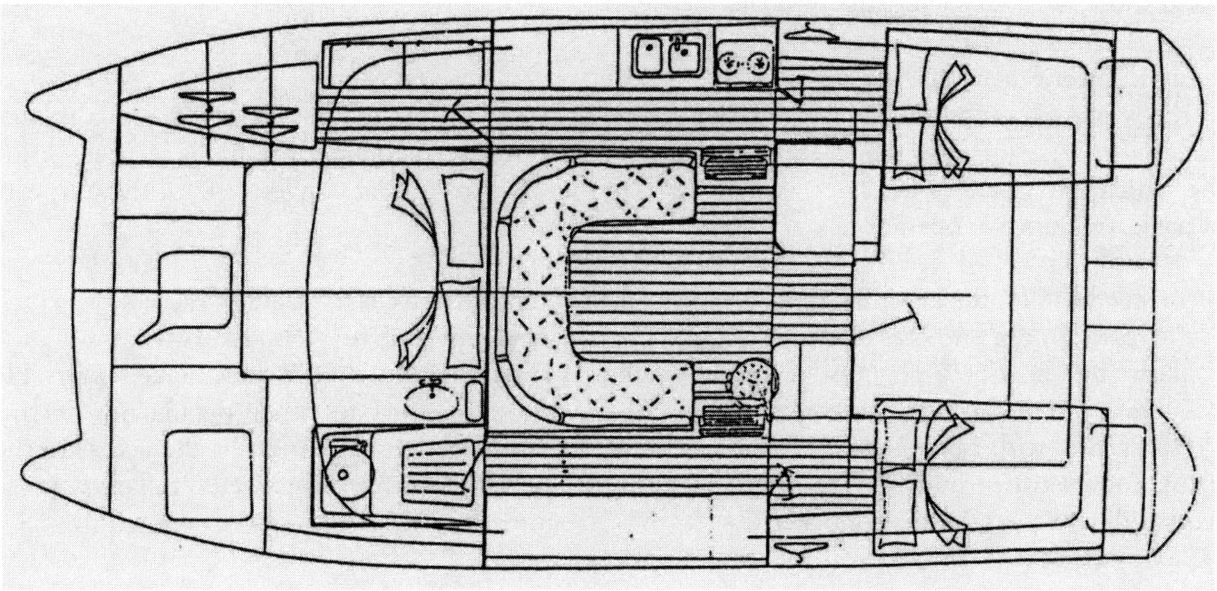

Bunk being defined properly as crew or total ships complement. 1000 pounds might seem like a lot at first blush, but it really isn't. Think how much your suitcases weigh on even a short vacation.

Cubic Volume
Here we have a developing problem and one of the disadvantages of catamarans. Many cruising catamarans are cavernous inside, but have little carrying capacity. The temptation to fill the boat up with "stuff" is close to overwhelming, but you must refrain. To be sure, it will also impede your performance on a comparable monohull. However, it is not quite so significant nor does a comparable monohull have anywhere near the interior volume as most cruising cats.

> **While on the subject of cubic volume, I should explain that the "Gross Tons" note on your vessel documentation is a volume measurement, not a weight measurement. There is an archaic derivation. It is the old English "tun." The "tun" is a 232-gallon wine cask. The original measurement standard dates to the English harbor dues of 1347 and incorporated into American maritime customs regulations during the Whiskey Rebellion of 1784. Some time after that, the spelling changed. My guess: a bureaucratic oversight! (Occasionally seen spelled tunne.)*

Gestalt
No question about it, we have a minor dilemma. Small is beautiful, yet bigger is better. It is imperative that you choose wisely for your own needs.

Classifying Multihulls

Why a Classification System?
A proper classification system promotes understanding of the various nuances between boats. If an auto salesperson tried to define a Chevy Corvette© as a family car, you would understand that his concept of "family" car was quite different from yours and perhaps being hungry for a sale made him stretch reality. At the end of this article, you will have enough practiced insight to intuitively understand those same characteristics in boats.

Stylists, Enter Stage Left!
In the world of compromise, we must keep clearly in perspective that a boat is not like an airplane, a boat is not like an automobile. In both of those conveyances, the operator remains securely inside the vehicle. An operator does not walk around on the exterior of said conveyance for the purposes of navigation and/or operation.

A cruising boat, on the other hand is quite different. An operator needs to be able to access every inch of the vessel for the purposes of navigation and operation. He must tie to docks, fend off from pilings, fend off other boats, etc. He must be able to weigh and lower anchors, change sails, make repairs. He must take stores aboard, raise and lower his dinghy and a myriad other details not even a consideration with cars or planes. Many automobiles are judged as utilitarian by the access to the trunk compartment and how many doors they have. Boats too may be judged similarly, by the accessibility of their decks and lockers.

You can streamline an airplane to the very limits of science and technology with little negative effect upon its utility. An automobile is streamlined within the limits of stability and legal statute. It is possible to obtain close to ideal streamlining with those vehicles. Modern computer programs can diagram airflow over the vehicle or aircraft to an unprecedented state of efficiency.
You cannot realistically or safely achieve an ideal degree of streamlining on a cruising boat, though the technology to achieve it, exists.

Every time some part of a boat is streamlined to make it more aerodynamic, it loses a certain amount of utility. The user loses a certain amount of on deck mobility. When it finally gets completely streamlined, it is untenable as a boat and unsafe on deck. Long before it reaches that point, it has ceased to be a viable cruising boat. Extreme streamlining for the sake of higher performance may be an acceptable compromise on a racing craft, but is unacceptable on a true cruiser. A cruising vessel must be a secure, safe environment for those operating it. It must be safe and secure without any extraordinary means, but simply by its design. A modern vessel I would consider over the line of acceptable streamline compromise for a cruising boat is the South African built *Wildcat* because of extreme rounded side decks and severely sloped fore and aft decks.

A true cruising sailboat design subsumes

- Total perimeter lifelines.
- Reasonably flat, accessible decks, the configuration of which allow safe access to every square inch of the deck.
- A good toe rail to protect you and your gear from slipping overboard.
- Sturdy and secure handholds, properly located. There should be no place on deck where you cannot reach a handhold.
- Properly sized deck cleats, large enough to accommodate three lines on each cleat. Preferably without the need to lead off boat through chocks.
- A comfortable, secure, protected helm, out of the sun, wind and rain.
- Good visibility from the helm and everywhere else. On a catamaran you should be able to see all four corners of the boat while comfortably seated at the helm. On a trimaran or monohull you should have reasonably unobstructed vision forward without the necessity of a bow man lookout.
- An appropriate engine or engines, located for convenient service, preferably not accessed through living space.
- A well protected drive train and rudder or rudders.
- And without doubt, a good, substantial, well placed rub-rail.

Secondary priorities

- Lots and lots of lockers, especially deck lockers.
- Adequate weight carrying capacity.
- Adequate fuel, water and stores storage capacity for the intended usage.

Those are the basic attributes of a vessel designed for cruising. All those attributes lead to compromises with certain aspects of performance. It is up to you to determine which, if any, compromises you are willing to make for the sake of that elusive high performance.

As the vessel gets smaller, the extent of the compromise gets greater!
When you get down into the 30 to 35 foot size range, it becomes virtually impossible to have a really high performance cruising catamaran and especially a cruising trimaran. You can have good performance, satisfactory performance, but not superior performance in a true cruising boat. A true cruising boat being one that meets all the above requirements. Minimum cruising life requirements are, food, crockery, personal and health items, recreational equipment, clothing, safety gear. Theses items remain virtually the same no matter what size the vessel.

> *A remark was once passed about multihulls that: "If you don't have speed, you don't have anything." I suggest that perhaps, as far as a cruising boat is concerned: "if you can't carry at least one thousand pounds per bunk, you don't have anything."

Symposium theme
One symposium theme at International Multihull Week, Newport, Rhode Island, brought out loud and clear the need for a measurable classification system for multihulls. We even joked about defining what categorized a boat as a cruising boat and tongue-in-cheek decided that: "Any boat that does not win the race is a cruising boat."

The purpose
The purpose of this classification system is to help prospective purchasers understand the nature of the different boats. There are similarities and differences between cruising catamarans, trimarans, monohulls and power cruisers. This classification system describes what goes into making up a class, the similarities and the differences and where and how to draw the line.

All boats are compromises!
All boats are compromises. Dick Newick, a popular designer, once said: "You can have high performance, lots of amenities and low price. Pick any two of the three." That truism being the crux of my argument for a rational classification system that any individual, especially a novice individual, can readily understand.

To define the various sailboats by the intended use, I have devised the following 1-6 classification schedule. There will be arguments about where the dividing lines might be. That is simply the nature of the sailing industry.

Perhaps, if we compare the classification differences of sailboats, (both multihull and monohull) to something far more familiar to us, it might be easier to visualize the subtle variations. We recognize the differences with our automobiles. Perhaps the following is a valid comparison:

Markers
I have designed six automotive symbols markers that I will use to denote the classification I believe that each particular boat ought to have. In most instances, when I discuss a particular vessel, I will add that symbol marker in the heading.

The following system describes in elementary terms, the nature and intent of the compromises necessary in classifying a multihull sailboat. It is for you, the reader, to be able to see through the blizzard of hype, armed with a system to compare vessel to vessel. It is for you to compare your discreet requirements to those offered by any particular boat, no matter how many hulls it has.

Automobiles		Sailboats
1. Grand Prix racing cars	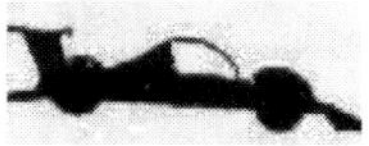	1.Experimental/high performance.
2. Sports Car		2. Racing
3. Coupes		3. Racing/sport-cruisers
4. Sedans		4. Cruising/racing
5. Station wagon		5. Cruising
6. Vans & RV's		6. Residential/utility

This is a comparative assessment of type, one through six, of both multihulls and many more popular monohulls to help you grasp the essence of your comparisons. Far too many people will try to compare a pure cruising multihull (class 5) say a Victory 35 to a racing/cruising monohull (class 3) say a Beneteau 365. This is essentially comparing apples to oranges. The tables will help you in making relevant comparisons. All too many salespersons have enthralled their customers with fabulous tales of high speed and extraordinary pointing ability of their distended looking cruising boats. There are many disappointed former owners who, perhaps should have had enough intuition to realize that if it looks like a summer cottage with a mast stuck on it, it probably will sail like a summer cottage with a mast stuck on it! The biggest tragedy of that sales approach has been to create a lack of credibility within the industry. This criticism applies equally to monohulls!

A second tragedy of at least as great a proportion, is that those summer cottages with a mast stuck on do sail quite well. They sail better yet compared to the monohull equivalent of a sailing summer cottage or a sailing museum replica. Not only that, but those summer cottages make some of the best true cruising vessels ever built.

Class 1: **Experimental / High performance**
(Grand Prix Race)

Experimental/High Performance means that the sole reason for the boats' existence is to go as fast as it is possible to make a sailboat go. Boats fitting this category most specifically are the newsworthy sponsored boats. Others are the "one of's" designed and built for various races and speed trials. Others, are those in which every compromise, every sacrifice is for better performance. Special spars, rigging and fittings are the usual rather than the exception for those who can possibly afford it. However, this category could and should also include inexpensive home-built boats if the intent was to make it go as fast as possible without compromise.

Examples of Class I, Experimental/High Performance are "C" Class, or "Formula 40" experimental/high performance, by definition of this formula. Minor restrictions might apply, such as overall length. Rudy Choy's incredibly successful *Aikane X5*, Tony Bullimore's *Spirit of Apricot* and Philip Poupon's prize winning *Fleury Michon* trimaran with its ten day C'STAR Atlantic crossing record. (To windward, of course). Most recently vessels like Cogito and Playstation. (Perhaps "C" class and Formula 40 might be classed 1.5 because they are designed within certain parameters)

Class 2: **Racing** (Sports Cars)

Racing boats differ from Experimental/High Performance in that there are more compromises in the design to facilitate production and the creation of class or level racing. It also can mean compromises made to conform to a racing or rating rule. This subsumes using some off the shelf parts and accessories, the intent of the boat design being to go as fast as possible within the confines of a rating rule, a production facility or a price range. Examples of pure Class 2, Racing Boats, might be: F 24C, F 31, Hobie 18, Tornado, NACRA, etc.

Class 3: **Racing /sport cruiser** (Coupes)

Racing/Cruising category means, that the purpose for the design is to achieve maximum performance and/or to conform to class rules or to achieve a more favorable rating. Some amenities exist, possibly by class rule. Technically, the addition of a head and a galley allow you to use the word cruising in the description of your vessel. The International Offshore Multihull Rule (IOMR), the rule which so many Multihull associations use for racing, describes these conditions.

Boats like *Gougeon 32, Condor 40, Dragonfly* 600, 800, 1000, *F24, F27, F31, Stiletto, Jarcat, Turissimo* fit well into the top of this category. Perhaps even delineating a class of 2.5. At the bottom of the category it is possible to fit most of the MacAlpine-Downie production boats, depending upon how they were fitted out. Many Shuttleworth boats such as *Damiana* and most Crowther, Kelsall, Choy,

Cross and Woods boats and Chris White's *Juniper* also fit this category. Possibly classifying a 3.5, but so much for hair-splitting.

It must be noted, that in this class the intent, and *intent* is the key word, is to make the boat sail as well or as fast as possible with only minimal compromise for cruising.

Class 4: **Cruising / Racing** (Sedans)

Cruising/Racing boats get a little harder to define. However, the essential element to consider is, when compromises in the performance potential of a given design exist to add amenities, it will most likely fit into this class.

For instance, a Gemini, a Seawind 1000 or an Iroquois could be either class 3 or class 4 depending on it's equipment and the type of auxiliary power. A stripped out Gemini with an outboard and some exotic sails could be a class 3 while a Gemini with air conditioning, RADAR, dinghy in davits etc. is a definite class four, or perhaps 4.5. Other examples might be Spronk catamarans which though they are speedsters, must carry all the weight of first class cruising boats for the charter trade. A partial rule of thumb: if the boat has a center-board it is probably class 3 or 4, if it has fixed keels it is probably class 4 or 5. There are always exceptions. Many European built class 3, Racer/Cruisers have fixed, low aspect ratio, keels. These keels are now coming into the forefront of design development.

> *Fortunately or unfortunately, we are being sold the virtues of these keels. However, by adding these keels, we lose a most important aspect of a multihull for cruising: shallow draft. Keels cost less to produce, simplify design requirements and do not intrude into interior space and in some cases add a degree of protection to the bottom. Therefore, the hard sell. Time will tell if the new multihull producers will make the same mistake that the monohull producers did with their bolted on fin keels and diminish their own market.*
>
> *The sailboat market is inversely proportional to vessel draft. The deeper the draft, the smaller the market. On the other hand, low aspect ratio keels actually work better than daggerboards at low speed because high aspect ratio daggerboards tend to stall at low speed. Since most cruising boats spend more time at lower speeds this may not be a disadvantage if the keels are not too deep.*

The addition of twin diesel engines to a class 3 or class 4 boat, will change it at least one class. That is the very nature of compromise. Few people interested in racing are willing to accept either the weight or the drag of these engines. However, consider that as the boats get bigger, the proportionate weight and drag become smaller. What is significant weight and drag on a 30-foot boat, may be insignificant on a 60-foot boat.

Class 5: **Cruising** (Station Wagon)

Attention to livability as its design priority defines a cruising boat. Ease of handling, simplified rigging systems, and the lack of physically taxing go fast gadgets outweigh performance under sail, except, unfortunately, in their advertising. They usually have larger engine installations, either inboard or outboard. They compare to the so-called motor sailor in monohull circles. Often, they are designed, built and fitted out for long term cruisers, liveaboards and those who adventure under sail. Home like interiors and appliances are the usual items.

Typically, performance is reasonable despite the design orientation, usually more so than monohulls with the same compromises. For instance: *Westsail*, *Morgan OI*, most Taiwan teak plated cruising boats, *Irwin 37*, *Whitby 42*, and many others in the size range, can be compared to the *Snow Goose* 34 and 37 by Prout, and the *Solaris* 33, Lagoon 380 and 410 are examples of roughly equivalent class multihulls that routinely out sail, on all points of sail, those listed monohulls. Check the results of the ARC (Atlantic Race for Cruisers). In these races, a group of *Snow Goose* 34's outperformed many larger cruising boats on equal terms. Those terms were the essential terms faced by the lion's share of cruisers in that race.

Those terms are: overweight, foul bottom, shorthanded crew, conservative sailing outlook. (Possibly low skill, small children and lack of total commitment by one or more crew members)

I suggest that the Catalac catamarans in the older boats and the Victory 35 or Manta 42 in the new offerings represent the classic class five multihulls.. Catalac (with over 600 units built and sailing) this occasionally maligned vessel has probably brought as many hours of happy, comfortable, safe boating to more people as any other. It is hard to find any comparable production vessel that has so well achieved its design objectives. One that comes close, I think, is the monohull, Morgan, Out Island series, the most popular cruising boats ever .

If you think a *Catalac* does not sail well, I suggest you go sail a *Watkins* 27, or an Irwin "Wide Body" 27 or one of those similar and comparable monohull cruising boats. Then come back and tell me the *Catalac* 27 doesn't sail! Other more modern and better sailing class 5 vessels might be the Victory 35, the PDQ 32, the Prout 45 and many of the French catamarans.

Judging by the formula, "how fast it sells as opposed to how fast it sails," these obviously successful vessels will be still sailing when some of its most strident detractors are just a faint memory.

Class 6: **Residential** / Utility (Van/RV)

Residential/utility is a catch-all category to include all those floating multihulls that do not fit into any of the five categories above. There are many creative, useful, entertaining vessels floating around that satisfy the requirements and dreams of their owners. There is no criticism implied or intended unless someone ventures beyond safe limits with such a vessel. With any residential vessel no matter how many hulls it has, the owner is still obliged to know and follow the rules of the road and to understand and practice good seamanship.

Cruising catamaran features (Ergonomics)

Overview

There are several features that make a sailboat, any sailboat, fun and safe to handle or conversely, make it difficult and even dangerous. While this chapter is about cruising (class 4 and 5) catamarans, the principles involved really apply to all cruising boats, regardless of how many hulls they have or how they are powered.

Streamlining and aesthetics

At the risk of being redundant, I must remind cruisers and aspiring cruisers that there are practical limits to streamlining a sailboat. Conveyances such as aircraft and automobiles can be streamlined to a maximum degree because they are operated from *inside* the conveyance. Cruising sailboats, on the other hand, are operated from *outside* the conveyance. Thus safety and practicality overrule the minimal performance gain potential of extreme streamlining. (And at five knots, the gain is minimal) Aesthetic standards for catamarans are different from the classic vessels of yore. Yet style and fashion must not be allowed to interfere with the basic necessities of vessel operation.

Deck Layout (Macro)

Deck layout, cabin height and side-deck accessibility. They are important items when choosing a cruising sailboat. It is necessary to climb on cabin tops to handle sails, sail-covers and make adjustments or repairs to sail lifting and handling apparatus, or just to get a better view. Moving from the cockpit to the foredeck safely and easily must be considered. The width of the side-decks, the slope of the cabin sides and the hand-holds available are important items for cruisers. Cabins with extreme slopes often are difficult to maneuver around or climb upon.

Secure access to the mast and its controls that is easy and safe. The most user-friendly cruising sailboats have good access and secure footing at the mast base. Many extremely sloped cabins also let in excessive sunlight, thus requiring special curtains to prevent it, negating one of the cruising catamarans best attributes: its panoramic view from the interior. These sloped windows also turn the main saloon into a steaming hothouse. Many of the Privilege and older Lagoon models exhibit hothouse characteristics.

Overly high cabin tops interfere with visibility from the cockpit. This must be judged on an individual basis, depending upon the height of the users. Aesthetics also suffer from overly high cabins. Certain vessels have no forward visibility at all from the cockpit. Vessels like the *Dufour Nautitech* and certain *Privilege* models, have a continuous Bimini top from the roof line thus precluding any forward observation whatsoever from inside the cockpit. In my opinion, this not only negates enjoyment of the cockpit and limits proper ventilation of the cockpit, but the excessively limited visibility creates a dangerous navigation and piloting situation.

In my opinion, the very best deck layout of the trampoline type boats is the now out of production, Fountaine Pajot, *Tobago 35*. It has every feature I described in my first book, *Cruising On More Than One Hull*. It has relatively flat decks, wide unencumbered side decks and good handholds all around. The top deck treatment is superb. It has enough overhang on the upper deck to shade and protect the moderately sloped windows and this overhang has a molded integral full length hand grip that also functions as a rain gutter. Other boats by this manufacturer have similar features. Other manufacturers exhibit these features to greater or lesser extent. However, there are some vessels that clearly go beyond the bounds of safety and security. For example, the *Wildcat*, manufactured in South Africa is a boat I use as an example of having decks far too rounded and sloped for security and the *Mayotte 500* is an example of a boat with insufficient handholds, no rub rail and no toe rail. Imagine two people bringing a fifty foot long by 30 foot wide boat up to a fuel dock in rough or windy conditions with no rub rail to protect the hull and no toe rail to protect them?

Deck layout (micro)

Deck hardware. The layout of deck hardware is of considerable importance. For instance:

Turning blocks. If you are leading lines back to the cockpit, the type of turning blocks that are being used? Are they fixed blocks bolted through the deck or individual blocks shackled to the mast base? Do the lines lead fair? Do the tails of the lines have a reasonable storage area? (With all lines led back to the cockpit, an enormous spaghetti pile of line tails will pile up, often right over the top of the steering wheel, engine controls, etc. Are there provisions to avoid this?)

Winches. Are the winches installed in such a way as to be user friendly? Can you swing a winch handle a full 360 degrees without banging into something? Do you have to kneel on something and stretch way over to operate them? Are they usable by the smallest, weakest, least committed crew member?

Traveler. Is the traveler situated and arranged for easy use? Does it have multiple purchase and an endless control line or does it have individual port and starboard lines which add yet more spaghetti to the cockpit? Does it have manual stops that endanger your fingers every time you move them?

Mainsheet. How is the mainsheet arranged? Does it conveniently lead to a single point, usually at a winch or rope clutch or do you have to chase it around the cockpit on the traveler, as you would with a day sailor?

Genoa tracks. Are the Genoa tracks installed in a convenient location? Do they appear to be sighted for proper leads?

Spinnaker blocks. Are they in a convenient location? Will spinnaker sheets in use, cross your cockpit?

Pad eyes. Do you have pad-eyes where you will need them and are they installed with real backing plates? Pad eyes are used for attaching snatch blocks, life lines, jacklines, etc.

Hatch gutters. Are the cockpit hatch cover gutters deep and lead properly so that water drains and items in the lazarettes or seat lockers stay dry?

Hatch openings. Do hatch openings face in the proper direction? Hatches low and forward should optimally open towards the stern thus putting their strongest edge towards wind and seas. An accidentally opened forward facing hatch in bad weather can be a disaster.

Cleats. Deck cleats should be large enough to handle three (3) lines of a size normally used on that size vessel. The best installations lead fair from the cleat without the need for chocks.

Chocks. Most chocks are skene chocks and are unsuitable for all but benign conditions. Truly good anchor or dockline leads, lead fair to the cleat to avoid the requirement for a chock. In some cases where it is not practical to lead directly to a cleat, roller-chocks can be used. On boats with bulwarks and hawse pipes, the best ones have internal rollers to avoid chafe. The reason being that it is close to impossible to prevent lines from chafing at chocks.

Lifelines and Stanchions. Lifelines should have gates at convenient locations. Stanchion bases should have proper support. There should not be any "crackle" when you pull on them.

Bow rollers . These are very important fittings for anchor handling. These should be easily reached and properly engineered for strength and utility. (Covered in greater detail in Book II)

Anchor lockers. Many cruising catamaran manufacturers are being seduced into trying to hide the dirty old anchor and have come up with many seemingly clever ways of doing it. Some moved the anchor back to the top hamper of the underdeck, just where the main deck joins it. This location leaves a gaping hole in the most vulnerable area of a catamaran. This hole allows sea water, rain water, fog or anything else that comes along to drive directly into the interior lockers of the vessel. The windlass, an item with a vulnerable electric motor, is right in the line of dousing. In order to lower or raise anchors from this location, you will need to work below deck level in small spaces with heavy, clumsy items. This is an unnecessary danger. Imagine, when (not if, but when) the windlass fails? Consider how are you going to raise or lower the anchor from that location? On charter boats this is less of a problem, you simply call the charter base on your VHF and soon a person will show up to fix it.

Fuel fill pipes. These should be located so that any spillage does not run down into lockers. Certain boats have diesel fill pipes inside lockers. On the surface this seems like a good idea but a minor spill can create a major problem with fuel running over stored gear and/or trickling down into inaccessible areas under the tank that cannot be easily cleaned thus consigning your boat to noxious fuel odors forever. Fuel is often decanted from jugs in rough conditions making spillage an unavoidable reality.

Boarding ladder. Most modern cats have them leading into the stepped transoms. In my opinion, the very best arrangement ever built was the old Prout *Snowgoose* where the ladder, complete with handrails, is an integral part of the stern pulpit arrangement and led directly into the stern platform. In any case, the ladder should have close secure handholds. There are times in less than ideal conditions you will be thankful for that consideration.

Mast step. The mast step should be accessible for reasonable mast access. You should not have to balance yourself in precarious positions to operate any winches, cleat lines or sort out the tails of lines. On many offshore yachts, a mast pulpit is a great convenience and safety factor.

Boom access and sail access at the goose neck is important. Consider the need on certain vessels to balance yourself on a mast step or worse, on a winch, to attach a main halyard or sort out tangled lazy-jacks.

Roller furling lines. The convenience of furling headsails is an often overlooked item. You should be able to conveniently haul on the furling line from the shelter of the cockpit. In most boats, furling lines require a winch if for no other reason than to dampen the jerking on the line caused by the flogging headsail as it is furled.

Mainsail reefing, furling and stowage are very important items. There are all sorts of modern conveniences alleged to make mainsail handling easier. Some do some don't. Lazy jacks are a mixed blessing. In most cases you are going to need to get to the boom and sails to make adjustments, repairs, attachments, etc. Ability to conveniently reach the boom and mainsail are important items.

Foredecks. There are two types of foredeck treatments, solid deck to the bows and trampolines of various types. The ability to manufacture a proper cruising boat with a solid foredeck is limited by maximum beam (BMAX). When the overall beam of the boat exceeds one-half the waterline length, the trampoline becomes a necessity rather than an option. Standard beam vessels with solid foredecks have the advantage of the solid footing for anchoring, dingy handling, etc. Head stay attachment is simplified, as is the construction and application of anchor handling equipment. Solid foredecks provide a drier ride in rough going. However, in many cases, the solid foredeck has greater under-deck slamming.* In certain cases the solid deck arrangement also provides enormous additional buoyancy when vessels try to drive their bows under. There is a full description of these items in Book II under: Foredecks and Trampolines.

**Most often the tendency to slam the under-deck is exacerbated by two factors: overall overloading and too much weight in the bows.*

Freeboard

Cruising catamaran design since inception, has had to face the conundrum of standing headroom versus excessive freeboard. To have a properly proportioned vessel, with both sufficient bridgedeck clearance and standing headroom, a minimum of forty feet in length was required in order to retain the classic sailboat aesthetics. In earlier years, this conundrum was faced by either not even attempting standing headroom (except in the hulls) or by building a nacelle in the center. Most recently, aesthetic prerequisites have been relaxed to include vessels of less than forty feet with considerable freeboard. Excessive freeboard in a small vessel creates a host of problems, including handling the vessel around docks and piers. Low docks and floating docks present a serious access problem. Getting on and off the vessel while docking and especially clearing the dock, is difficult.

Victory 35, Primo, does what most cruisers do in Boot Key Harbor, Marathon, Fabulous Florida Keys. Despite what might be considered excessive freeboard, the fine graphics cut it down to size.

Next, is excessive windage. Multihulls with high windage share the same wind related maneuvering problems as houseboats and top-heavy motor yachts.

Then there is the problem of excessively high center of gravity. This problem requires the need for extreme attention to where tankage is placed and how and where heavy items are stored. Care must be taken to offset the designed high center of gravity. A too-high center of gravity may create excessive and undesirable pitching as well as a potential safety hazard.

Fountaine Pajot very good underdeck clearance

Freeboard out of proportion to length

- Creates windage.
- Makes access to and from docks difficult.
- Raises center of gravity.
- Increases pitching.
- Increases difficulty in docking.
- Changes the aesthetics, usually for the worse.

Catfisher 28. An example of a really slick quality vessel that just has a pitching problem because of the high center of gravity and the canoe sterns.

Bridgedeck clearance

Underdeck slamming (pounding) is the bane of many catamarans. As a generality, the catamarans with the most trampoline area slam the least. Higher bridge deck clearance generally means less under-deck slamming. However, there are certain caveats to that. First, according to information supplied by world class designer Gino Morrelli, accepted practice for offshore boats is that clearance should be at least 6 % to 7% of hull-center to hull-center beam. (Not BMAX) The wider the boat, the higher the clearance needs to be. That is why boats like Gemini, 14 foot beam, get away with minimal clearance and many Fountain Pajot boats sport high clearance and have little complaint of slamming. Some boats, mainly from South Africa, are extra wide, extra low and slam badly.

Endeavour 44 has low clearance and a flat forward facing section that will prove to be annoying in heavy seas.

Underdeck, as used in this book, is the bottom side of the bridgedeck. It is differentiated from bridgedeck because of other references to bridgedeck that might be confusing. Bridgedeck/underdeck, slamming/pounding. Slamming and pounding in many cases can be interchangeable terms. I use them differently because there are technical differences between the wave slapping or slamming a bridgedeck receives and the pounding of a monohull or powerboat.)

I consider one inch of bridgedeck clearance for each foot of center to center beam as a ballpark guideline. Therefore, a cat with a sixteen foot beam would have a minimum of 16 inches by that method or 13.44 inches as 7% of beam method.

There are many other factors involved in underdeck slamming, such as extremely fine bows with fat sterns, causing buoyancy distribution problems, forward facing flat sections of the underdeck, the slope of forward end of the underdeck, the shape of the hull and the fineness ratio of the hull. Power boats are using a "wave breaker" between the bows, some cats are using the high-speed ferry proven delta pod bow.

PDQ 32: Note good bridgedeck clearance. Only the pods for the motors show on the underdeck

The cruising trimarans have an enormous advantage in that department as the classic designs of Brown, Cross, Piver and others had exceptional underdeck clearance and almost never slam even in very severe conditions.

Certain catamarans take the opposite tack and design bridge-decks almost at the waterline. Examples would be the Endeavour Cat and its predecessor the Intercat. Interestingly, these boats have been quite successful despite the low bridgedeck. Many people, especially those who use powerboats, do not consider the noise from under-deck slamming as significant. My experience with low bridgedeck cats has shown me fewer than expected structural problems because of it.

The nacelle is another method used to increase internal headroom without raising external height. The early Prout *Snowgoose* is a classic successful example of this technique. The nacelle not only got more internal headroom, it also lowered the center of gravity and created an ideal location for the power plant and outdrive.

Piver trimaran has unassailable bridgedeck clearance

Rub-rails and toe rails

Rub-rails

Prout Snowgoose underbody showing nacelle and clean design. Note the extremely strong headstay attachment.

A rub-rail is an essential part of a cruising boat. Rub-rails are designed to protect the sides of the boat from damage when docking, undocking, rafting or otherwise coming in contact with solid objects or other vessels. The size, placement and composition of the rub-rail is determined by the size and weight of the boat. The fewer the crew, the more important the rub-rail. Some geographic areas, such as New England or the Caribbean, rub-rails are of lesser importance because of the universal use of moorings. Other major sailing areas have no moorings at all. All boats are at docks. In these areas, rub-rails are critical.

There are two schools of thought on rub-rails. One, uses a rub-rail sacrificial facing of soft, pliant material such as vinyl or rope. The concept is to absorb shock. Another school uses a metal sacrificial rub-strip, in order that the dock or piling take the wear, rather than the rub-strip. In both systems the sacrificial strip is easily and inexpensively replaced. Either system will help prevent topside damage.

Toe-rails

Toe rails are at the outer edge of the deck and get their name from their primary purpose of stopping your toes (feet attached, of course) from going overboard. Toe-rails play an important safety role, not only in preventing you from going overboard, but often your equipment, such as accidently dropped winch handles or sunglasses. Well designed toe-rails also act as gutters for water, often channeling rain water to scuppers. With rain water handled that way, it lessens the dirty black streaks so often seen on the topsides of boats without them. Some are easily converted to fresh-water catchment systems.

One excellent modern way to create both a rub-rail and a toe-rail is the use of a perforated aluminum extrusion at the hull-deck joint. These extrusions give you the opportunity to attach accessories to them as well as acting as both toe-rail and rub-rail. Snatch blocks,* fenders, child retaining netting, spinnaker poles are but a few of the items often seen attached to perforated rails.

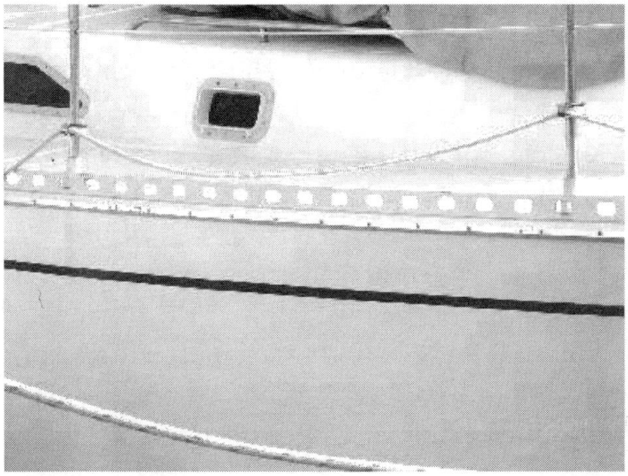

*Most perforated rails are aluminum and not strong enough to handle the direct load from a Genoa sheet in a snatch block. A way around that is to use a short piece of line going through at least two perforations and attaching the snatch block to it.

Helm. The location, layout and accessibility of the helm is critical to your safe enjoyment of your vessel. In a cruising boat, the very best helm locations have the following features:

Visibility. The helmsman should be able to see all four corners of the vessel while seated at the helm station. Forward or navigational visibility should be unencumbered by low sails, overturned dinghies, or other obstructions.

Comfort. The helmsman must be able to concentrate on steering without the distraction of discomfort, awkward seating, uncomfortable back rests or poor visibility of instruments.

Accessibility. The helm station should have easy access and egress. You should not have to climb over stumbling blocks or around coamings. Changing helmsmen should be smooth, uncomplicated. Some vessels have made excellent seating accommodations for two people to sit at the helm in a double seat.

Protection from the elements. This critical need will often determine whether or not your cruise will be an adventure or an ordeal. Little recognized is the fact that many folks simply get worn out on long passages by constant exposure to sun and glare. In some boats this is exacerbated by sitting sideways in the cockpit thus constantly looking over your shoulder. You need to be comfortably out of the sun, the rain, the wind, the cold at the times you want.

Steering. You should not have to do battle with the wheel. Steering should be light and comfortable on all points of sail.

Uncluttered. Clutter around a helm is both distracting and potentially dangerous. Some vessels have a sheet winch directly in front of the helm, with all the lines leading to it through rope clutches, ostensibly, so the helmsman can operate it while seated. The problem is, so many of these installations have no provision for the tails of sheets and other lines so they fall across the wheel, engine controls and instruments in a big tangle we refer to as spaghetti.

Ergonomic Features

Questions you should ask of such arrangements include

1. If a rope clutch is accidently released, will the line running out snag some delicate instruments?
2. Can a line being loosened or tightened snag the throttle controls?
3. If there are no serious provisions for the tails of various control lines, can they snag the wheel? Can they get underfoot creating a tangle that may require sorting out at a difficult time?

Often you find that many of the "conveniences" sold to boat owners actually create more problems than they solve. Leading all lines back to the cockpit, especially if the tails fall over the instrument panel, can be one of these instances. I believe that leading all lines back to the cockpit is a throwback to monohull racing and given a promotional life of its own by hardware manufacturers. In many catamarans, leading lines back to the cockpit simply transforms a simple one-person job to a complex two-person job.

● CAUTION: Certain vessels purport to be cruising boats yet have the helm stations out on the aft corners. One of the models actually has the helm appearing to be an afterthought stuck out on the top step of the transom *outside* of the cockpit! Even the best of these aft locations are not really suitable for true cruising comfort. Generally, these locations have very poor visibility, are completely exposed to the elements, require switching helm stations while docking or maneuvering and in some cases are actually dangerous. In addition, they do not conform with ABYC standards for visibility from the helm. (See Helms, Good, Bad and Ugly in book II, see page 35)

The location, layout and accessibility of the helm is critical to the safe enjoyment of your vessel. The true test of helm location is

1. You can see all four corners of your vessel while comfortably seated at the helm.
2. Your forward view is not impaired by sails.
3. You can access and exit the helm in a simple and safe manor.
4. You can see all the controls with little or no difficulty.
5. You can operate the vessel without creating hazards such as: sheets getting caught under or in the spokes of steering wheels, no space for coffee cups or writing materials and little ability to operate the sail, and/or engine controls that are lead back to the steering station.
6. You have no protection from the elements or no comfortable, dry environment safely inside the cockpit, protected from sun, wind, rain and following seas.

If you rate helm comfort, convenience and safety as described above on a 1-10 scale, the description would be a "10." The following features detract from that "10" all the way down to a Zero which is represented by certain boats with helm stations actually out on the transoms with not only disastrously poor visibility but extreme lack of ordinary comfort and a major safety hazard.

Subtract points accordingly for each deficiency noted

- Visibility obscured by sails.
- Unable to see all four corners of boat while seated.
- Unable to see all four corners even while standing!

- Difficult to get to helm station, harder to leave quickly.
- No comfortable seating arrangement.
- Awkward placement of controls or need to contort body to view instruments.
- No protection from the elements.
- Difficult for second person to share your view thus add to your ability to make decisions or simply enjoy the sailing.
- No appropriate area to securely accommodate a cup of coffee.

Examples: ideal "10" helm stations: *Packet Cat*, *Victory*, *Island Cat 48*, The ZERO helm station: Dufour 395 (The Rambo helm)

"9" helm stations: Seawind 1000, PDQ 32

"8" helm stations: Tobago 35,

"4 or 5" helms: many tiller steered boats

Unsatisfactory helm stations for a true cruising boat, Dufour 395, Privilege, most models. Catana, most models. Some Catana models have helm stations high enough for the helmsman to see over the cabin top increasing visibility slightly but putting him or her right in the path of the highest apparent winds with no protection whatsoever from the elements and possibly in the path of the boom. There is also the question of the massive helm stations on the decks creating sufficient windage to negate a significant part of the advertised performance factor. (Helms: good, bad, ugly. Book II, page 34) (this critique does not include sport trimarans as they do not fall into class 4 or 5 cruising boats)

Beam (BMAX)

Catamaran maximum beam is always a consideration.

Beam (BC)(centerline)

Centerline beam is measured between the functional centers of the hulls. Thus two catamarans with the same BMAX but one has Asymmetrical hulls while the other has symmetrical hulls will have different centerline beam measurements. Some designer's claim that it is the centerline beam that determines the vessels righting moment. I suggest that is not accurate. None other than legendary catamaran designer Rudy Choy, a man with many credits in the multihull field, has disproved that assumption.

Hull beam (HB)

This important ratio will be explained in greater detail in a following section. It is the ratio of the waterline length to the waterline width of one hull of a multihull.

Cockpit

The following cockpit consideration should be a major item in the choice of a cruising boat.

Cabin entry

How easy or difficult is it to go between the cockpit and the interior? Most catamarans have some type of doorway. Those with large sliding glass doors really bring the inside to the outside.

You will hear whispered concerns about the safety of this arrangement but from my historical perspective I have not seen any problems. My greatest safety concern is retaining the door in the open position to prevent accidental slides either open or closed in rough beam seas.

More important, are there any ankle twisters or head knockers on the way inside or out? Some boats, the old irrepressible Iroquois for example, have an entry where you need to duck your head and lift your foot over a coaming at the same time. Ouch! Sometimes I think they ought to issue hard-hats with those boats. Some boats have partially hidden step-down areas at the entrance. One elegant cat even had a step-down area in the galley! What an ankle-twisting surprise that was for show goers.

Cockpit seats
Often do not get much attention, however, that is the place where you will spend most of your time. Test them for comfort and convenience. Some boats have separate tables in the cockpit. Some are excellent and some only look good but have no foot room underneath nor enough room for any full size person to slide around the seat.

Hatches and storage space
Does the boat you are considering have enough convenient accessible locker space on deck to house your fenders, sails, swimming gear, cleaning equipment, dinghy motor and fuel, fish cleaning and catching gear, PFDs, bicycles, water toys, spare jugs and all the ancillary parts and equipment that go along with the fore mentioned items?

Cockpit entry
How do you get from the decks into the cockpit? Is it fraught with peril? If you have to climb over steep sided coamings, are there appropriately placed hand holds? Do you need to step down on the cockpit cushions? Certain new boats have really taken advice from users and made significant change in ergonomic design. One of these leaders is Robertson Caine in South Africa, manufacturer of the Leopard and the Moorings 3800. These boats have excellent access to the cockpit.

Unabashedly class 5, cruising, the PDQ 36 shows a combination perforated rub rail-toe rail, and traveler on the hard-top. The traveler on top takes both the traveler and the mainsheet out of the cockpit. Note the strong bracing at the stern pulpit corners and note the easy access on flat decks from the stern sections to the cockpit. The helm station has good visibility, protection from the elements and is inside the social area.

Foredecks and Trampolines

Fully decked foredecks versus trampoline foredecks, will the real answers please show up!

Parameters

Cruising catamarans can be divided into two groups, those with trampolines and those with solid foredecks. There are also a few you might consider hermaphrodite, that is showing the characteristics of both.

First, let us set up the parameters of the theory and practice behind the systems and then, what it means to you, the cruising sailor. We will confine this to cruising catamarans first, then trimarans since all the day sailing multihulls are trampoline boats generically.

To a certain extent, the ability to carry a full deck, both bow and stern, is related to maximum beam.(BMAX) The engineering data is beyond the scope of this book. Suffice it to say that once a cruising catamaran design goes beyond the length/beam ratio of: BMAX=WL/2 (waterline) you must shorten the bridgedeck proportionately. For instance, the boats now in production that have full decks are represented by; Gemini 105M, 34 x 14, Victory, 35 x 16, Packet Cat, 35 x 16, Prout *Snow Goose*, 37 x 16, and Endeavour Cat 30, 30 x 14. Older boats with solid bridgedecks are Catalac, Prout, Aristocat, Oceanic, Intercat, etc.

Next, note that the really wide boats, the ultimate racing boats, have no bridge deck at all, but are held together by aluminum or composite tubes and have full or almost full, trampolines.

The others, hastily described as the "French style" boats, all have trampolines forward and sterns that extend beyond the end of the bridgedeck. The Fountaine Pajot series of boats are the most ubiquitous in our area. Other common French boats are Privilege and Lagoon. For the sake of this discussion I will refer to boats showing the characteristics of both as hermaphrodite. They are represented by the Manta, the Privilege series, Prout 45, Island Cat, and Renaissance 35.

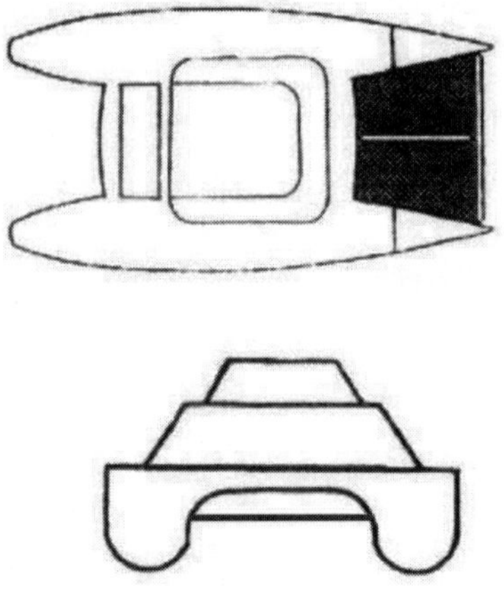

Full cabin, trampoline foredeck

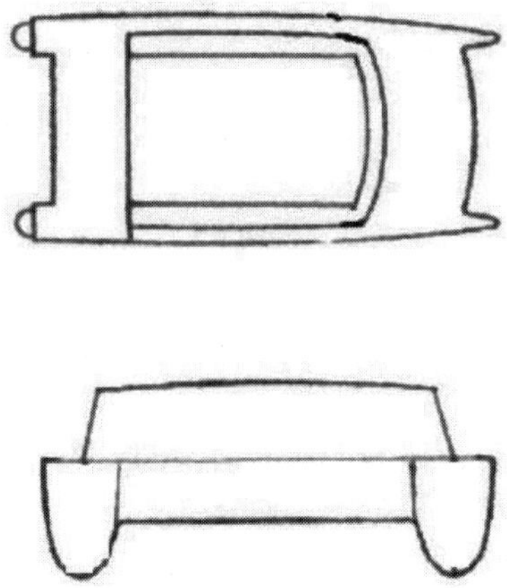

Full cabin, Full foredeck

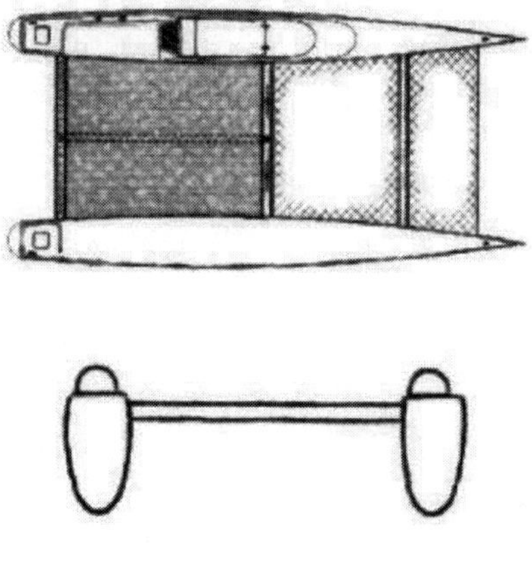

Open deck, no cabin, trampoline

Full foredeck advantages
Dryer sailing, lower maintenance, less expense, elimination of crossbeam truss construction and expensive high maintenance trampoline, deck space, sail lockers, and simple headstay attachment points are the salient advantages. The disadvantages are more weight forward, the tendency to overload the bows, and a lot of nonsensical rumors and just plain false information.

It is very important at this juncture to put a screeching halt on the totally false rumors spread by people with hidden agendas. Catamarans with solid foredecks do not blow over backwards from wind under the bridgedeck. Catamarans of this design have been manufactured since the 1950's and there has never, I repeat, never, been a documented case of this happening. Not only that, there are well documented cases of boats not burying their bows in survival storms running downwind.

If anything, from a scientific point of view, quite the opposite would be true. Wider, higher, trampoline boats would have considerably more wind resistance. Thus, as was discovered with the wide beam open wing trimarans, which did indeed capsize with a greater frequency than the older, narrow-beam solid-wing-deck models, they would be the more likely candidates for such a catastrophe. In either case, this never does happen except in some ignorant or malicious people's imagination. Again, I ask people with any doubts on this subject to read Marchaj for the full scientific and mathematical answers to these questions and to check with the complete statistical data available through MULTIHULLS Magazine. (The entire issue of capsize is a giant red herring anyway and is fully explained in Book 2 under How to sail, Beating a Dead Horse.)

Trampoline foredeck advantages
Nothing beats the thrill of sailing along, lying on the tramp, watching the water, the dolphins and feeling the breeze. There is no comparable exhilarating experience on any other type of vessel. These nets, as trampolines are often called, fill in space between hulls that otherwise would not be used. They lighten up the forward end of the boat and some claim that they increase the air flow in the jib because air is forced into it from between the hulls up through the nets. (personally, I have never been able to prove that claim)

There is also the condition of underdeck slamming. Fully decked over cats tend to slam, depending upon loading, more than their open foredeck cousins. According to Peter Spronk, a legendary Caribbean multihull designer, the further back from the bows the bridgedeck starts, the less the boat will slam and that distance is even more important than bridgedeck clearance height off the water. For those familiar with the Iroquois, 30 x 14 or the extended version, 32 x 14, it has an open trampoline foredeck but only moderate underdeck clearance yet it seldom slams, even highly loaded.

The disadvantages, however, are considerable. First, trampolines, or nets are very expensive and a high maintenance item. The sun, acid rain, air pollution and constant stressing wear out even the finest nets. Some of the boats come with trampolines made from fishnet. If you like to walk around barefoot, this is definitely not for you. If the netting has wide spacing it is a danger to walk upon especially for young children.

Foredecks and Trampolines

Trampoline boats have another unfamiliar construction, the forward crossbeam. Some forward crossbeams do not have a truss arrangement but a simple wire bridle to support the headstay across a tube between the hulls. The cross tube is only in compression and there are no lateral bending stresses because the bridle is in the same geometric plane as the headstay. This is the lightest, strongest and most satisfactory arrangement but the least used because it places the jib tack higher off the deck thus losing some sail area and the "end-plate effect." Note, this particular carryover from the speedophiles gives marginally better performance but in many cases devastates visibility. *For the cruiser, visibility is far more important that the perceived gain in performance.*

To keep the jib tack at the lowest possible level, a strengthening truss is created out of the cross beam. This construction by its very nature must be extremely well engineered since it takes tension, compression and torque stresses from different directions. On most applications, the truss is not in the same plane as the headstay thus transferring bending loads towards the bridgedeck. Compensation must be made for these loads and on many boats a central compression member that does double duty as a walkway or anchor chain support and runs from the center of the crossbeam to the center of the bridgedeck splitting the trampoline area in two. Obviously, crossbeams and trampolines are expensive.

Both worlds

The hermaphrodite foredeck shows the best of both worlds. The finest examples of this is the *Manta* and *Ocean Cat* which have a pre-stressed curved beams molded in the hulls forward and a central molded compression strut that divides the trampoline area into two. Another example is the *Renaissance* which has a circular section cut out of an otherwise solid foredeck. The *Privilege* boats and the new *Prout* 45 have a sort of central gondola that splits the foredeck area which makes them like almost trimaran-like in appearance.

Trimarans

All the modern, folding sport-trimarans have nets between the hulls. Those boats are represented by the F27, F24 F28 and F31, Dragonfly 800 and 1000, Contour 30, 35 and 40 and the Elan. The exception to this rule is the Telstar, which has high buoyancy floats and hinged solid wing-decks.

While some 1500 Telstars were produced, they are currently not in production and only available on the used market.

Catamaran lovers trampoline update, or:

What you need to know about that snazzy trampoline from the perspective of a surveyor.

The majority of production catamarans are built with trampolines as a foredeck. This represents a sea change from the earlier models where most catamarans had a solid foredeck. The topic of which system has what advantages is explored elsewhere in this book. I must convey trampoline safety findings to you that are becoming apparent as the fleet ages.

Many cruising catamaran purchasers do not realize that it cost more to build a trampoline and forward cross beam, often with a complex gull-striker than a solid deck. Therefore, it is to be expected that manufacturers will look to off-the-shelf items and standard technologies to ameliorate costs and increase production efficiency. Many have simply borrowed techniques and material used by day sailors such as Hobie, Tornado or NACRA, others are using inexpensive and marginal materials. My findings show that the materials and techniques that work for small boats may not work as well for big cruising catamarans.

How they are used

First, we must review how trampolines are used. With a day sailor, most of its sailing life is spent with one or two people sitting on the tramp. They roll around on it and seldom stand or jump on it with only bare feet. It's average storage life is under cover or rolled up and stored away.

Conversely, on a cruising cat, trampolines have constant use by people standing up and walking on it, thus you have many more people, often of great individual weight, giving it higher point loadings over considerably greater spans. Jumping on it just for fun is not restricted to children! When not being used it stays exposed to the elements, lying horizontal in the blazing sun and exposed to various sorts of air pollution especially in urban areas and near airports.

A nicely used trampoline

My observations lead me to believe there are three inherent weaknesses in the trampolines I see constructed. There are simply too many people falling though trampolines not to bring up this vital safety issue.

Rule of thumb

A trampoline should be strong enough to support twice the weight of the entire crew.

First item is geographic location. The further south you go, the more the sun affects your trampoline.

Boats stored in the tropics deteriorate at an alarming rate from the sun and because trampolines are horizontal, they are especially vulnerable.

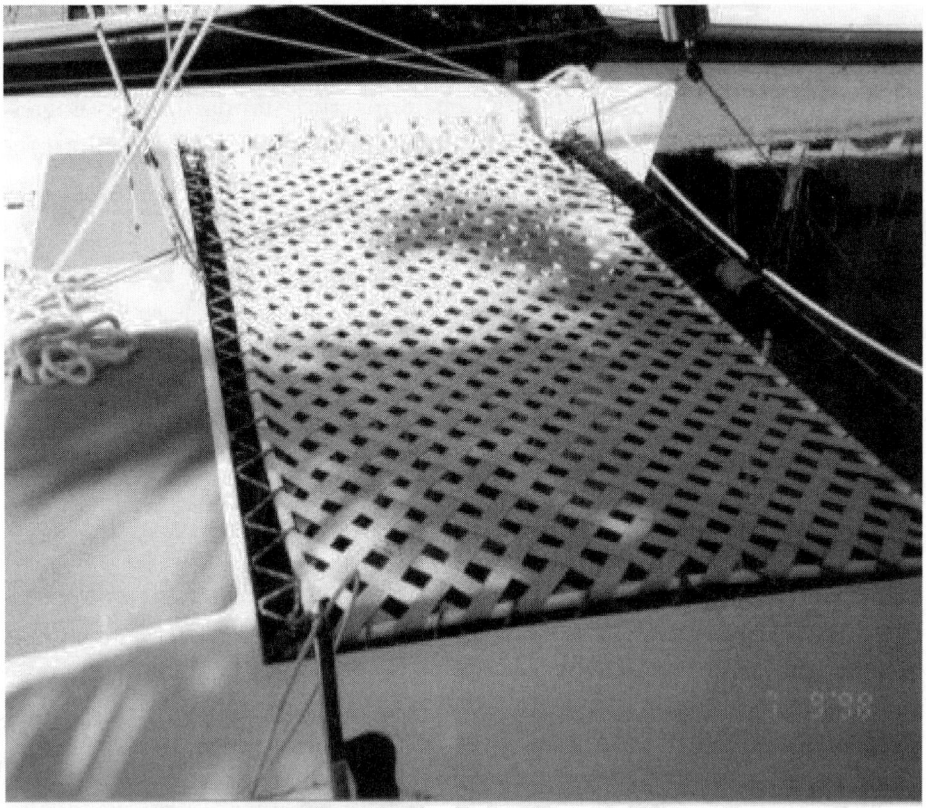

La Forza's trampoline is made of a single piece of belting, has PVC pipe end bars and is center lashed with 3/8 line.

Second, is the manner in which trampolines are connected to the vessel. Many are connected by lashings or shackles to ordinary plastic sail slides. White sail slides are not a good connector though they look snazzy and do not chafe either the metal extrusions or the lashing lines. They are simply not strong enough for the shock loads and are susceptible to sun damage.

The lashings themselves must be oversized, not for strength but for protection from sun damage. Lines smaller than quarter-inch are theoretically sufficiently strong, but strength disappears quickly under the savage tropical sun because of the line's small diameter. On the trampolines themselves, metal grommets pressed through material eventually create stress points that distort the material and pull out, leaving the lashings pulling directly on the material.

Catana trampoline past its prime

Third, and perhaps most important, are materials and methods of construction. I suggest that any trampoline that relies on stitching alone for its integrity is not going to give you satisfactory service. My observation over the past two decades is that trampolines that are constructed by folding back the materials around a reinforcement of some type that distributes the load and has several rows of stitching over a wide area of material will give reasonable service. That reinforcement can be rope or solid material such as stainless or aluminum bars or PVC pipe.

I feel that the very best trampolines have the fewest parts or (theKISS (keep it simple system) principle. From my observation, the most fool proof trampolines of all are cargo nets woven from a single piece of two-inch belting. That way, there is only one joint in the entire trampoline and that is usually joined in the center of the trampoline. After weaving and shaping, the best cargo net constructed tramps are then hot-glued and sewn with UV resistant thread at each crossing. This construction is far from the cheapest but it is not the most expensive either.

The heavier the materials, the longer they last. Two-inch wide belting materials stand up better than thin mesh. Knotted fish net was meant for catching fish. It is overly stretchy, uncomfortable to walk on barefoot and does not last very long. However, it is easy to attach strongly and cheap to replace.

If you must drag anchors, anchor chains or a dingy across your trampoline, it must be made of material rugged enough to handle it. The point loading from your feet while handling those items is enormous. For that type of usage, a cargo net construction is almost a must.

Maine Cat 30. Good installation, center promenade, groove for anchor chain, split tramp is always stronger.

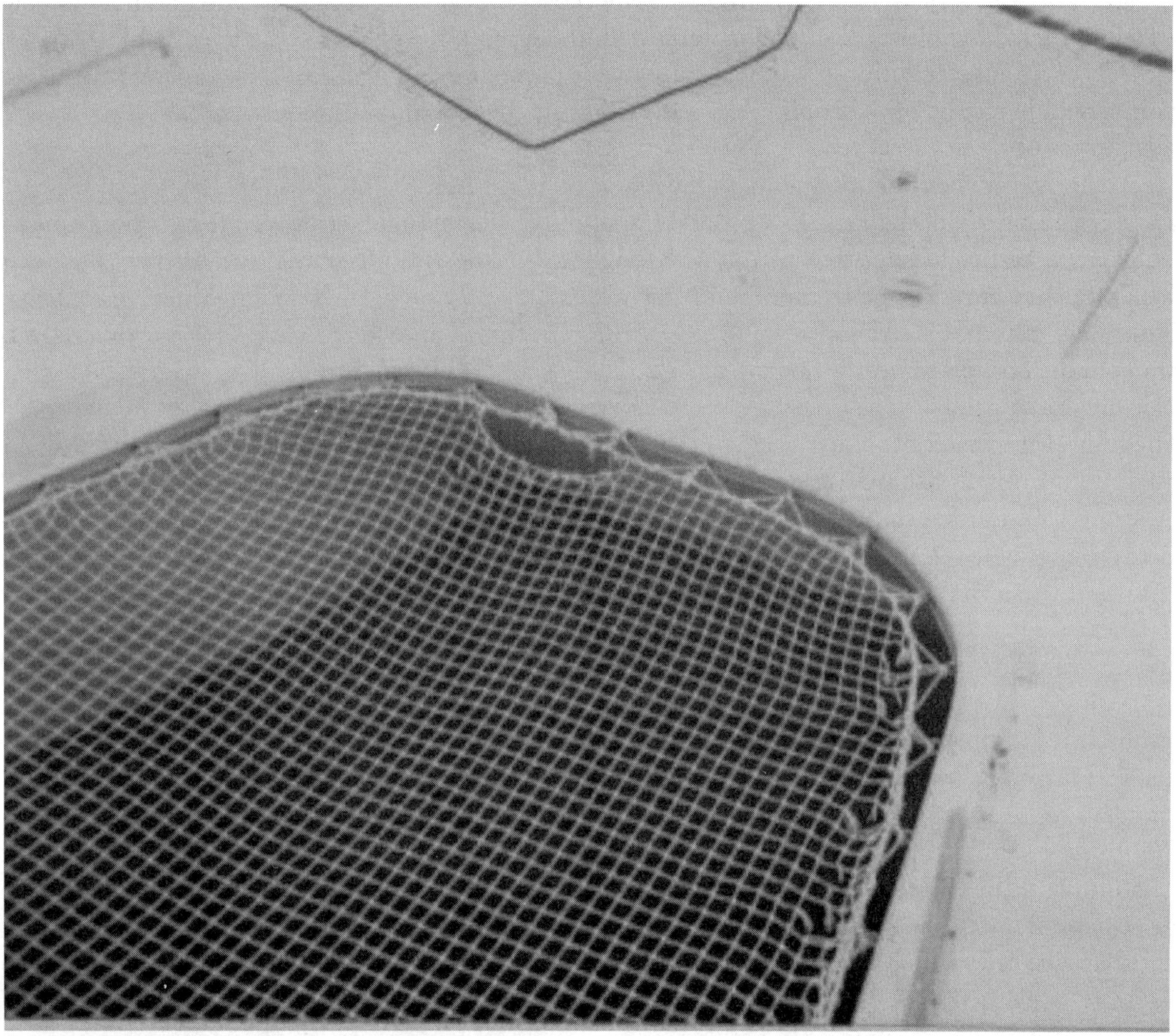

Privilege charter boat ready for a serious law suit.

I examined several failed trampolines, all with the same MO. That MO was that the owners failed to observe any failure until there was catastrophic failure. Racer-sailor-author Rob James was lost at sea when he fell through a trampoline. I have had a serious accident myself when I fell through a trampoline in Marsh Harbor, Abaco, Bahamas while approaching a dock on the last leg of a delivery from Jamaica. I have outlined a couple of procedures for all you owners and especially charterers to do the next time you board a trampoline catamaran. These procedures will require you to use a magnifying glass and get on your hands and knees, but it is worth the exertion in the end.

Connections first. Take your glass and examine the sail slides or other connecting devices be they wooden strips with drilled holes or eye-bolts, U-bolts or whatever. If bolted through the glass, check for cracking and crazing at the penetrations. Check, not only the fittings, but the lashing rope. Look for fraying in the rope. Small stuff, unless almost new, is not up to the job. Quarter-inch line is the smallest diameter you should tolerate. This has nothing to do with theoretical breaking strength but with susceptibility to both chafe and sun damage. Cored line is susceptible to internal damage that is not readily noticeable. Perhaps I am a bit old fashioned, but dear old three strand rope has the advantages of having wear quickly and easily noticeable, and is inexpensive to replace.

How is a trampoline strung? Individual lashings or shackles are the least vulnerable to catastrophic failure, whereas windings or continuous line is the most vulnerable. If you are using a continuous line, it should be secured and started in the middle of each side and spiraled in opposite directions. That way, if there is a failure, only one half the span is lost rather than the entire side of the trampoline. For instance, you are on the leading edge of the tramp, the one lashed to the crossbeam. You are changing sails or weighing anchor or some other activity that has your full attention. Suddenly a lashing snaps under the load. If it is properly lashed from the center, only one half the span will fail perhaps saving you from injury.

Sewn tramp with grommets. Check the grommets to see if they are stretching the fabric enough to unseat themselves. A little stretching is normal. If the holes are elongated enough to have space behind the grommet, that is signal of the end of useful life. Many tramps have several rows of sewing. Check this sewing with your glass to assure yourself that the threads are still intact and contiguous. A dead giveaway are the ends of stitches sticking up looking like a row of mini-prairie dogs. Again, at or close too the end of useful life. The material itself can generally be gauged by comparing the underside to the top for the difference in deterioration. Any broken strands are a dead (and I do mean dead) giveaway that you should not be cavorting on said trampoline.

Examining trampolines, we find three basic types, fishnet, woven strapping and sewn fine mesh materials. Each has its advantages.

Fishnet is painful to walk upon barefoot, uncomfortable to lie upon or attempt to sleep upon, has very poor longevity but is dirt cheap and easily replaced. There are different types, sizes and grades of fishnet. Some, like the coated knot-less variety, more suitable than just plain knotted net. In fact some knot-less coated nets actually are very good, however, they still have the disadvantages of excessive stretch and toe catching hole size.

Woven strapping is by far the strongest and has the best potential longevity. The larger the straps the greater the longevity. Two-inch Nylon or polypropylene is the most common size. They can be either lashed directly or with end-bars such as seen in photo of La Forza's trampoline or fastened in other ways.

Fine mesh materials can take many forms. There is the ultra-fine polypropylene mesh and the various grades and textures common to day sailors. The materials themselves are excellent. They are the most comfortable and have reasonable wear characteristics though they are the most expensive. With these trampolines, construction is everything. All of these trampolines rely upon stitching to maintain integrity. If they are built so that they rely solely on the stitching, they may be dangerous from the get-go. You can tell if this is the construction method by inspecting the joint between the edging material and the base trampoline material. If the trampoline material ends abruptly and the edge material begins and is separate and distinct, then the trampoline is relying only on threads and may be questionably constructed no matter how many rows of stitching there are. Sometimes this construction is difficult to observe because the edging is a folded over piece of material with the grommets through it. The fold is then slipped over the raw edge of the base material forming a sandwich with the base material in the middle and then sewn. This method looks very strong at the outset, but it still relies solely on stitching.

If they are built so that the base material is a direct part of the sewn edging, most likely they will give you good service. That infers the base material is inside the folded sandwich and any grommets or fastenings go through all the layers including the base material.

Some multihull owners voice concerns that the fine mesh materials do not rid themselves of water fast enough and will not allow a buried bow to shake itself out sufficiently fast and/or keep the bow buried excessively. Depending upon the boat, the loading and the usage there is a certain validity to this concern.

Viability. Trampolines should be protected from rough use and excess wear. Some vessels are built so the trampolines must have the boat's anchors, anchor chain or even dinghies dragged across them. Other designs carefully avoid this wear by providing central utility walkways often referred to as promenades or catwalks, for that purpose. The central walkway also provides a structural component for larger boats. I often see bights of chain, flukes from anchors and other damaging items hanging through trampolines This is detrimental to safety, ease of use and longevity.

Example of a foredeck built with only production efficiency in mind

- fishnet trampoline
- no supporting promenade
- inadequate anchor roller
- much too wide spacing of lashings
- lashings too light
- open ends of crossbeam
- headstay fitting pulling on welds

Foredecks and Trampolines

Forward Cross-beams

On trampoline boats there will be a forward cross-beam. This unit has several functions, all to do with handling compression, tension and torque loads. While handling compression between the hulls and trampoline attachment, its main function is to distribute the headstay tension load. There are two basic designs for this purpose, the simple bridle arrangement and the "gull striker" truss arrangement.

The bridle is the simplest, lightest and most effective. It's greatest disadvantage is that it raises the tack of the headsail off the deck, thus reducing sail area and the end plate effect of the deck on the sail. On the other hand, raising the sail up from the deck increases visibility from the helm. In modern, fractional-rig boats, minor differences in performance are of lesser importance and visibility and simplicity are of greater importance.

Bridle

A bridle transfers the head stay tension to the bows using the tube as its compression member. This is structurally a superior arrangement. It is also possible to keep a tighter head stay with this arrangement because the head stay itself is shorter, there are fewer connections and the loads are distributed more directly to the strong points at the bows. It also eliminates the need for a central compression structure (usually incorporated in a promenade) since the head stay and the bridle are in the same plane, eliminating the inward pull of the head stay on the resulting torque on the crossbeam.

PDQ 36 with no gull-striker just a simple bridle for headstay. Note there are no provisions for anchor handling in the center of the beam. Anchors are properly handled from the bows.

Gull- striker

The gull-striker is essentially a truss arrangement that allows the tack of the jib to be placed on the cross-beam itself, thus allowing the maximum possible length of head stay and sail luff. The upward tension created by the head stay is counterbalanced by the central compression strut. The wire which is used as the tension members, runs from the ends of the beam across the top of the central strut thus forming the classic triangular truss.

There is considerable torque and wracking (twisting, bending) in any cross-beam arrangement. Therefore design is critical. The best crossbeam designs are all bolted rather than welded. It is not good engineering practice to pull directly on a weld, especially an aluminum weld. Cross-beams will also have anchor rode handling equipment of various kinds installed. This too puts wracking and bending strains on the beam. How the beams are attached to the hulls is also of critical importance. The simplest, strongest, lightest and most easily repaired are the glassed in structural tubes that are part of the hull structure and encompass both sides and the deck. However, this construction is also the most difficult because of the need to cut and refit gel coat and to assemble the striker (if any) after the tube has been put in place. The various types that are bolted or clamped on can be prefabricated as a unit thus simplifying the production process.

Norseman 430. An ideal photo that says almost everything. Gull striker, baby stay, split trampoline, promenade, web trampoline but no chain groove. It has the chain hidden in the tube and the windlass below deck. Good for marketing but not good for cruisers.

Cross-beams that are free-standing, such as those used in a bridle arrangement, have neither gull-striker nor promenade and are usually not strong enough to anchor from the center in severe conditions. Since you are putting enormous force in the center of the tube with nothing to counteract it, it is best after anchoring, to use a bridle from the hull deck cleats. This is especially true with all-chain rodes which generate enormous shock loads in heavy weather, making it more important than ever to use a shock absorbing bridle. Some vessels built without gull strikers have their anchor handling tackle on the individual bows. (See PDQ 36 photo)

Manta 42 curved arch composite forward crossbeam. Extremely strong design that eliminates all the metal parts and puts the jib tack down on deck.

Occasionally you will see some excellent fiberglass composite cross-beams. An interesting and effective approach is the curved arch, as used on the Manta 42, Island Cat and others.

Cruising Catamaran Features

The Cruising Catamaran Seakindly Galley Advantage

Introduction

There are certain things about galley placement aboard small to moderate sized cruising monohull sailboats (26 - 45 ft.) that limit potential layout and maximum ergonomic design. Those three items are the core galley items of sink, stove and refrigerator. Their layout is virtually preordained simply because in a monohull, you must account for heeling whereas, in a cruising catamaran that is not a consideration. Certain sterling advantages of cruising catamarans do not receive their fair share of attention.

The sea kindly galley

How do I really know catamaran cruising boats are more sea kindly? Thus far in my experience, I have been on a significant number of ocean going, globe girdling catamarans and I have never seen a rubber galley apron or a galley harness installed at a stove. Next time you are checking out offshore cruising boats at boat shows or better yet, previously owned ocean greyhounds at brokers, check their galley equipment and see what they consider necessary to operate their galleys. Then, check the cruising catamarans. You will be convinced just by the dramatically different approach to the galley of the inherent catamaran sea kindliness.

On your boat show or broker tour agenda, check out the ergonomic layout of galleys on the other type of cruising vessels. You will note an amazing similarity in layout and in galley position between monohull cruising vessels without regard to size or manufacture. Note that all the galley sinks are at or close to the center of the boat. The reason for this remarkable similarity is because those boats heel over and if the sink is off center it may be below the waterline while heeled thus water could come back up through the drain.

Note the layout of gimbaled stoves with ovens. You will observe stoves are only installed parallel to the centerline length of the boat and close to the outside of the hull, no matter how inconvenient it makes it to use them. Stoves on monohulls need to be gimbaled meaning they pivot to remain vertical when the boat is heeled. (However, the counters upon which you prepare food or place hot pots are not gimbaled) They are up against the outside because there is no other realistic place for them that would not consume excessive floor space.Those items noted, you can readily understand the design and engineering constraints of the other type of cruising vessel. According to modern kitchen ergonomic theory, a kitchen (galley) should have the sink in the center between the other two major appliances. A horseshoe shape being the optimum. This is based upon universal time-distance studies showing that arrangement provides the least motion and shortest distances in preparation and cleanup among many other advantages such as separating the refrigerator from the stove. Since the stove is hot and the refrigerator cold, separation is an important item.

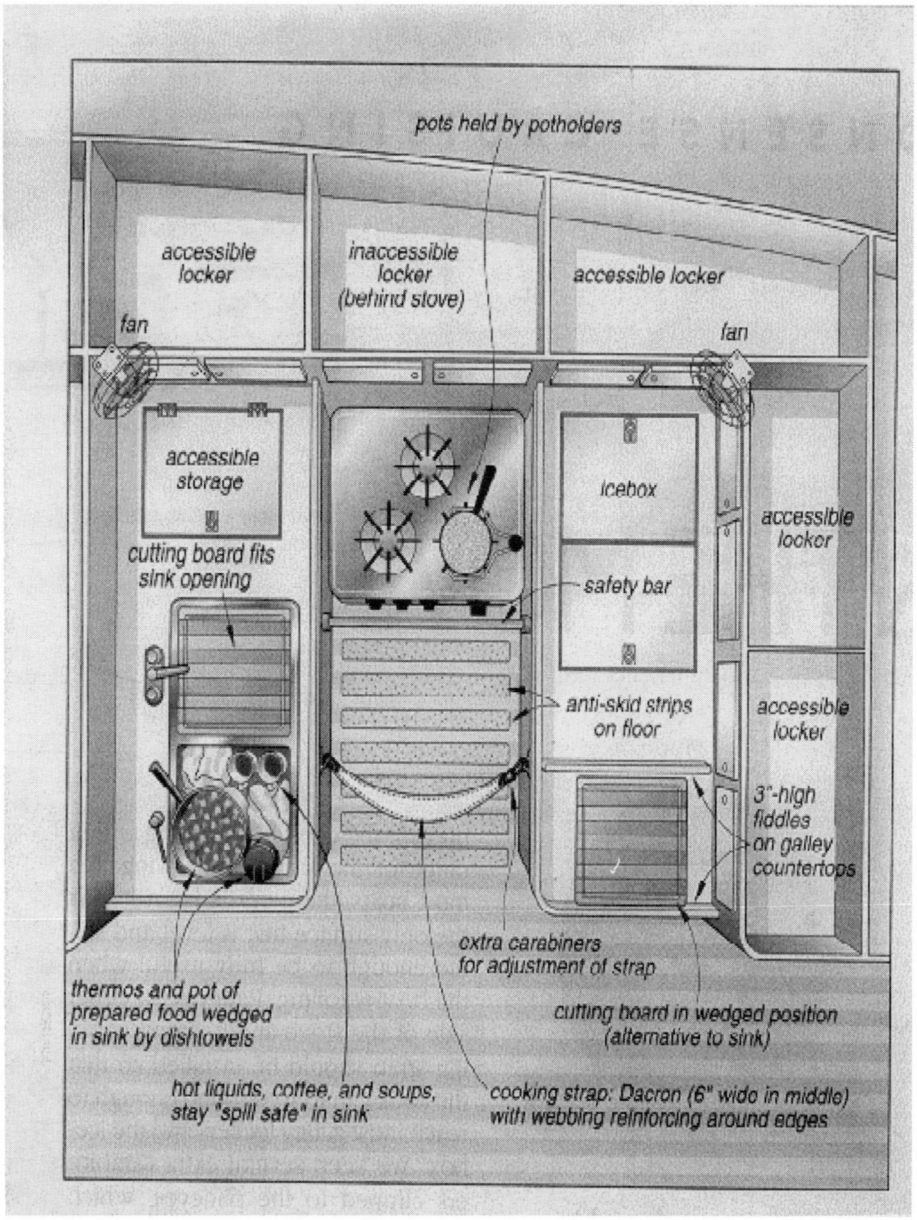

The following diagram is a galley plan from a famous cruiser's forty-seven (47) foot monohull. The diagram is reproduced from an article that appeared in a recent issue of a popular magazine. The title of the article was: *Cooking On The Move.* This particular galley is put forward as being a good, well laid out galley. As far as cruising monohulls are concerned, it is more or less average but certainly not as poorly organized as the galley in Calypso, the Practical Sailor boat. Please note that the stove in the diagram is so situated that it is adjacent to the refrigerator, obviously not a good location for several reasons one of which is heat transfer. In addition, if you need something from the fridge you must reach across the stove to get it. Not good practice in any event, but could be deadly if the stove is in use.

Note the fact that there is no counter space on the left side of the sink and that some of the cabinets marked "accessible" are really not because you have to reach across a stove in use to get into them. The design also calls for two inch fiddles around the counter tops. Much of the counter top is actually the top of the refrigerator and a storage locker. This is an undesirable practice for two important reasons.

First, because spilled liquids find their way down the seams. This is especially important for the locker next to the sink where various types of spills and cleaning water will trickle down into the locker possibly contaminating foods or ruining other stores.

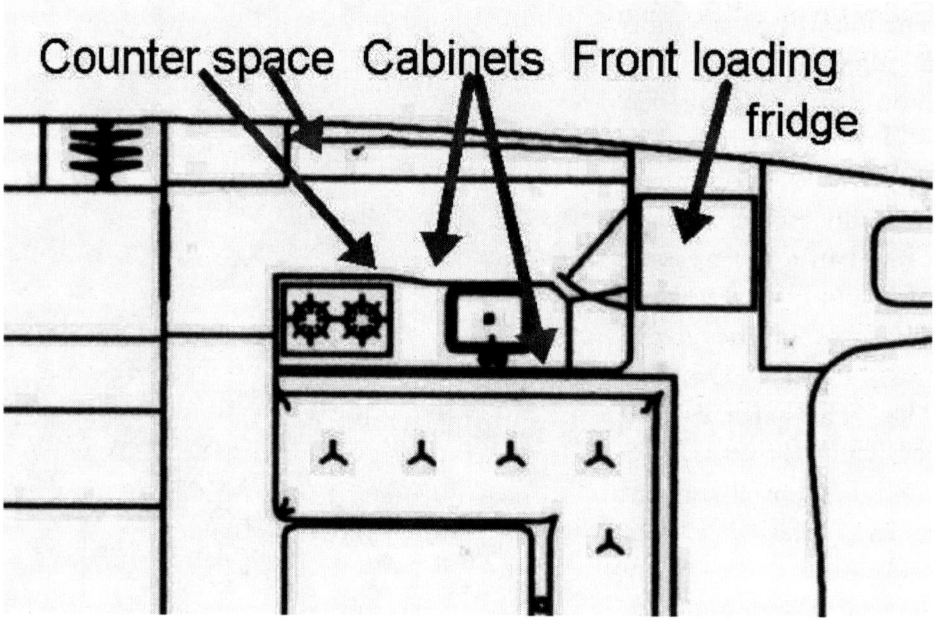

Second, because of the gross inconvenience of needing to clear the top while cooking to obtain items from the locker. Also note that this boat has the galley crash bar and harness. Note further, this is essentially a one person galley.

The Alternative

Now consider the galley down arrangement from a PDQ 32 in the next diagram. (A boat eleven feet shorter) Note that none of the deficiencies noted above apply to this galley. In addition, the front loading refrigerator is up at eye level and the visibility both to the outside and to the saloon is excellent. The refrigerator is completely separated from the stove. The sink has copious counter space on both sides and there is an opening port directly opposite the counter next to the stove.

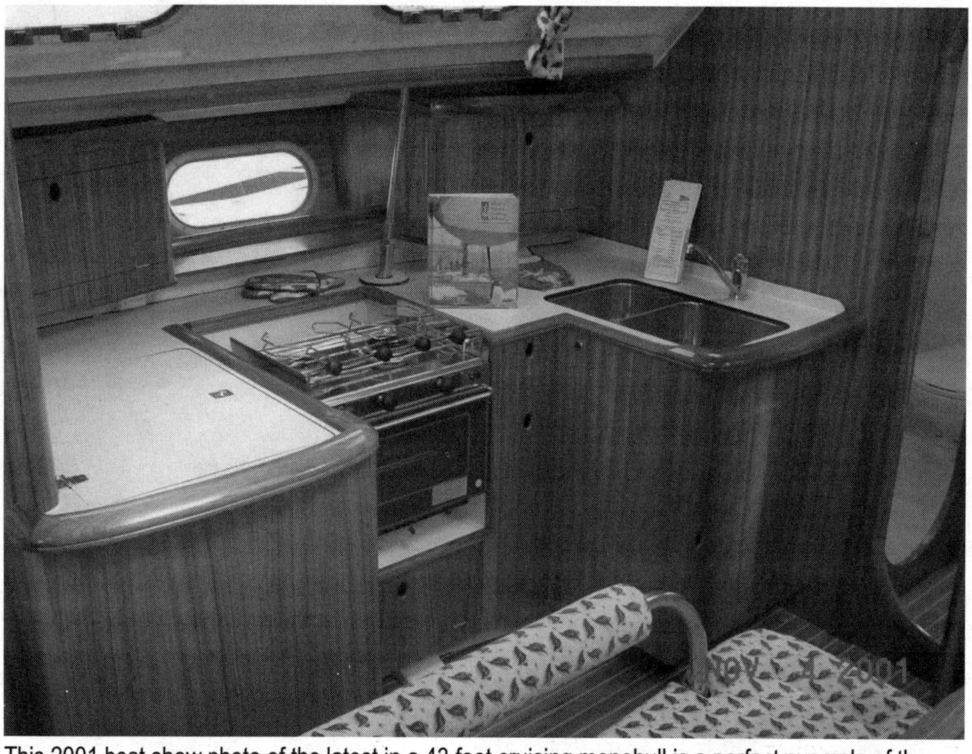

This 2001 boat show photo of the latest in a 43 foot cruising monohull is a perfect example of the major differences in the galley. Note, the top-loading refrigerator next to the stove, no counter space next to the sink, miniature cabinets above the counter and virtually no storage space under. Note. With the stove tucked under the deck, there is no combustion protection over it and the curtains represent an additional fire hazard.

Galley up or galley down?

With catamaran cruising boats there are two basic options for galley placement, up, on the bridgedeck or down, in the hull. Each has its own advantages, disadvantages and lifestyle choices.

Galley-Up

Galley up is seldom seen on smaller cats because of the need to keep the saloon area a reasonable size and the difficulty in achieving standing headroom on the bridge deck. In addition, because it is usually only two steps down into the hull, a person operating in a galley down environment can still be physically part of the social activity.

The greatest appeal of galley-up is the social aspect. It is the "country kitchen" of the sailboat world. It is an informal arrangement and a lifestyle choice. There are disadvantages. The smaller cats that have galley up that come to my mind are the Packet Cat, Victory and Tobago 35. With many smaller cats it is very difficult to achieve sufficient cabinet space in the galley up configuration. There are several different popular layouts in the galley-up style. For instance, there is the galley at the back of the main cabin, as in the Manta 38, 40, 42, the Maldives 32, some Catana models and some custom design vessels. Then there is the side of the cabin arrangement as seen in the Venezia 42 and the Kennex 420, 445 and many Fountaine Pajot models. There is even a central arrangement as seen in the Packet Cat 35. All these arrangements tend to support the theme that the galley should be an attraction in and of itself and be a central part in the entertainment or nurturing center.

Galley-Down

The basic galley-down arrangement means that the galley is configured in one of the hulls. The Gemini 105, Prout *Snowgoose*, Privilege and PDQ, Lagoon 42 are examples of this technique. They have the galley layout in a straight line or two lines back to back with an isle through the middle.

Galley-down location has many sterling qualities. First, the ability to have yards of counter space in a small vessel. In larger vessels it separates the galley activity from the other activities and could be considered a more formal lifestyle, if anything on a cruising boat could be considered formal. Some boats, like certain Privilege models, have galley down but are still reasonably open to the main saloon. Galley down has considerably less traffic through the galley area. The galley, being separate, allows more open space in the main cabin. Traffic patterns are better. Galley-down arrangements are often proportionately larger. They usually also have considerably more counter space and often have eye-level front loading refrigerators.

On certain small boats the galley-down is not only in the most ergonomic layout but a joy to use because the windows are right at eye level. The PDQ 36 is an example. In the older Catalacs for instance, the excellent opening ports are right over the galley sink. In almost every galley- down there are lots of convenient cabinets and spaces for dishes. Prout and Gemini, both galley- down boats that lose little of the social aspect of galley up but share some of the virtues of galley- down such as masking piles of dishes awaiting washing.

In most galley-down cruising catamarans, the head is in the opposite hull from the galley, thereby eliminating that cross-traffic. Galley-down is the favorite of charter operations because it keeps food preparation away from the main saloon where people congregate, many of whom were strangers prior to their charter. In many cases of charter, food preparation is not a prime concern because many or most evenings the charterers dine ashore thus leaving the galley mainly for snacks and breakfasts.

Counter space and then some

Lots of counter space mark catamaran galleys, especially those with galley- down. Boats like the *PDQ 32, PDQ 36* and the *Gemini 34* and *105M*, Prout *Snowgoose*, most *Privilege* models and many of the *Lagoon* series. The hallmark of the galley-down is exceptional amounts of good, clear galley counter space. Often, this is more than your home, more than many condos and far more than any comparable monohull, even monohulls up to twice the length. With rare exceptions, this counter space is clear of those troublesome top loading locker hatches.

Even the smallest catamaran galley is judged by a different standard than a monohull galley. An industrial kitchen designer is freed from the requirements and fixed locations of monohull galley design and designs a catamaran galley based upon the ergonomic science used in home kitchens. For instance, the fact that you have counter space on both sides of every appliance, the traffic pattern from refrigerator, range and sink, trash handling facilities and other amenities simply not possible on a monohull. Almost without exception, cruising catamaran galleys have double sinks. Double sinks are a delight. For those of you who are new to cruising, a few of the advantages, over and above the standard double sink advantages are described below.

The galley sink

First, let us discuss sink placement. A sink in a monohull must be placed on or very close to the vessel centerline. The reason is so that the sink will not back siphon when the vessel is heeled, placing the sink below the waterline. Because of this strategic layout requirement, many monohulls have either very small, shallow sinks, single sinks rather than double sinks, or sinks that are both shallow and single. As you peruse vessels in your quest for a reasonable cruising boat, visualize washing your favorite pot or pressure cooker in any sink installed in a boat you might consider buying.

The double sink delight

A double sink, especially an extra deep one, is fraught with advantages:

- Use the second sink as a dish-drainer. That way you can rinse dishes and let them drain dry without splashing water all over the counter. It allows you to have dishes draining while using the other side to prepare meals. Often, cruisers simply leave them there, safely protected for the next use.
- The second sink makes a convenient rough-weather temporary storage for items most often left on the counter top. A handy place to stow your thermos, your coffee cup or other beverage when the going may get rough.
- Allows a second person to clear and safely stack while the first person washes.
- Many people like to have one sink section filled with soapy wash water and spray rinse the draining

dishes in the other. This beats having a separate drainer up on a drainboard both for efficiency of washing and efficiency of counter space usage.

- Have room for salt-water taps, soap dispensers, spray heads and other accessories.

In rough weather your deep double sink gives you a protective place to stow liquids such as the hot soup you just removed from the stove while not compromising its use as a sink. A cutting board grooved to fit one sink with a corner cut out makes the most ideal vegetable handling area. Cut and peel, push seeds & skins through hole, rinse in the other side. Double sinks have so many advantages it is hard to list them all.

(The original Manta 38 Galley)

Note: This is not a boat show picture. We took these photos after a week's cruise. Notice the eggs, canister full of kitchen utensils and the glass coffee pot. They would stay right there no matter what the weather conditions.

Notice the ventilation, light, space and layout. Notice all the cabinet space under the counter which is clear of seams. I consider this an almost ideal galley, afloat or any place else.

Later models have improved upon this arrangement, adding more cabinets and larger, double extra deep square sink basins. The refrigerator and freezer are top loading and under the chart table opposite the galley. (I get complaints about top loading refrigerators)

The Galley Stove

Next, let's look at the stove. Stoves on cruising monohulls are most often gimbaled. A gimbaled stove must be parallel to the fore and aft axis of the vessel, so its pendulum motion mimics the angle of heel. That limits the possible installation location. No such limits exist on a cruising catamaran where stoves need not be gimbaled. Stoves most often are placed in a satisfactory ergonomic position. Most stoves are fueled by LPG, (propane) which has over the years has proven to be the best fuel for the job.

Stoves also require combustion protection. A stove up against a bulkhead or under a deck should have protective covering over any flammable trim materials. Care must be taken in the installation of curtains or other materials that could ignite. Fire aboard a vessel is considered the greatest safety hazard. Provisions must be made to protect the vessel from flame ups from ignited cooking such as bacon grease or stir-fry.

Pot clamps, so necessary with gimbaled stoves, are seldom used on catamarans, because of previously explained reasons, things do not slide around, although exterior rails or fences are seen occasionally.. Ventilation is an important item. Most catamarans use LPG stoves. See note about fuels in the chapter on safety.

Stove operating advantages

No need for gimbals on stoves. For anyone who has used a stove on gimbals, you will truly understand. Once, just once, getting a pot of hot soup dumped in your lap because of a rotating stove, and you will be a believer forever. (That is why they make protective leather galley aprons)

Corinne wrote the best selling nautical cookbook yet published; The *Galley K.I.S.S. Cookbook* : She did it while living aboard and cruising on *La Forza*, a simple boat, a K.I.S.S. boat, a *keep it simple system* boat. Based on our comparative experience aboard other boats, our catamaran was the great facilitator.

Less movement

Less tossing things about, less spilled beverage. While not perfectly still, a cruising catamaran is the closest thing to it. Setting down your coffee anyplace you happen to be, is routine. Beverage containers often adorn every level surface. It takes a really serious powerboat wake to upset one. Regular waves seldom do it.

The Galley Refrigerator

Last but far from least are refrigerators. Popular folk-lore states that top loaders are much more efficient than front loaders and that is the reason for their almost universal installation on cruising monohull sailboats. What is not said is that front loading refrigerators are extremely difficult to manage on monohulls because monohulls heel! Items slide back and forth on shelves, things become dislodged and worse, if the refrigerator is not mounted athwart ships, you cannot open the door when heeled because everything might fall out. It is difficult to manage a front loading refrigerator when heeled over. With modern efficient technology and easy access to electric energy, the difference in cooling efficiency is but a minor consideration.

Refrigeration

Much has transpired in the past five years. Most notable, are dramatic advances in refrigeration technology and installation. Gone are the old days of demanding top loading refrigerators because they were more conservative of precious cold. The new breed of front loaders are actually just as efficient as the old style top loaders. If they are used judiciously and not opened too often, and kept full, they come within reach of top loading efficiency. If cold conserving techniques such as a front air-brake are installed, they probably equal the actual energy consumption of best top loader. The basic reason is that despite the fact top loaders are inherently more efficient, they rapidly lose their efficiency edge when large amounts of stored foods must be removed to find a needed item at the bottom and you consider the time they must be kept open to clean the materials dripped down through the spaces between the refrigerator body and the top opening hatch.

The Search Fridge Position

Creighton Henry

Check out the sketch of a petite female cruiser in the most undignified positions ever assumed by the distaff set, that is "the search-fridge position." As you are most likely aware, the usual procedure with a top loading refrigerator is to :

1. Remove all the assorted stuff from the top of the fridge that in so many boats is essential counter space and find a place to temporarily stow it.
2. Open the top, if removable, search for an unencumbered location to put it, most likely balanced on the stuff you just took off of it.
3. Since whatever it is that you want is on the bottom (naturally) you need to lift out the stuff that is on top of it and find a place for that, probably balanced on the precariously balanced lid which is on top of the stuff you took off the top in the first place.
4. Now you must assume the position, that is bent double with your topsides down in the cold and your bottom sides waving in the air whilst your feet desperately look for firm footing.

The following three accommodation plans show three different approaches to galley layout. Two galley up and one galley down.

Voyage 370 galley up. Note U shape forward facing galley.

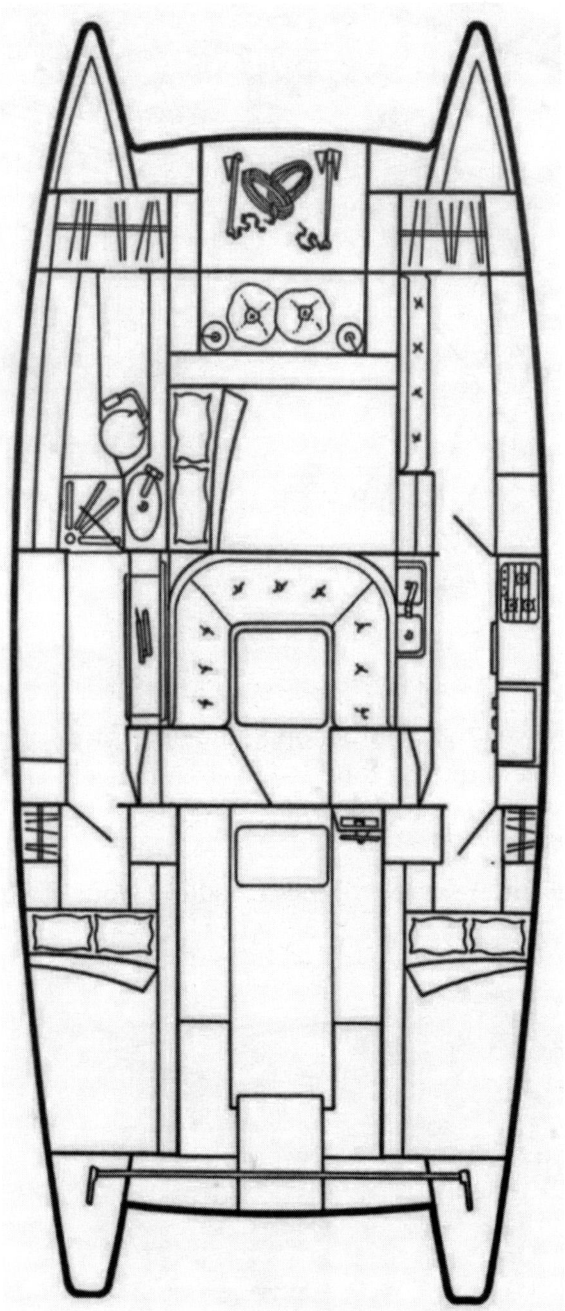

Prout 34 Galley down. Note counter space, double sink.

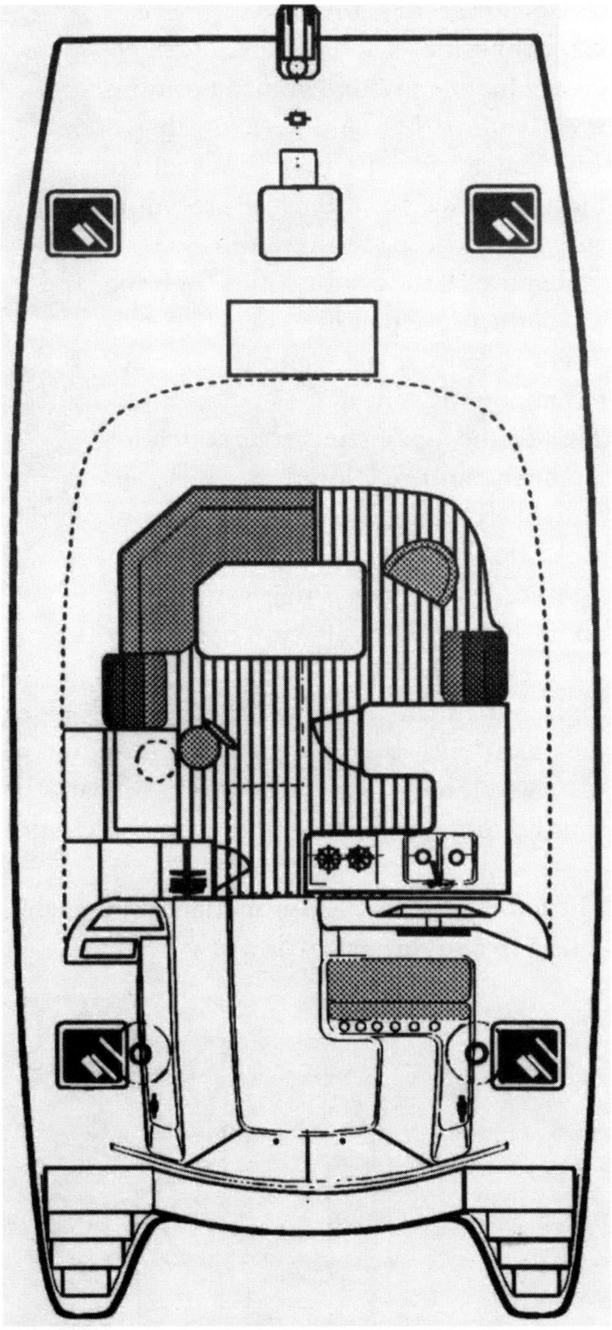

Victory 35, Galley up. Note U shape, eye-level front opening fridge opposite double sink.

Catamaran galleys have room for creative shapes

● In *La Forza's* galley, we have only foot pumps. One for fresh water and one for sea water. We have a soap dispenser. This inexpensive little aid saves having a detergent bottle as a constant companion. We use regular dinette chairs. They are light, handy, inexpensive and, best of all, comfortable.

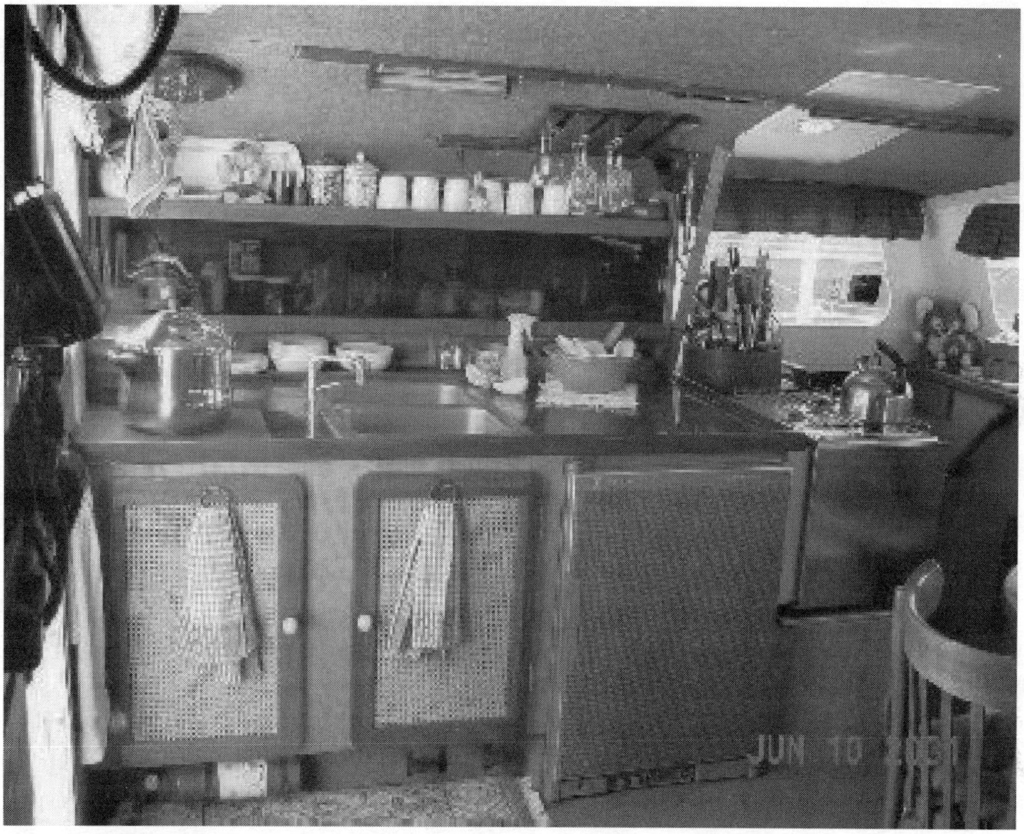

Corinne's bears look on as I snap this shot. Note the orientation of the sink, the forward facing three burner stove to better utilize the space, the LPG fridge, the sliding door spice cabinet, the crystal wine glass rack, metal combustion protection and ventilator over the stove, the dinette chair, the fire extinguisher just inside the door under the sink, the carousel for cooking implements and the great visibility.

Refrigeration options

● DC motor driven compressor, water cooled or air cooled.
● AC motor driven compressor, water cooled or air cooled.
● Engine driven compressor, water cooled or air cooled, driven either by generator or other auxiliary engine or main engine.
● AC driven option either from inverter, shore power, generator or any combination.
● Hybrid systems utilizing combinations of the options above.
● LPG.

That is a lot of options. Full discussion of these options is beyond the scope of this book, however, the various advantages and disadvantages of each system are important items for those actually using them. For instance, an engine driven system must be run every day, thus creating a definite schedule for the owner. Tough if you want to be off the boat for a few days! The most automatic systems, are the DC systems. Yet, they too have their problems because of the basic inefficiency of running low voltage motors require large amounts of electricity. If you use wind generation to supply this electricity, you must not only monitor it for excess wind, but listen to the thing running all the time which is another minor irritant. On days when there is no wind? Many users find they can generate sufficient power with solar panels but most of them are in the tropics. If you are a full time live aboard virtually always plugged into shore power, the question is moot.

LPG (propane) Refrigerators

Catamarans allow successful use of propane refrigeration. Propane refrigerators were almost universal aboard English cruising catamarans for decades. The largest producer of cruising catamarans in the United States features LPG refrigeration. I have successfully used a propane refrigerator aboard my own catamarans since 1974. The advantages are compelling. Those advantages are: low cost, light-weight, front opening doors, low fuel cost, noiseless operation, no maintenance, no moving parts, no daily chore to operate, no greed for large amounts of electricity.

Household type double door models are available, as are multi-fuel options. Having had numerous opportunities to compare refrigerators under actual conditions during deliveries, I find interestingly, all the propane refrigerators, no matter what size, seem to use the same amount of fuel. I am not sure of why this is true, but I suspect that it has to do with the ratio of cubic volume to external area. The larger the box, the lesser the exposed side area as a ratio of the two. For instance, a one cubic foot box would have six square feet of exposed side area. A two cubic foot box will have ten square feet of exposed area. As you can readily ascertain, the one cubic foot box has a ratio of 1:6. The two cubic foot box has a ratio of 1:5. an eight cubic foot box has a ratio of 1:4, a considerable drop.

Three other important considerations are: cost, weight and complexity. Ask most any cruiser. They will tell you that refrigeration is the most chronic repair and maintenance headache aboard their boat. Besides the complexity of the compressor unit itself, you have the support machinery of power generation and compressor cooling. Water cooled units are forever clogging with debris and jellyfish and air-cooled units suffer from ventilation problems, unwanted heat in cabins and excess noise.

Power generation for electric refrigeration can take the form of solar panels, wind generator, auxiliary gen-set, high output alternator on a main engine or compressors belted directly to the engine. In any event, they are expensive, heavy and/or require manual operation, fuel and cause noise and fumes.

Most propane refrigerators have front-opening doors and are just like household refrigerators. They are economical to use, burning about one pound of fuel per day, seemingly regardless of their size. The larger and more sophisticated units have a computerized brain that automatically seeks out the most economical fuel of three available: propane, 110 Volt AC or 12 Volt DC. Therefore, if you motor for a while, your refrigerator will automatically switch from propane to 12 Volt DC. When you shut down your engine and resume sailing, your refrigeration switches to propane. Arriving at your dock and plugging in, your refrigerator seeks out the line voltage automatically.

We all know that excess weight aboard a catamaran is detrimental. A six cubic foot propane refrigerator weighs about 40 pounds. Two, 20 pound aluminum tanks, both full, weigh about 60 pounds, including hoses, regulator, piping and fastenings. Total system weight, including maximum fuel: 100 pounds. Total cost, less fuel: $950.00 *retail!* (Add $250.00 cost of aluminum tanks, otherwise ordinary steel tanks go for around $20.00 each)

Compare that to any other system. Weights might be: compressor and cold plate unit, 75 pounds. Two batteries: 150 pounds. Box with lid 50 pounds. Windbugger (if feasible) 50 pounds. Generator, 50 pounds (and up). Fuel: 6 gallons @ 7 pounds per gallon, 42 pounds. Total weights guesstimate: 250-350 pounds. Total costs guestimate: $2,000.00-$4,000.00.

However, as I write this book, new and far more efficient units are being built. New companies sporting new and vastly improved technology are coming on the market. It may not be long before the LPG refrigerator loses any advantage over electric except cost.

The disadvantages of propane refrigeration are: safety (although this seems more theoretical than statistical) difficulty in keeping full cold in high ambient temperature environments (although the bigger they are, the better they work) difficulties with ventilation depending upon location and not being constructed of marine materials. Careful installation, extra insulation and careful attention to ventilation, will minimize the disadvantages. There is a little known aid to refrigeration efficiency, a circulating fan that goes inside the unit. Pick one up at any camping store.

The disadvantage of losing cold from a front opening door is more than made up by the convenience of using it. Minimize front opening door cold air loss by keeping the refrigerator chock full and employ any of several aids such as air brakes. If you have spare space in your refrigerator, fill it with plastic bottles of water. If you do not have shelves on the door, cover the front with an air brake constructed of strips of clear vinyl suspended from the top edge. (just like the air brakes you see at Home Depot or the local butcher)

There are some down-sides to propane. First and most important is that the equipment was designed and built for recreational vehicles and country homes, not the marine environment. Therefore, longevity may suffer due to heavy service and salt water exposure. Installation is critical. The open flame must be placed so that there is no possibility of any accidentally escaping propane reaching it. Since most propane refrigerator installations are on the bridge-deck, this is not difficult and drainage is directly overboard. Bilges, if any, that may collect gases, are widely separated and most have no spark creating machinery.

More important down-sides are temperature controls. In very hot climates, there is the possibility of not keeping the box sufficiently cold, especially if you add large quantities of warm foods. Again, this will vary with size, location, installation, attention to detail and extra insulation.

Rigging

The why and wherefore of the catamaran fractional rig

Note that the majority of modern multihulls sport a fractional sloop rig with only three wires holding up the mast. This is essentially the same rig that has appeared on day-sailing cats for decades and the reasons for its popularity on day-sailors are quite similar to those on cruising boats. There are modifications to this system. For instance, the Privilege has three additional stays that intersect at mid mast, but still no backstays The new Catana 475 and the PDQ have no spreader rigs.

Strength

The first question I usually get about those rigs is: are they strong enough? The simple answer to that is yes, but the real answer actually is that the system is actually stronger. It is not the number of wires holding up the mast that matters, its the quality and location and distribution of load. Remember the maxim: “if you ain't got it, it can't break!” Thus it is with rigging.

Function

The function of the three stays is to hold the mast up. Far simpler than masthead rigs that require esoteric tightening systems and have multiple spreaders, backstay adjusters, etc. The headstay, as a bar-tight attachment for a jib is not a primary rig function in a fractional rig as it is on a masthead rig. The headstay, and thus the jib, gets bar-tight when the mainsheet is tightened to its maximum.

Trimming

In a manner of speaking, the fractional rig is the automatic transmission of the rigging world. When the main sheet is slacked, the mast relaxes allowing a degree of slack thus increasing the camber of both main and jib. When the traveler is center-lined and the main tightened to go to windward, the rig is tensed, the sails flattened and the camber moved forward.

The traveler is a priority trimming tool on a multihull. The mainsheet tension sets the camber and twist of the mainsail and the traveler sets the angle of attack. Many multihulls have a self-tacking jib that does not overlap the mast. Several are using the highly sophisticated Beirg Camberspar® jib which not only is self tacking, but self-shaping with what amounts to half a wishbone boom in a pocket in the sail.

Sails

Many modern multihulls have full-batten mainsails and roller-furling jibs. The mains have considerable roach giving them a far better planform (shape) than can be achieved on a rig with permanent backstays. Approaching an ideal planform is simplified by not having to worry about interfering with backstays and/or making compromises in both shape and area to accommodate them. Full-battens allow not only better sail shape, better control of camber and but also simpler handling. Once you have sailed using a full batten main, its pretty tough to go back to all that flogging. Exactly the same can be said for jibs. (The ideal planform is the old WW II Spitfire wing)

Large, overlapping jibs are usually the bane of a cruiser's existence. If not properly handled, they flog, they droop overboard, or they flog the jib-sheets into impossible balls of spaghetti. Smallish, non-overlapping jibs ameliorate many of those problems. Self-tacking jibs eliminate them to the maximum.

Tuning

In tuning the fractional rig, the single most important thing is not to over tighten it. I have seen boats literally ripped apart by people trying to get the headstay bar-tight. Not only can it not be done, it should not be done. The stays should be just tight enough to keep the mast from jumping, and no tighter. That translates to about the tension you would have on the lower shrouds on a masthead rig. Under sail, the leeward stay should slacken but not flop.

The mast should have some rake, usually about five degrees, however, many people like considerably more. The more rake you have, the tighter the headstay will be simply because the angle between the headstay and the mast gets smaller thus increases the leverage on the stay. Most objections to mast-rake come from racers who see it as a lowering of the masthead thus reducing the effects of wind gradient and decreasing projected area. Perhaps these are mathematically correct observations, but from the point of view of the cruiser, there is far more to gain and less to lose from additional mast rake.

A good example of prudent use of mast rake are the Chesapeake Bugeyes and Skipjacks where mast rake is an art and a science. The extreme rake eliminates the need for backstays and also allows the use of the main halyards as a tackle to handle cargo. The old Hudson River sloops were also vessels that relied upon mast rake and had no backstays. (Is it true there is nothing new under the sun?)

Stepping masts

Many monohull rigs rely upon a keel stepped mast. There are two main advantages to that. First, because the mast is essentially a cantilevered beam, a lighter section can be used than would be needed on a deck stepped mast. Next is the facility of layout and interior design. Not needing a major bulkhead under the mast, the placement of which severely restricts accommodation planning, is a huge asset to designers. The catamaran fractional rig uses a considerably larger mast section than an equivalent monohull. Masts are stepped on deck in rugged base plates (steps) that are placed over either a major bulkhead or a beam construction. In many cases, the beam is an enormous box beam that is integrated into the main transverse bulkhead. Some of these constructions are truly astoundingly strong.

Mast design and construction

Visiting the boat shows and looking at the different approach to mast rigging can be mind-boggling. You can see virtually any combination of spreaders and diamonds varying from a dead simple single spreader and diamond stay, through multiple interlocking tri-directional spreaders and diamonds. Much of this has to do with the size and wall thickness of the extrusion and much has to do with the outlook and paranoia quotient of the designer and builder. Suffice it to say that the history of these rigs is excellent. You see far more mast failure on the esoteric racing craft trying to stay their noodle thin sticks than on the very rugged cruising catamarans, but then, that is as it should be.

As with any mast design, parasitic drag from rigging wire is extremely important. Therefore, it is important to ask the sellers of these vessels about the design and construction of the staying system. If they are going to a fractional rig to clean up the rig and the drag, but then add numerous spreaders, struts, jumpers etc. have they really improved anything? Would they be better off with and ordinary mast-head sloop rig? There are some cruising catamaran manufacturers using the tried and true masthead rig. It is still unparalleled for providing the most sail area for the least weight and expense. In addition, many people claim the masthead sloop is actually better downwind than the fractional sloop.

At the other end of the spectrum, we have the new and highly innovative Walker Wing Sail and the unstayed rotating Aero rig both of these show great promise. We will watch them carefully and report back to you any experiences we have with them. My experience with Aero rig thus far is encouraging.

Other rigs

Occasionally you will note some different or unusual rigs on catamarans. I will just mention and describe briefly since none are used on production boats but are mostly used by avant-garde thinkers on class 1, 2 and 3 vessels. Nevertheless, these rigs find there way to class 3 through 5 occasionally.

Aero rig

An unstayed rig with the boom as a fixed member. The fixed boom also projects forward to support the jib. The entire assembly rotates to trim the sails. This rig is the ultimate in detuning in a blow and can actually let you sail away from a dock with reasonable aplomb because of the excellent control available by being able to feather the entire rig into the wind.

Rotating wing masts

A wing shaped mast that uses the its shape to decrease the mast interference drag. They require several additional controls, put more weight aloft and in most cases they are expensive, have high maintenance costs and do not really provide any advantage for a cruiser. Almost all the small day sailing and sport boats use them, however, scaling them up to cruising boat sizes is not really economical.

Bi-plane rig

The bi-plane rig features a mast in each hull. In some cases, they are unstayed masts as in the *Freedom* or *Nonsuch* monohulls. There have been successful racing boats using this rig. The French catamaran JAZ being one of them. I am aware of at least a dozen catamarans with a bi-plane rig in existence. Walter Castle of Big Pine Key, Florida, built several between 1970 and 1980.

Una rig

A una rig is a single sail rig, often called a cat rig. The *Nonsuch* series of monohulls and the *Intercat 28* are boats featuring this rig. Despite its inherent efficiency, in larger sizes it simply too unwieldy to be a good choice for a cruiser.

A-frame mast

A rig that has the mast as an A-frame rather than a single central mast. Harken built *Amoco Procyon*, an experimental boat featuring such a mast and Hunter has a mast with some of the better features of an A-frame by having struts transfer the load to the edges of the vessel more or less turning the deck into the tension member of a truss. Essentially, the A-frame is stronger because it utilizes the outside edges of the vessel for its support which are the strongest parts of the vessel, clears away the problems of a mast penetrating the interior and a host of other advantages. So far, no major builder has fully developed this innovative concept.

(Chinese) Junk rig

The most universal of all rigs, probably on more sailing vessels than any other or perhaps even all others combined. There have been some attempts by Westerners to use it, Blondie Hasler, James Wharram, Tom Colvin to mention the most prominent of them. I can not explain why this rig which has so many advantages has not been fully developed.

Gallant rig

This rig design was an attempt to streamline and modernize the junk rig by doubling the cloth thus putting all the battens and control lines inside the sail. The luff is open to fill the sail with wind thus turning it into a virtual aircraft wing. There is tremendous unfulfilled potential here.

Lateen rig

The Sunfish rig. A rig used for millennium by Indian Ocean traders. Variations of this rig are found in native cultures worldwide.The **Crab Claw** of the South Pacific being one of them. Perhaps the most efficient rig ever produced was the old *Aqua Cat* rig which was a lateen sail hung from an "A" frame.

Sliding Gunter

This is a type of telescoping mast. Has lots of potential for those who are constricted by bridges.

Cutter

Variations and corruptions of this rig feature two headstays or a headstay and staysail. By definition, a boat with the mast stepped at 50% of the waterline length. Offshore boats favor this rig because they can drop the Genoa entirely which moves the CE aft. (center of effort)

Schooner

A two-masted rig with the main mast aft. It never really became popular with recreational vessels.

Ketch, Yawl, Gaff, Lug, definitions in glossary. Little application to modern multihulls.

Traditional masthead sloop (Bermuda rig)

Still the rig of choice for most recreational sailing vessels. Flies the most sail for the least money. Infinite variations on the theme, most production monohulls, Gemini, PDQ and other catamarans.

Catamaran Performance Factors

Seakindly motion, energy and momentum

Back in thirds school grade science, our teachers taught us the law of conservation of energy. We learned about kinetic and potential energy and momentum. Our teacher used a model airplane with a rubber band motor for demonstration. When we were winding the rubber band we were storing potential energy. When we released the propellor and the plane took off, we were converting potential energy to kinetic energy. Thus it is with the motion of our catamarans and the simple explanation of the seemingly mystical characteristic that even in fairly rough conditions things do not fly around the cabin but stay put where you left them.

The most concise description of this phenomenon is that when a monohull is heeled by a wave the force that rights it is gravity acting on the keel. This is a lesser force, thus there is energy left over since this is a zero-sum game. With a catamaran, the wave that lifts the hull is the same wave that counteracts it by lifting the other hull, thus there is no energy left over. This may sound a little hokey, but when you experience it for yourself you will fully understand.

Putting down your beverage on a flat surface, going to do something else, returning and finding your beverage in the same place you left it is one of the most desirable characteristic of the cruising catamaran. Think of the implications of that characteristic when preparing a meal.

(I remember as though it were yesterday, Bud Jenkins, the five-times navigator on the Bermuda racer, *Dragonlady*, just staring at the little restaurant table candle burning as our chart reading light on his first Atlantic passage in my 24 foot Hirondelle and muttering, "I can't believe it just stays there!")

At sea, a catamaran is the most seakindly of all vessels. Its roll and pitch will be in tune with the waves. Some people describe this as a short jerky motion but what they fail to realize is that this type of motion helps retard the onset of seasickness, does not fling things about the cabin, allows normally habituated reflexes to function (meaning you don't need to develop sea legs) and you can work on deck, raise and lower sails on the level)

Comparing cruising multihull design features

This is a complex subject that I will deal with in three separate parts First, let us discuss the shapes that make the boat go fast or slow, carry plenty of weight or be weight critical and affect the boat's performance.

Underbody shapes: keels and centerboards

Every boat, without exception, is a series of compromises. In today's industry, we see two distinct trends in multihull cruising boat underbody design. Some of these design compromises are for simplicity and cost, others are for performance and draft.

Comparing the compromises in underbody configuration, the catamarans with fixed low-aspect ratio keels and spade rudders have the following attributes

- Protection from grounding damage to main hull and rudders.
- Simplicity. No moving parts, no mechanical gear, no pivot-point to leak.
- Easier to sail, no board to trim.
- No trunk to clog with debris or to leak.
- More interior room.
- Less expensive construction.
- Simpler to design, no trunk to plan into accommodation.
- Many vessels have easier steering and better tracking.

Catamarans and sport trimarans with centerboards, daggerboards and either transom hung or kick-up rudders have the following attributes

- Very shallow draft.
- Better windward performance (deeper draft board down.)
- Better overall sailing performance (less wetted surface.)
- Ease of rudder maintenance and repair.

Beam, Length/Hull-beam ratio. (L/HB)

For the individual hulls, the ratio of the length (L) to beam is the clue to how high performance the boat will be. Hull-beam (HB) is measured at the static waterline. Thinner is faster. A high performance boat will have hull-beam ratios of around 12:1, (L/HB) often greater. Dennis Conner's 62 foot Stars & Stripes of America's Cup fame has a L/HB ratio of: 16:1. That means a 32' racing catamaran with a 12:1 L/HB ratio will have a maximum hull-beam width of 32". The higher the performance expectations, the higher the L/HB ratio. The trade-off in L/HB ratio is carrying capacity. Therefore, as boats are designed closer to cruising specifications. they will have much lower L/HB ratios. Many cruising catamarans and trimarans are in or around the 8:1 L/HB category. While this L/HB ratio still projects above average speed, it does subsume lower potential top speeds.

The common hull speed limitation factor ($1.4\sqrt{WL}$) so often expressed in monohull circles, changes as the HB ratio changes. For instance, a hull of 10:1 LB ratio will have a hull speed (K factor) factor of ($2.8\sqrt{WL}$) thus greatly increasing the potential top speed. This principle is the reason why catamarans are faster than comparable monohulls. However, they only are potentially faster given the right conditions. (See graph)

Beam overall (BMAX) Overall wide Beam versus Standard Beam.

It must be noted that the standard length to beam overall ratio (L/BMAX) for cruising catamarans over the preceding 40 years has been: Beam = ½ waterline length. (BMAX = ½ WL)This standard is still adhered to by many present catamaran manufacturers including Performance Cruising (*Gemini*), *Victory* and *Prout*. This beam ratio will be considered the standard beam for the sake of the following descriptions. It must be further noted that this beam has been proven beyond any doubt to provide appropriate initial, athwart ships, fore & aft and transverse stability. This LB ratio is the same as used by the early Polynesian colonizers and is scientifically established as providing equal stability in all directions.

Wide beam in high-performance craft

- Allows the carrying of significantly taller masts.
- Allows carrying full sail area into higher wind strengths with certain technical limitations.
- Eliminates or greatly reduces the possibility of hull-wave interference
- Increases athwart ships stability and proportionally decreases fore and aft and transverse stability.

Wide Beam in Cruising Craft

- Allows enormous amounts of interior volume and huge saloons.
- Creates huge deck areas.
- Allows creation of luxurious interior and exterior layouts.
- Allows more headroom in central bridgedeck cabins with fewer aesthetic penalties.
- Requires foredeck trampolines.

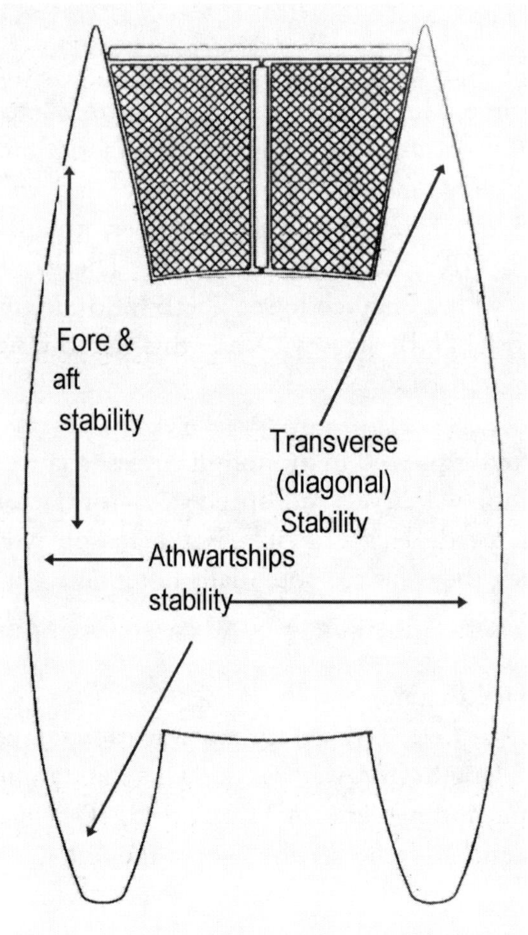

Standard Beam in Cruising Craft

- In most cases, easier to handle and maneuver.
- Allows access to marina and repair facilities through a greater length range of vessels.
- Allows full-decks, both fore and aft, thus eliminating the need for a trampoline.
- Fully decked catamarans have a drier ride, less maintenance, a simpler head-stay attachment.
- Have better utility for anchoring and mooring.
- Have a better ride.
- Can be designed and built much lower with less freeboard and lower center of gravity, resulting in less windage, less tendency to pitch and more traditional appearance in sizes forty feet and over.

Stability

Beam has a great effect upon stability. A well proportioned catamaran will have equal stability in all directions. Athwart-ships, fore and aft and transverse stability as measured in foot-pounds should be close enough to equal to not create a worry for a cruising catamaran. Excessively wide beam, coupled with very fine bows and fat sterns does indeed create a worry.

This stability diagram defines the relationships of the different directions that are important in judging catamaran stability. Change in any of these relationships changes them all. Some three-thousand years ago the Polynesians figured out the relationships and that has not changed. However, the marketing has changed and introduced some erroneous concepts as "facts."

The new thinking about wide beam

"Over square" defines excessively wide beam for catamarans and over length defines it for trimarans. "Over square" refers to the traditional concept of: beam equals one half waterline length for a catamaran (BMAX=WL/2) and beam equals 85% of length for a trimaran.

Over square beam in a catamaran changes the ratio of stability between fore and aft, beam to beam and transverse, giving an illusion of more sail carrying ability. It often greatly increases the danger of pitchpoling, and increases the danger of driving the lee bow under. This trait is more pronounced in designs featuring big wide transoms in the stern.

On a class 1, experimental/high performance, or class 2, racing, vessel, taking a chance like that may be the only way to win a race. On a cruising vessel it may just undo years of great safety records.

For a comparison, consider the stability problems certain four wheel drive vehicle are having on the highway. They *capsize* because of their short wheelbase. In other words, when the proven wheelbase (length) to track (width) changes, other factors change also. Certain All Terrain, style vehicles, are a highway menace because of instability. They compare favorably to a certain 22-foot catamaran with a 20-foot beam!

Bet On The Champ Until He Loses!

One must take into account about 50 years and tens of thousands of production catamarans. These catamarans have countless ocean crossings to their credit and they sport an incredible safety record. A *safety record far better than any other class of boat!* To deny this in favor of the unproven benefits of modern thinking is to do exactly what the monohull industry has done with excessive draft: create vessels you cannot sell because *they do not fit into the reality of cruising life.*

Occasionally, you will hear salespersons vociferously promulgating their catamarans as safer because of their wide beam. Of course, the innuendo is that the narrower cats are not as safe because of their narrow beam. Many of these wide beam boats have not sold! They are comparable to the 28-foot monohull with a 5.5 foot fin keel that has not sold. However, in both cases, the salesman's prattle has damaged the potential sale for *any sailboat.*

In addition, a cat with over square beam is often more difficult to tack than a catamaran with conventional beam. Over square beam increases the need for bridgedeck height over the water. Extra height is self-defeating of both the aesthetic and sailing requirements. Over square beam exacts a high price for the extra amenities it produces.

Excessively wide beam creates a host of problems, the aggregate of which are worse than the simple problem the "wide beam" was trying to cure.

On the older, "narrow beam" cats, the formula was: beam = one half the waterline length (BMAX=LWL/2). Bridgedeck clearance was approximately one inch per foot of beam. In other words, boats like Catalac, which have 14 ft. beam, had about 14 inches of clearance. As you get wider, the relationship changes on a curving scale. When you get to boats like mine, which are built to the formula: Beam = 1/2 overall length, (BMAX=LOA/2), you need to have a clearance factor of 1.25 inches per foot of beam. When you begin to get over square, you begin to get clearances like two inches per foot of beam.

As the beam gets wider and the bridgedeck height increases, the center of gravity also goes higher. We all know that a lower center of gravity is more desirable not only for safety reasons but for comfort and ride, which also affect performance. A higher center of gravity disturbs the pitching equilibrium as well as the rolling equilibrium.

Wide beam may have only minor or inconsequential effects on the center of gravity on class 1 and class 2 racing boats because deck structures seldom exist, Clearly, you can see the devastating effects of wide beam on any boat with a bridgedeck cabin. If you build the bridge deck the correct height, and try to maintain a proportionally sized cabin, you have little usable cabin space. You cannot expect anything approaching standing headroom, until you get over 40 feet in length in a wide beam boat. If you lower the bridge deck, you make the boat pound unmercifully. There have been catamarans built to both extremes. Some, have a reputation for excessive pounding, (Norseman) some appear like a huge looming colossus due to their extraordinary freeboard.(Catana)

An example of excessive pitching was the old *Catfisher*, 28. It pitched miserably partially because of the high center of gravity . Redesigned to include four feet to the sterns, it corrected the problem. (Note, the canoe sterns and heavy deckhouse also contribute to the pitching problem)

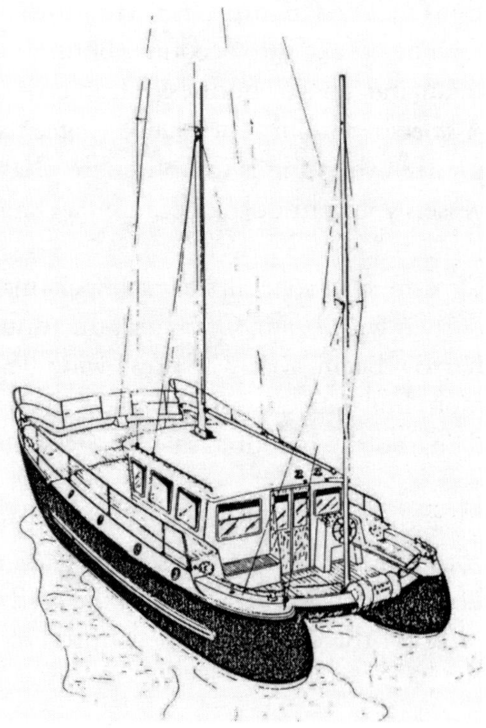

Remaining are the several problems of vessel utility. Note the helicopter picture of *La Forza* in the canal behind my house. *La Forza* is 16 feet wide by 32 feet long. Do you think my neighbors would let me get away with anything wider? The questions of: "where do you keep it?" and "where can I go with it?" are the cogent, important questions of vessel utility.

No *illusion of performance* can offset *lack of utility*.

Beam is the limiting factor for multihull sailboat utility.
Draft is the limiting factor for monohull sailboats utility, designers must take these limits into account.

> *There is a positive argument in favor of wider beam: hull wave interference. However, this is insignificant for cruising boats with expected lower top speeds.*

Thus far, I have not had any problem at all getting into marina slips with *La Forza*'s 16 foot beam. Sometimes the slips have been a tight squeeze, but I have always managed. The reason is simple. A 45-foot Hatteras is 16 feet wide and most marinas designs accommodate them. Being wider by just a few inches begins to seriously limit using marina slips as well as travel lifts.

The world is more crowded and more affluent. The face slips once the darlings of the wide body set, have disappeared from general usage. The competition for them is increasing. Most likely, you will find a huge motor yacht at the face slip you once took for granted. Alas, that is cruising reality circa 2002.

I believe that lack of customer acceptance of the wide beam is a causal factor in recent fledgling multihull production failures. The question I pose to designers of these wide beam extravaganzas is: "*To whom are you going to sell them?*"

Decks

Then there is the decking requirement. Once beam exceeds one half the waterline length, you need to fill the forward and possibly aft areas between the hulls with netting. (The mathematics of this process are beyond the scope of this book). On the traditional boat, say the Prout line or a Gemini, the area between the hulls is solid deck. This has enormous advantages for cruising. Nets allow the entry of water, require high maintenance, (deadly dangerous if not properly maintained) and are not as comfortable to work upon as solid decks. (One of the things I miss most from my twenty-four foot Hirondelle is the incredible foredeck). On the plus side, however, is the fact many people enjoy sleeping or lounging in the nets.

Ride

There is yet another consideration and that is ride. The traditional beam boats have a more comfortable motion. As the beam gets wider in relation to length, the ride gets choppier and more abrupt. A Gemini, 34 x 14 has a better ride than *La Forza*, 32 x 16, and a MacGregor 36, which is 36 x 18.

Waterplane shape

Waterplane shape and its distribution of buoyancy and displacement also has a great effect upon stability. Increasing hull beam increases reserve buoyancy; often referred to in inches of immersion. The distribution of buoyancy in the hulls is a major factor in stability, pitching moment and overall ride. So many designers sacrifice buoyancy forward for knife-like bows. I always ask, "if fine entry is so vital, how come the monohulls, which certainly have less fine entry than cats, go to windward so well?" I never get a reasonable answer.
In this diagram are three sample waterplane shapes, all quite different yet with the same basic length/beam relationships. You can see the distribution of buoyancy and the displacement.

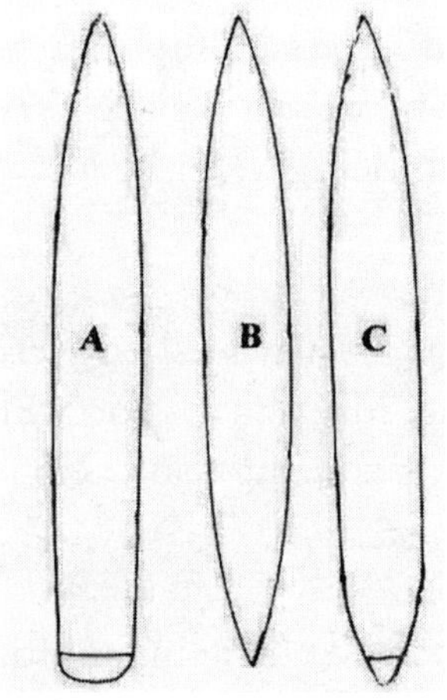

A. The most total buoyancy of the three and considerable buoyancy aft.

B. Classic double ended design, has the least buoyancy and has it centered.

C. Closest to modern hull shape has buoyancy in the mid to aft section of the vessel.

Displacement

There are several aspects to displacement. The first, is designed displacement. Designed displacement is the estimated weight of the vessel with the water and fuel tank half-full and all necessary gear installed. Installed gear would include sails, rigging.

The total displacement is the weight of the vessel as it is sailing, including the crew, stores, dingy, etc. Total displacement is the important figure. In any boat, just like any other vessel or vehicle there are total displacement figures that you ought not exceed. You do not expect your compact car to haul a palette of bricks and like your car, you should not expect your boat, any boat, to exceed appropriate displacement figures.

The difference between designed displacement and total displacement is the carrying capacity of the vessel. Designers and builders vary on how they arrive at that figure, nevertheless, that is a number you need to know. You need that number on any boat, power or sail, monohull or multihull. That is the number that dictates how much stuff you can put on board. Note the placards required on trucks: Gross-Tare-Net. Gross being the entire vehicle weight, Net being the unladen or designed weight and Tare being the allowable load. Your boat has the same limitations and restrictions, though there are no highway patrol scales nor potential summonses if you exceed the limits. Exceeding weight limits does the same thing to both vessels and land vehicles. It make them poor performers and possibly dangerous.

It is important for you to obtain from the manufacturer of your vessel a realistic load capacity figure. Note, on large commercial vessels you see clearly marked, bow and stern, what are called "Plimsol lines." These notations are there by law to delineate the load carried. While yachts are not required to have such lines, nevertheless, they are still subject to the same laws of nature and should only be loaded within proper engineering parameters.

Hull fineness ratio (HFR)
Since hull fineness ratio is a main determinant of overall potential speed, many designers will agree that the hull fineness ratio (hull length/hull beam) is the determinant of whether a boat is a cruising boat or a racing boat. The skinnier the hull the faster it might go but the less it can carry. Hull fineness ratios of 16:1 would definitely be racing boats. A ratio of 10:1 is in the "could go either way" zone, and 8:1 would definitely be cruising. (Monohulls in cruising class usually run in the 4:1 range) There are always exceptions based upon designer intent. There are other numbers and ratios that enter into this equation. Some will be discussed in other sections of this book. Others are left to the theoretical mathematicians among us and are mentioned in the glossary.

Multihull performance is determined the same way as monohull performance
1. Horsepower (sail area) to weight ratio.
2. Waterline length.
3. Stability. (righting moment)
4. Wetted surface.
5. *The Critical difference*; length / beam ratio. (HFR)

(As with all explanations in this book, the scientific names, theories and equations have been deliberately omitted for the sake of brevity. Technical explanations and mathematical proofs for these assumptions can be found in several good references in the Bibliography)

The effects of hull-length to hull-beam ratio. (HFR)
The "K" Factor concept, courtesy Bill Roberts

The comparison of the 27 foot sailboats with beams ranging from 16 inches to 72 inches, considering their respective displacements, produce the listed K factors. The K factor actually being derived from the hull/beam ratio. It is a refinement of the speed/length ratio found in text books.

LWL	BMAX	DISP	Fineness Ratio	MAX Speed	K Factor
27'	16"	650	20	27 Kn	**5.2**
27'	23"	1300	14.3	20 Kn	**3.8**
27'	28"	1950	11.7	17 Kn	**3.3**
27'	32"	2600	10.2	15 Kn	**2.8**
27'	72"	7500	4.5	7.3 Kn	**1.4**

Listed below are a few examples of how the K factor acts in practical terms.

60'0"	18'3"	71,000	3.90	10.8 Kn (Alden 72)
53'6"	16'3"	50,200	3.60	10.7 Kn (Bluewater 60)
42.0"	15'7"	43,700	3.30	9.07 Kn (Bristol 51.1)
34'10"	13'6"	18,000	2.59	8.2 Kn (Catalina 400)
28'7"	11'6"	13,500	2.50	7.4 Kn (Island Packet 32)

$$VMAX = K\sqrt{LWL}$$

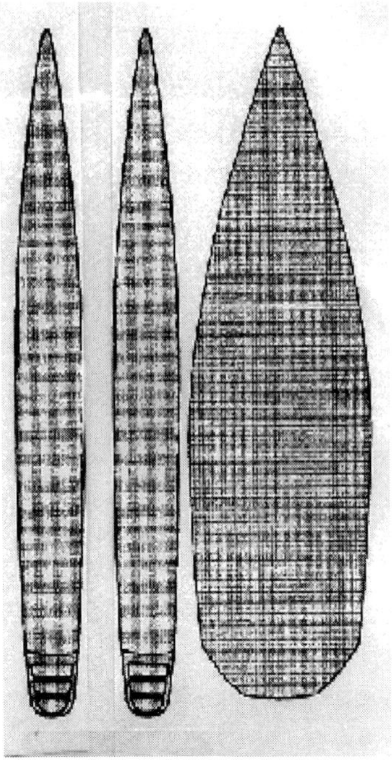

This diagram of the waterplane (footprint) of two, thirty-two foot vessels, one a popular catamaran and the other a popular monohull should graphically describe just why skinny hulls go faster than fat ones.

The monohull has the typical K factor of 1.4 because of its 32 foot length and its 8.5 foot beam.

The catamaran has a K factor of 2.8 based upon its 32 foot length and its 3.5 foot beam.

Looking at this concept gives you the general idea of why catamarans are potentially faster than monohulls. In practice, displacement and the other factors mentions play an increasing important role in actual speeds attained under sail.

Comparing Catamarans to Trimarans

Occasionally, I stand accused of bias against trimarans. I don't think that is either fair or accurate. It is true that I believe for long-term cruising or adventuring under sail the catamaran has all the advantages over the trimaran. Nevertheless, there are some areas in which the trimaran clearly excels.
First let us look at some generalize arguments.

Catamarans
- Have greater weight carrying ability.
- The majority fit into marina slips.
- May be lifted by conventional travel lifts.
- Generally have more interior room.
- They are structurally simpler.
- Have fully enclosed accessible decks and all other space covered by trampoline or nets and enclosed by lifelines.
- Drier to sail.
- More generally accepted by the public.
- Safer in serious offshore weather.

Trimarans
- They are faster in light air.
- They are easier to judge when over pressed.
- Tack more easily and securely (always exceptions.)
- Have most of the favorable sailing characteristics of monohulls, with the fewer of the disadvantages usually associated with catamarans, such as difficulty in tacking or headstay tension.
- They are much more successful as trailer/sailors and pocket cruisers.

Some Empirical Realities About Using Trimarans for Cruising

See-Saw Syndrome
See-saw syndrome, occasionally referred to as elevator syndrome, is that process by which the trimaran toggles back and forth between amas (outer hulls) while at dockside or at anchor. In the more modern, class 1 and class 2 trimarans, one hull is always out of the water while at rest. In some designs both outer hulls are just kissing the water at rest.

In either event, the slightest shifting of weight, either by occupants inside the vessel, people entering the vessel or boat wakes changes the balance of the boat. Just as people would react on a see-saw, the vessels suddenly bolts from one at rest position to the opposite at rest position. The greater the dihedral, (the angle between horizontal and actual) the greater the acceleration and concomitant shuddering stop it makes when the opposite hull hits the water.

at rest port **moving** **at rest starboard**

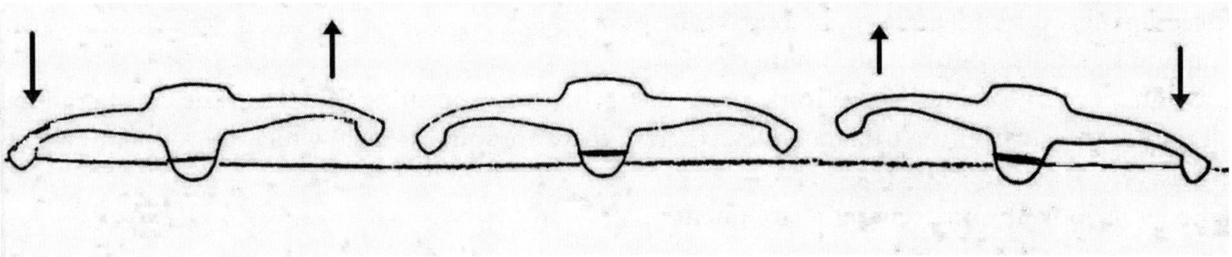

The results of this syndrome create a user unfriendly atmosphere. First, the vessel is *never level at rest!* Try to do any gourmet cooking on that one! Second, the vessel is subject to the slightest movements of both wind and wave. Strong gusts of wind, powerboat wakes, people shifting positions within the boat, will set off the ama elevator ride. Third, it is difficult to raft up. See-saw up and down movement precludes keeping fenders in place. The same problem virtually prevents you from docking for provisions. Every time you walk across to the center to put down the provisions, the ama rides up. I have yet to see a modern, open wing trimaran with dihedral in the amas, which has been cruising, that has remained undamaged from this syndrome. Couple this problem with lack of lifelines and other handholds and you not only have a destructive problem, but a bodily dangerous one as well.
The older or traditional designs are not subject to this syndrome, nor are several of the latest offerings.

Access to the interior
In the open wing trimaran, getting provisions on board is always a tedious, difficult task, far out of proportion from other vessels, including monohulls. Groceries, ice and everything else, must be carried across the nets. Many closed wing trimarans have really nifty little pens under the wing deck. These spaces just fit a dinghy. A pen of this type, alleviates the problem of boarding, both people and provisions. (Aboard *La Forza*, we call that our garage)

In the open wing versions, most users face a conventional sliding main hatch and companionway ladder arrangement. The same system used on a monohull. That means handling washboards, going down backwards and all the other little petty annoyances of the style.

Limited visibility
From down below, visibility out of portholes or windows may be just like or worse than a monohull. In many trimarans, looking out the ports gives you a good view of the inside of the amas. The ability to have a panoramic view is a very important feature of the best of cruising boats.

Wet sailing
In open wing trimarans, any spray at all comes straight back to the cockpit. This is a serious problem, even on very large, high freeboard boats. Note the popular cover picture on *The Cruising Multihull* of Chris White's *Juniper* and you will see special spray guards installed on the flanks of his boat. That boat is 54 feet long, just think what a mere 40 footer must be like?

Narrow cockpit

Many trimarans have small, monohull like cockpits where you are consigned forever to sit sideways. This is a type of cockpit where your knees must compete with the tiller for room. On some models, you must climb completely out of the cockpit, onto the main deck in order reach the main hatch and get below.

Limited usable deck space

This is especially true of the open wing variety. While the closed wing trimarans have decks approaching the size of tennis courts, open wing varieties have practically no usable deck space. Strict limits exist as to what you can store in or on nets. Even an upside down dinghy, lashed onto the nets can take a terrific beating or become a dangerous hammer against the nets in serious weather.

Engine installations

Outboards, except on the smallest models, vary from difficult to impossible to properly install on a trimaran. There are several systems in use, with varying levels of success. Perhaps the *Condor* with its hydraulic raising and lowering mechanism is the most successful in a larger boat, to date.

Charleyhorse, a beautiful custom built trimaran is a virtual mini cruise ship

Trimarans with inboards, have all the same disadvantages as inboards in monohulls: through hull fittings, smell, noise, mess, danger of explosion in the bilge and latent heat. Latent heat is something hardly ever addressed.

After you have been motoring all day with an inboard, the engine and everything in the engine room area is hot. When you anchor for the night, if your bunks are over the engine, you sleep on and live with all this extra heat and smell! Phew!

Summary

The design of the above seven considerations is to be descriptive of actual user processes. They were not designed to be any type of condemnation of trimarans. Not at all, as long as you are dealing with class 1 and class 2 vessels. The seven thoughts are strictly meant to very graphically point out the extent of the compromises you will need to make in order to use a vessel of that design as a *cruising* (class 5) boat! *Cruising*, being understood as defined in Chapter 1. I should probably add that these characteristics are seldom if ever directly addressed in print. The reason is the previous nature of the industry and the clientele it served. That has now dramatically changed. Sailing is no longer an industry dominated by *speedophiles*. Speedophiles view such things as trivia not worthy of discussion. Sailing is an industry dominated by the majority sailors, who mostly buy production boats and to whom such things are of primary importance.

For comfort and convenience, the catamaran configuration is clearly superior to any other. I think that is a fair summary. That configuration describes the user friendly and cruiser friendly requirements. At this moment in time, I believe that to be correct, yet change is inevitable.

I must confess that trimarans represent a certain perplexing dilemma for me since I believe them to be a specific class of *monohull* as well as a multihull. If you carefully examine trimarans, you find a single hulled boat with a positive form of ballast instead of the monohull negative ballast, either internal or external. Ballast being defined as the device, counterpoise, or system that holds the boat upright. The stability system. All other problems, characteristics and attributes remain much the same as with other monohulls. Many will not agree with this concept. Yet I suggest to you who are in search of more knowledge, at least play with the idea. The concept answers many questions for you and gives you a perspective from which to start when searching for techniques about handling and sailing trimarans.

This book is a primer for people newly introduced to cruising catamarans and trimarans (multihulls) or to sailing itself. It is an introductory manual and intended to critique designers on Class 3, Class 4, Class 5 and Class 6 designs. Class 1 and 2 sailboats will be critiqued another time in another volume. However, certain traits and characteristics trickle down into the designs of Class 3, Class 4, and Class 5 vessels. These traits affect newcomers or the less sophisticated. I feel it obligatory upon the spirit of this book to point them out.

Over the years, I have been a constant critic of the trend toward deeper and deeper keels on monohull sailboats. They are the antithesis of the needs of cruisers. I think comparative sales have proven me correct. I am also a critic of what I consider user unfriendly or unsafe trends in multihulls, both sail and power. The most important of these trends is extra wide beam.

Trimarans have proven their worth on many race courses. Except in areas of the world where you have dependably strong winds, the lower wetted surface, heeled, trimaran dominates. Generally, they are better in light air, and enjoy essentially easier or better handling. (As a generality only) Since most races are run in light air, they have a distinct advantage.

Catamarans lend themselves to a different level of extremism. For instance, "C" Class, Formula 40, ProSail or our infamous Stars & Stripes of The Americas' Cup, are radical. These boats are more weatherly than any, but obviously in a different category and at this moment in history, bear little relationship to cruising boats.

Cruising In Catamarans

Of course, it is possible to cruise on any boat. People do it all the time. The page 2 photo is a picture of people on a Hobie Cat, cruising up the Intra-Coastal waterway. However, that does not make a Hobie a cruising boat any more than the guys from Route 66 made a Chevy Corvette® a station wagon. Of course you can cruise on anything you like, as long as you are willing to pay the price. However, just because you are cruising on it, it does not, ipso facto, transform into a cruising *boat.*

In a certain respect, we are beating a dead horse. There are so few cruising trimarans around that they are insignificant. On the other hand, if you take a trimaran and seriously overload it, you stand the chance of turning it into a dangerous dog. Overloading is a reality in various types of cruising.

A modern open-wing trimaran cannot handle serious overloading, first, and most important, it degrades the rotational momentum of inertia. In other words, you assume more characteristics of a ballasted monohull. A catamaran does not suffer from this particular problem. Weight distribution favors the opposite outsides of the boat, increasing the moment of inertia by the amount of added weight. In other words an overloaded catamaran is less likely to capsize, an overloaded trimaran is more likely to capsize. (The best technical explanation of this phenomenon is contained in *The Cruising Multihull*, by Chris White)

The second serious problem with overloading a trimaran has to do with immersion. You degrade the bridge deck clearance to the point of danger by overloading. With depressed wing decks you stand the chance of swamping or submerging the leeward float. Obviously, not a good condition.

On a catamaran, while overloading will turn most any cat into a dog, it has not proven to increase the danger. On the contrary, it apparently increases initial stability proportional to the weight. However, this must not be construed as an endorsement of overloading, it is not. This is especially true if you have one of the wide beam extravaganzas like so many of the South African cats. A cat with over square beam, a full bridgedeck cabin and minimal bridgedeck clearance can be unmanageable when overloaded.

On a trimaran, excessive beam greatly increases the lever arm of the parasitic hull. Not only does this increase the centrifugal force and capsizing moment of inertia, but it also increases the windage. Consider the windward ama as a pendulum. The longer the fulcrum the greater the force. (For a full mathematical analysis see Marchaj, The *Aero-Hydronamics of Sailing*). Add to that force the additional *windage* from the wind gradient. This gradient increases rapidly the higher off the water the ama rises. You can understand there is a major rotational force that the leeward ama counteracts. The leeward ama has *significantly reduced* buoyancy and size, which was the reason for going so wide in the first place!

Couple that to the fact that the leeward ama is in the counter rotating portion of the wave and you can readily understand what Alain Colas meant when he said; "To broach is to die."

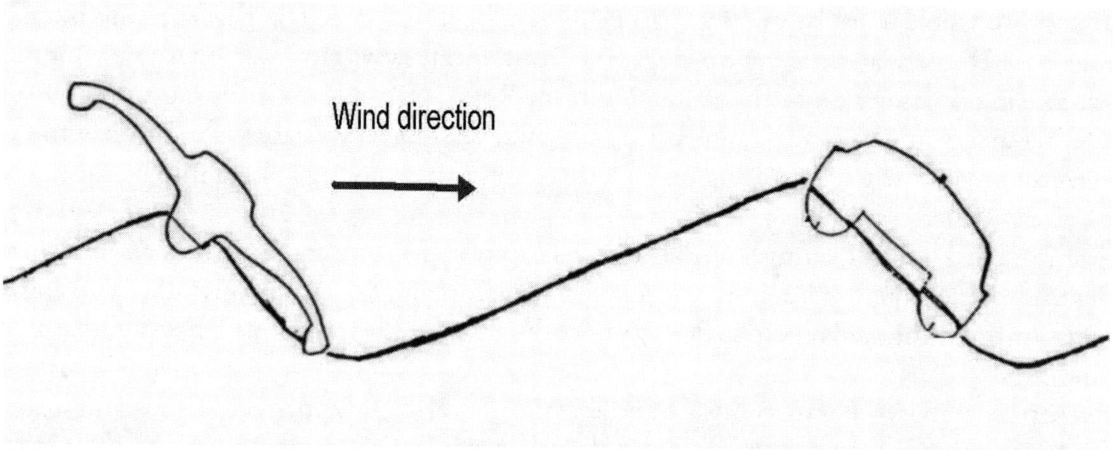

Diagram shows modern wide beam open wing trimaran and traditional cruising catamaran on a sea with 20° slope and 45° face. Aside from the windward hull of the trimaran being higher into the wind gradient and the fulcrum of that hull against the less buoyant ama, Marchaj suggests that the wind blowing around the ama cause an airfoil effect creating more drag than if it were a solid deck.

Borrowing again from Marchaj, the diagram leads to the thesis that the vented (open) trimaran wing decks might possibly be even more dangerous than the closed (solid) wing decks of the past designs. Judging by statistics, trimarans of the open wing type positively have the worst safety (capsize) record, Far worse than even the self-designed, home built boats of decades past. (See *The Capsize Bugaboo,* by Charles Chiodi, for some statistics, although dated.)

When Walter Greene made his now famous (or perhaps infamous) statement that trimarans are unsafe offshore, it was precisely that type of craft he was referring to. Present state of the art, however, appears to have eliminated much of the danger. Yet I would not recommend that type of craft to anyone as an offshore cruising boat.

In March, 2001, an article appeared in *Soundings* Magazine concerning an experienced sailor who was capsized and lost sailing an F31 trimaran from Florida to Europe. A tragic accident, but indicative of the themes I have stressed in this book about using technology not truly suitable for the job at hand. Previous passages aside, single-handing, overweight and inappropriate technology increase the risks beyond a reasonable level.

Comparing Catamarans to Monohulls

Compare by size

- **Length.** You pay by the foot for dockage, survey, haulout, etc. Catamarans give you close to double the boat in the same overall length thus represent a bargain in most jurisdictions.
- **Beam.** Over 16 feet can be a limiting factor in some cases. Between 16 and twenty feet options diminish and over 20 feet you are in a different game.
- **Sail area.** Varies with displacement. Since less displacement usually requires less sail area, cats usually have smaller sail plans than monohulls of the same length.
- **Volume.** About forty percent more in a cat compared to a monohull of equal length.
- **Displacement.** Averages about two thirds of an equal length monohull despite the cat being about forty percent larger.
- **Displacement/Sail area ratio.** About the same.
- **Draft:** Shallow draft is a main feature.

Compare by motion

- **Heeling angle.** No cat normally heels more than five degrees.
- **Pitching:** About the same as other boats depending upon length and displacement.
- **Rolling at anchor or at docks.** An easy comparison you can make yourself anytime you pass a marina that is open to swell. Note that not only do cats roll much less, but they have a quicker motion.
- **Attitude in waves.** Cats are the most sea kindly of all vessels, thus their popularity in such a wide range of applications.

It is an interesting note that virtually the entire Australian dive industry is using catamarans for their dive boats. The motion is the primary reason although there are others. In Key West, Florida, the major vessels taking tourists to the Dry Tortugas are catamarans as are most of the day trip charter boats. In fact, many of the famous race boats have found a home chartering in Key West.

Compare by cost

- Length alone is deceiving. (see explanations elsewhere in this book)
- High tech costs more in any boat, power or sail.
- Type compared to type, costs are roughly similar.
- There is one 34 foot catamaran that is still the best value for the money in any cruising boat. (See Gemini)
- It is possible to have a lower cost seaworthy cruising catamaran than an equivalent monohull sailboat (See Wharram Catamarans, 88, 102)
- Based upon showroom prices, catamarans are only slightly more costly than roughly equivalent monohulls.

Compare by power options (Complete coverage in book two)

Myths and Realities about Cruising Catamaran Performance

The horsepower race

Separating fact from fantasy and good intentions from reality in the advertising and promotion of catamarans for cruising.

Remember the horsepower race in automobiles? After a certain point, it no longer fulfilled any useful function. We are faced with the same sort of mentality today in the sailboat industry, and most specifically in the cruising catamaran industry. Thus the following list of questions for you to ponder.

- How fast is fast?
- How fast can your cruising catamaran really go?
- How fast can you push your cruising catamaran and still derive the pleasure you expect from it?
- What are the parameters of fast?
- Fast compared to whom?

Certain facts emerge

As of circa 2002, no Class 4 or 5 production cruising catamaran is capable of any realistic speeds beyond true wind speed and even that is stretching a point.

Keep in the forefront of your mind that sailboats, in order to go fast you need wind. When it is windy it is usually rough. When it is rough you can get pretty beat up. Under conditions like that, most cruisers adjust their speed to the most comfortable rate. Usually, that rate is considerably faster than a comparable cruising monohull of the same waterline length but much slower than the potential speed of the boat. It is no different than adjusting the speed of your one-hundred plus mile-per-hour automobile to the road conditions. Not too many people drive at breakneck speeds on rough roads.

In the fifteen years we lived aboard and cruised on *La Forza*, we averaged seven (7) knots. The group of monohulls we cruised among and with averaged five (5) knots. That is a difference in average cruising speed of forty (40) percent. To add to that, the monohulls never would have averaged even close to that had they not motor-sailed a large percentage of the time. What it meant to us in practical terms is that very often we would sail right by monohulls motor-sailing at six knots while we were sailing at seven or eight in ten to twelve knots of wind. Could we go faster? Sure, and often we did. However, conditions had to be just right in order for us to do it.

I remember one special trip through the Bahamas where we were sailing a course of about 280 degrees from Chub Cay to Gun Cay across the Bahama Banks. We were buddy boating with dear friends on a Gulfstar 50 ketch. They have towering masts, a 100 horsepower engine and a waterline almost twenty feet longer than ours. We hauled anchor at six AM and left together. The wind was the typical winter pattern of ESE at 15-20 knots giving us a classic broad reach. Because the water is so shallow, there are no large waves to contend with and the two to three foot waves were coming right off the port quarter giving us a great boost.

By 10 a.m. the *Gulfstar* was completely hull-down on the horizon behind us. Based upon my guesstimate at the time we had made eight miles on them in four hours. We were sailing with only a full main and working jib and making a pretty steady 10 knots, occasional spurts to eleven and twelve. I wanted to fly my spinnaker but Corinne vetoed that out of hand. I know the boat is capable of a steady fifteen knots in those conditions, but we were cruising! Our goal was not to win a race or see how fast we could go but to make a comfortable and safe passage. Reflecting back on that passage, I believe the majority of the cruising catamarans I have sailed and reviewed these past ten years since my first book, *Cruising On More Than One Hull*, would have performed equally well and perhaps even better. This statement subsumes similar size catamarans. Obviously, much larger catamarans will sail proportionately faster.

In the fifteen years we lived aboard and cruised and in the other sixteen years of sailing and competitive racing, we have only had a handful of sailing experiences of that type. Therefore, I ask the potential purchaser of a catamaran for cruising if it is worth the required sacrifices in amenities and carrying capacity for the mythical and only occasional spectacular passage? That is the essential choice you must make.

Spinmeisters wax eloquent with tales of dreamy passages in the trade winds loafing along in gentle seas with the wind from the quarter and averaging 200 to 300 miles per day. I suggest that while there is considerable truth to this, ninety percent of your sailing is not going to be in those conditions. Even the intrepid adventuring sailor Eric Hiscock, stated that ninety percent of your time is spent in port. Thus, what I am suggesting is that the all around utility of your chosen catamaran is more important than its theoretical top speed. How easy is it to raise and lower sails? How does it tack? Do you need to start an engine to tack? How does the boat handle in close quarters, waterways, channels, around docks and pilings are characteristics needing more attention than top speed.

I am not suggesting for one moment that speed or good performance is not important. It is. I am pointing out that you must pay for that performance in one way or another and that you ought to be aware of the payment schedule. This process is identical to choosing the family automobile. Most of us would like a *Corvette* or a *Porsche*, but we settle for a sedan, station wagon or minivan. None of us expect to get the same performance from our minivan as we do from our sports car. Thus is it with catamarans.

Cruising boats vs Charter boats

Cruising Boat vs Charter Boat

Many people are enticed to purchase expensive catamarans with the idea that they can place the boat in charter, have charter fees pay for the boat and at the end of the period, usually five years, own the boat free and clear, or close to it. This is a profound concept that has in fact worked well for many people. Let us analyze the difference in approach to designing a charter boat as compared to a cruising boat as it impacts on your particular circumstances.

Charter boat requirements

- **Cabins.** As many cabins as possible with personal long-term comfort a minimal concern. Most charters are for a week or two. The folks who charter seldom analyze the layout features of the boat because they are only there for a short time and they are there strictly to have fun. By the time they begin to raise their consciousness, they are on a flight back to winter someplace.

- **Privacy.** This is one area in which no other type of vessel can even begin to compete. With a minimum of four smallish cabins or staterooms behind closed doors in the four corners of the vessel, and most with private heads, it is the most genteel and dignified way to group strangers together. Of course, families love this also when Mom and Dad are sequestered in their cabin and offspring are cavorting elsewhere.

- **Ease of maintenance.** Many boats are built of materials that are easily cleaned by indifferent maintenance workers. Some of the major charter companies will not install saildrives because of maintenance headaches, thus leaving no choice but conventional shafts and props. Engine layout is unimportant thus engines under bunks are perfectly acceptable. Charter clients are hardly expected to check oil or other items. These tasks are undertaken during turn around time by the maintenance department, thus it matters not where the engines are located. In fact, in some instances, they more remote and difficult the engine access, the less the curiosity seeking charter guest will pry. If an engine is under someone's bunk and that someone is using the bunk...you get the picture.

- **Minimal equipment.** Anchors, electronics, docklines, safety equipment etc. Short term usage in restricted environment does not call for much equipment and if the boat doesn't have it, the guests or workers can't break it or lose it.

- **The illusion of complete luxury.** Features like sliding liquor cabinets and hand rubbed burly walnut and such. Much of this cosmetic material is not repairable after five years of abuse.

- **Stylized interiors.** There might be but I have not seen a charter boat with any other arrangement in the saloon than the central table and surrounding bench seat. This is an inexpensive and expedient layout that is fashionable at the moment. It is of European origin and fits in with that life style as does the abomination of stern helms.

- **Stern helm locations.** These are essentially designed to separate the proletarian crew from the aristocratic charter guest. As a secondary attraction on charter boats, it allows free ranging fantasy to the guest, who, after all, usually only actually sail a few hours every day. (The Rambo helm)

● **Small galleys.** Big, well equipped galleys are not a requirement for a charter boat since most guests dine at restaurants the majority of the time. Occasional breakfast and lunches do not require major galley installations.

● **Storage space.** While most cats have enormous storage capacity just by virtue of their volume, charter boats are designed to accommodate transient storage such as suit cases, duffel bags and have less emphasis on groceries or food preparation and personal items.

Cruising boats
After reading the beginning of this chapter, you have a good vision as to the requirements of a cruising boat. Thus you have to make a decision whether or not a boat designed and built as described above will fit into your concept of cruising. In many cases it can. However, there are several considerations you should closely evaluate.

● **Bunks.** There is an enormous gulf between the reveler who charters for a week or two in the Caribbean or Pacific and a couple moving aboard for a lengthy period or a permanent home. Long term bunks must be easy to enter and leave, nocturnal interruptions should not unduly disturb sleeping partners, there should be very ample headroom over the bunk and engines under a bunk are not only a daily maintenance nuisance but have latent heat, odor and require workmen traipsing through your bedroom.

● **Helm stations.** (Review the chapter: Helms, good, bad, ugly.)

● **Heads.** Many charter boats have diminutive little heads with what appear to be miniature toilets and smaller yet, washbasins. Not a problem for a week or two in the tropics when you are swimming every day, but a major consideration in a temperate climate where you actually may use such facilities while wearing clothing.

Storage and galley layout are usually very important items to cruising boat purchasers. Of course, many charter boats do have the potential for such.

There is good opportunity here. If you can adjust your needs to the charter boat industry; your long range planning can accommodate the waiting period and you get to use the boat yourself occasionally; it can work out favorably for you.

Leasing boats from private owners to charter operators is a well established business principle and there are several reputable companies doing it. Check out the boats being represented by these companies and purchase the one that will come closest to your heart's desire. There is one other factor to consider. You can view the lack of equipment as a positive because it gives you the opportunity to fully equip with state-of-the-art equipment when you take over.

Defining Your Needs

A Systematic Approach

You've waded through the preceding pages and decided that you can no longer live without a real cruising boat. Now it is time to settle down, sharpen your pencil and make some plans. The first of these plans has to be an evaluation system for decision making.. Failing this, you will be forever coming back to square one.

Base your evaluation system on at least the following list of criteria

1. Budget. How much are you willing to spend (or perhaps invest) on a cruising boat? How much yearly maintenance, haulout, insurance, license, taxes and dockage can you afford? How much depreciation on your investment are you willing to accept?

EXPENSE	AMOUNT	REMARKS
Licenses and taxes		
Insurance		
Haulout and bottom coating		
Dockage and storage		
Maintenance		
Depreciation		
Interest		
Upgrades and new equipment		
Fuel and operating costs (crew?)		
Reserves for the unexpected		
Total		

2. Where are you really going to use this boat? Design this by latitude: Frigid, temperate, torrid, tropical, and by longitude: offshore, coastal, Atlantic, Pacific. This is a critical concern especially for first time buyers. Ever so many have unrealistic goals for themselves and wind up buying unsuitable boats. It is self defeating to buy a boat designed for circumnavigation and then never sail beyond your local cruising area.

3. What size boat do you really need?
Define this by how long you intend to stay on board any time, the distance between ports, your need for amenities, the number of people you plan to have onboard and your budget.

When we closed our nest, Corinne and I scaled down from a forty-foot catamaran to a thirty- two foot catamaran, and thus far have not regretted it.

Some say "buy the largest boat you can afford." I would suggest that if your motives are local or coastal cruising, buy the smallest boat you can live with. If you truly intend to do much open ocean passage making, buy the largest boat you can afford *and physically handle!*

4. How many people will I most often have on board? (pointless to have a boat fitted with nine bunks if it's just you and your spouse with an occasional guest)

5. Where do I fit on the scale of cruisers between the opposite poles of:

	1	2	3	4	5	
Performance						Amenities
Practical						Beautiful
Frugal						Decadent
Good sailing						Good motoring
Spartan						Opulent

6. How much racing will I do? (0 to 100%) Sailboat racing is really a lot of fun. Though you may feel now that you will never race, keep in mind that lots of people have felt just like you and later changed their minds. A boat that sails well is more fun and more satisfying. You will sail it instead of motoring it, and will sail it more often. This factor is important. It does not warrant short shrift. Good performance is a worthy objective and worth some compromise.

7. Where are you going to keep your dream boat when you are not using it? This is one item in which you must be both realistic and pragmatic. If you intend to keep your boat on a narrow canal, better check for deed restrictions first. Better check not only your permanent mooring facilities but also your haulout facilities as well.

8. When you do cruise, what lifestyle will you want? Will you spend most of your time at anchor, or most of your time at marinas? (Eric Hiscock, one of the most ambitious cruisers of all time, said: "even the most intrepid cruiser spends 90 percent of his time in port.") This is a most important consideration. It affects not only the vessel itself, but how you equip it and what priority you give to items like shore power, air conditioning or a water maker.

How to Acquire a Cruising Boat

There are several approaches to obtaining your dream boat. They are:

- New production boat.
- Used boat.
- Custom built.
- Home built.

Lets take the easiest one first: Home Built.

Home Built

My observation leads me to the distinct visceral subjective conclusions that there are builders and there are sailors and only rarely do the two come together. All too often those home building projects never seem to get completed. Often, the results are less than expected when completed. On the other hand, I have seen some outstanding home-built vessels, clearly head and shoulders above what are available from the production builders.

Before considering building your own boat, you ought to apply certain criteria to yourself.

1. How many boats have you built previously? (If the answer is none, don't even think about starting with a sophisticated vessel like a modern catamaran or trimaran). There are exceptions, of course. Plans for James Wharram catamarans, and Piver trimarans designed specifically for simple, easy building, are available.

2. How well did you do on your past projects? What finish level did your boat have upon completion?

3. How fast do you work? Fast enough not to destroy the rest of your life? Marriage? Job? Bankroll? Health?

4. What are your real objectives? Saving money? Knowing every detail? Getting your heart's desire?

5. Have you included the value of your labor?

6. Are you able to absorb the potential financial loss of a "one-of," home built boat? Unfortunately, home-built boats historically have the least resale value and are the most difficult to sell and there is always a resale time.

There was a time, back in the 1960 and 1970's, when you could save money by home building. Unfortunately, that historical period has ended. You must keep clearly in focus that the hull, that part where you can use significant amounts of your own labor, represents only 20 to 30 percent of the cost of the boat. The mercantile distribution system for the rest of the items you need are such that there is little money you can save. You will wind up paying retail or close to it for all your parts, hardware, mast, sails and accessories.

If you think that knowing each nook and cranny is important. It might be possible to haunt a custom builder or a production builder, carefully watching every step. Questioning an experienced builder about certain techniques, is more productive than doing it yourself and wondering why. Having a boat built under survey will help in this regard.

It is my observation that many home built boats are equipment and fitting-out starved. People spend inordinate amounts of time and money building the hull, then they get to the expensive part, fitting it out. About that time they have run out of not only money but patience. It is regrettable but a majority of the home built vessels I survey are poorly fitted out for the above reasons.

My suggestion is that unless you are clearly and strongly motivated to do it yourself, and you have the necessary skills and the necessary background, don't do it! In most marine communities there is a *Boatyard of Broken Dreams.* Go down, check some out. Ask the proprietor about the various projects that are underway. Ask some people there, building or refurbishing. You will get an earful, believe me!

Custom Built Boats

Combining a good builder with a designer that suits your needs is an excellent way of getting almost exactly what you want in a boat. Note the specific term: almost. I suggest that you not consider custom building until you are ready for your third cruising boat. By the time you have gotten to that point on the experience scale, you ought to have a good grasp on your own particular needs, desires and budget.

Do not expect to have fewer problems with a custom boat and fewer annoyances and dissatisfactions than you do with your production boat, only different ones. The simple tragedy of custom building is that you can only avoid the known mistakes, not the unknown ones.

Speaking of budgets, unfortunately, it seems that almost every custom boat finally tallies at between 20 and 50 percent over budget, and some far beyond that. It's really hard not to find a catalog of horror stories about custom building. Check some back issues of MULTIHULLS Magazine for specific examples.

One other facet of custom building is that often the sales persona of the designer is the strongest force in the entire process. Do not be lulled into the buying/building decision based on the designers' idea of cruising reality. Base that decision on your own definition of cruising.

Used Boats

Fortunately for the multihull enthusiast there is a plethora of good, used boats on the market. However, do not expect to save much money on the more popular models over the price of a new one. Demand has kept the prices of good production boat's high and still climbing. For instance, the ever popular *Iroquois* catamaran that originally sold for around 15K when imported in the 1970's now goes for upwards of 30K (good condition of course). This shows a gain of 100 percent in about twenty years (Adjusted for inflation, it still looks good!). Very few monohulls can state that gain in the present market. If you can adjust your cruising requirements to accommodate some of the ex-ocean competitors, you can buy them at a fraction of their original cost due to the low demand and poor suitability.

On the other hand, there are quite a few production boats scattered about that are in various stages of neglect and decay. This is a good place to invest your boat building energy and talent. If you do a good job, your reward will be to own a vessel with both pedigree and established value. However, do not underestimate the difficulty or cost to refurbish. If you are not a skilled person with lots of building savvy, you can be up to your armpits in alligators before you realize what happened!

I have had the very good fortune to sail, sell, own, survey and otherwise become familiar with a large majority of these production boats. For a full description of both my objective and subjective evaluations of these boats, see Book Three in this volume.

New boats

For the most complete and up to date list of new production catamarans, obtain a copy of SAILOR'S MULTIHULL GUIDE *To the World of Catamarans and Trimarans*. Presented in this book are only the boats with which I am intimately familiar by virtue of survey or delivery.

As a cautionary note, before you leave a deposit on a production boat, be absolutely sure the builder is capable of delivery. Most unfortunately, there have been people in boat building businesses who take deposits and don't deliver. If at all questionable, see a lawyer for an air tight escrow arrangement.

At the end of this chapter there are helpful guidelines and matrix charts to aid you in choosing the most satisfactory cruising boat for your purposes.

Owners Association

One item that should be very high on your list for your cruising boat is an Owners Association. An owner's association can provide you with enormous amounts of information, background and various approaches to problem solving. Most important, you gain a history, and perhaps a pedigree for your boat. With a one design or a home built, you may be on your own. Do not underestimate the importance of this moral support. Once you have purchased some designer creation, you may have only have your designer (if available) to commiserate with. If you have a problem, he probably won't be able to help you with it either. If he knew how to handle it, it would not have happened in the first place. Having sisterships and other owners at your level is a tremendous asset.

Value

One-of-a-kind boats only have value to the extent that there is demand and a pedigree. An unknown boat by an unknown builder has very questionable value. If it is self-designed as well, it is possibly close to worthless. Home built boats, built scrupulously to the designers' plans and scantlings that pass strict survey might be good bargains for people with realistic expectations. Some home built boats, upon which the builder has "improved" the designers' specifications, will earn the condemnation of all concerned parties, especially the designer! Changing designs, changing specifications or scantlings is risky business and you must be prepared to take the financial responsibility for your actions.

Most designers will deny the use of their name to vessels that have been modified beyond their approval, and rightly so. There are also some production builders that use famous designer's names because they had done the original design work decades ago, but have been reworked through the years by others until the original work is almost lost.

All around the country, home built boats exist in every stage of completion. A significant number of these builders will proudly assert that they have improved upon the design. They will tell you such things as: "I used 1/2 plywood where the design specified 3/8. I stretched the design two feet." Or, "I used heavier rigging, bigger frames," or whatever, believing that by second guessing their chosen designer, they have improved their boat. In many cases, they simply have turned their vessel into a dog with little or no value. If you think you know better than the designer you have hired at a significant retainer, why did you pay him in the first place? Save the money and do it yourself. Such vessels litter both coasts of The United States. They are the hulls of shattered dreams. Do yourself a favor, buy a production boat, unless you are one of the very rare few on whom all the cosmoses converge. Having a custom-designed boat is the province of a rare and fortunate few. You know who you are, there is no need for me to expound further into the subject.

Computing Value

Bargain and sale creates value. That is the way it works with real estate and to a large extent, is also the way it works with boats. However, boats will show a greater variation. An appraiser or a surveyor, can only estimate a value within a range determined by the sale of similar boats. Fortunately, or unfortunately, there are three agencies that monitor boat sale prices and are influential in deciding or more properly, reflecting value. They are: BUC, NADA and ABOS. They are reporting agencies that serve a function similar to Red Book agencies found with automobiles or real estate. Take into consideration that these agencies receive precious little information about multihulls thus rendering their services of little value.

Historically, the boats with the least retained value are the trendy racing boats. This is true from tiny one design, up to the class "A" ocean flyers. Once a design has become obsolete, either in technology or racing rule, it has little value to anyone. Even the day sailors suffer from this, as once a class is no longer racing, demand and, therefore, value, plummet.

Often, people approach me to find them an old ocean racing greyhound cat or tri, believing that it can be converted to the charter service. This is about like taking a sports car and converting it to a dump truck. Those who really know the charter business, appreciate the need for high quality, rugged, properly designed vessels. Vessels designed to take the dock crunching, feet stomping, beer spilling crowds day in and day out. The most successful charter boats, both monohull and multihull, were built that way from the drawing board up. Gold Coast Yachts in St. Croix is my prime example.

Every sale starts off with an asking price. To help arrive at some reasonable approximation of value, consider this quick rule of thumb. When buying an older boat, look at the cost of a similar new one. If the boat you are negotiating for is no longer in production, pick the boat that is the closest replacement to use for comparison. From that selling price, deduct 5 percent per year. Depreciation should slow or stop at 50 percent. However, very old boats will depreciate further. In other words, when it reaches ten years of age, the boat has depreciated to a plateau, or as far as it will go until the next decade. This system considers the currency devaluation that has taken place over the past three decades. This is not an absolute system. It is only a starting point to estimate the value of an older boat in modern currency.

Modern production catamarans have been manufactured since the 1950's.

Power Options for Catamarans and Trimarans

Catamarans have installed auxiliary power in five standard ways:

1. Outboard.
2. Inboard, twin screw (standard shaft).
3. Inboard, single engine with Sonic Drive leg.
4. Inboard, single engine with twin hydraulic drives.
5. Saildrive (twin or single engine installation).
6. Miscellaneous.

1. Outboard

Personally, one reason I own a catamaran cruising boat is that it is feasible to use an outboard for auxiliary power. A comparison looks like this:

Advantages

1. The engine is outside the boat! Therefore, you do not have to live with latent heat or engine room smell.
2. The danger of gasoline aboard is completely minimized because tanks and engines are outside and vented from below, as in an automobile.
3. The engine is easily removed for service and/or replacement, either at the dock or into a dinghy.
4. Exceptional service organizations are universally available. No waiting for parts to come from Europe or Asia.
5. No need to haulout for engine repairs or replacement.
6. Inexpensive! Usually less than 1/3 the cost of a comparable diesel installation.
7. Lightweight, usually less than half that of a comparable diesel.
8. The modern outboard is quite reliable and dependable.
9. Can be completely retracted from the water, thus lessening drag and galvanic corrosion.
10. In some installations, can be adjusted for depth.
11. Maneuverability. Most outboards are connected to the rudders, dramatically improving close quarters handling and backing.

Disadvantages

1. Fuel consumption: The rule of thumb for fuel consumption:

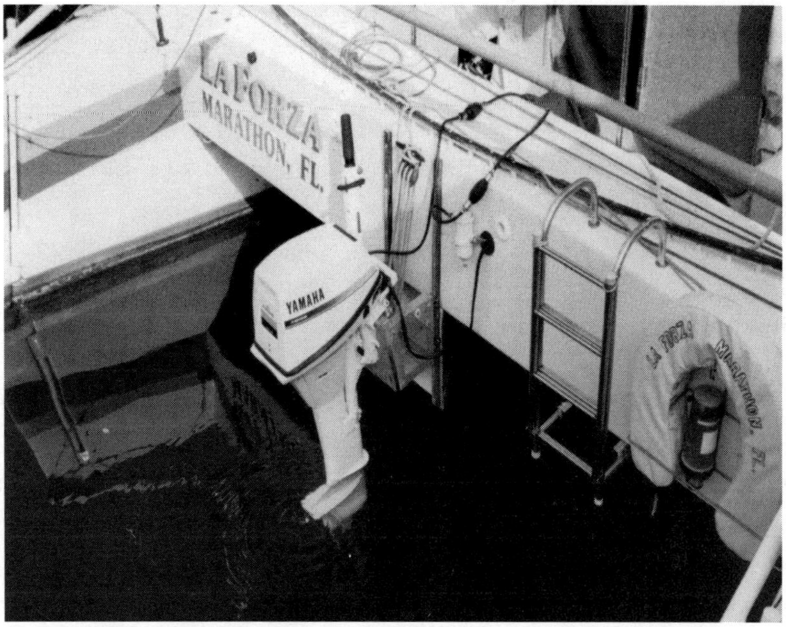

La Forza sliding vertical bracket. Note 6:1 purchase lifting tackle

- *Two cycle burns one gallon per ten horsepower per hour.*
- *Four Cycle burns one half gallon per ten horsepower per hour.*
- *Diesel burns one quarter gallon per ten horsepower per hour.*

2. Low electric output (except Yamaha 9.9)
3. Cavitation. Some vessels are more affected than others. My numerical descriptions will take this into account on the vessels with which I am familiar. Check Book III. Look up the individual vessel your are concerned about.
4. Ease of operation. There are very few outboard installations that are as user friendly as a nicely installed inboard, where you just touch a button and you are under way.

Gemini tilting "bucket. 4:1 purchase

5. User acceptance. There exists a large body of cruisers who will not accept outboard power simply because it is outboard power! To them, an outboard has negative connotations of jury rig or of not being truly seaworthy. No amount of persuasion will alter their concepts. This has implications at resale time, and there is always a resale time.

2. Inboard, Twin Engine

Advantages

1. Nothing, but nothing is as user friendly and as easy to handle around docks and piers than a well designed twin engine catamaran. It is more maneuverable than even a twin engine powerboat, because of the wide beam.

I can remember one particular incident when, at Yacht Haven Marina, St. Thomas, USVI, I inadvertently sailed down the wrong channel looking for my assigned slip. I was sailing a Cherokee 35 (*Panache*). All the docked boats were in their slips backwards. Their bowsprits loaded with anchors were sticking out into the channel, looking for all the world like WW II fortifications. It was a tight and dangerous squeeze. When I got to the end, there was no place to go! I slowly spun the boat around on its own axis and motored back out. All eyes were upon me, watching for what they believed would be the inevitable disaster. As I motored back out, there was a round of applause from the assembled cruisers. That is the beauty of twin engines.

2. Safety. (Redundancy) There is safety in numbers. If one fails, the second engine may get you back.

Disadvantages

1. You have the cost of two engines and their service parts.

2. You have the weight and complexity of two installed engines.

3. You have the same underwater protuberances and chronic problems of struts, shafts, through hulls, zincs, galvanic action etc. as your monohull counterpart, only you have it double.

3. Diesel with Sonic leg.

Many of these installations are in a nacelle between the hulls of the catamaran. They are comparable to the inboard/outboard of motorboat circles. That is, they raise, lower and turn, though few, if any are hooked directly to the rudders or steering systems in the manner of outboards. Perhaps for those that must have a diesel engine, for whatever reasons, this is the most practical setup, as has the most advantages and fewest disadvantages.

Advantages

1. Retractable leg, nothing in the water to drag, corrode or conduct electricity, when under sail or at dock.

2. Engine is outside the living quarters, therefore, less latent heat, noise, smell and mess.
3. Able to power accessories, refrigeration, alternators, hydraulics, etc.
4. Easy access for service and maintenance.

Disadvantages
1. Weight.
2. Cost.
3. Maintenance parts and service availability.

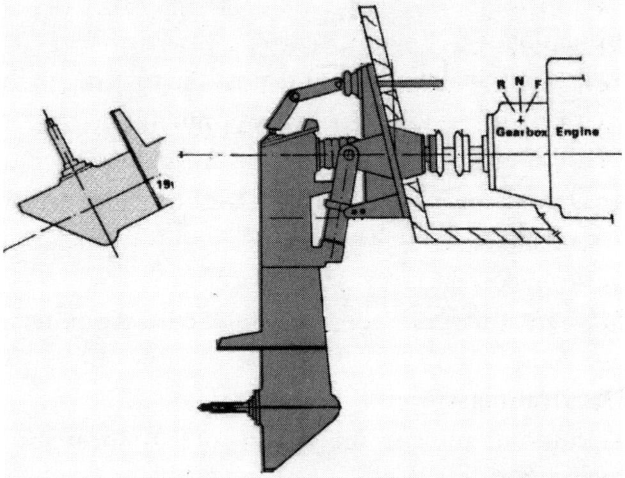

Sillette Ltd.
190/192 Garth Road
Morden, Surrey, SM4 4LU, UK
Tel: 01-337 8687

4. Inboard diesel with twin hydraulic drives

This once popular system has almost faded into oblivion. In the days when marine diesels were outrageously expensive and inordinately heavy, it made good sense to use a single engine and hook it to two propellers via hydraulic motors. With the coming of the lightweight high speed diesel, it has become an unnecessary, complicated arrangement.

This system has the additional disadvantage of introducing yet another expensive, esoteric, complicated system: the hydraulic system, into the boat. The engine is coupled to a large hydraulic pump instead of being hooked to a drive shaft. This pump in turn, feeds through a manifold and director valves into hydraulic motors. The two hydraulic motors turn the propellers. Try to get this one fixed in the boondocks!

For those in existence, they have many, but not all, of the advantages of twin screws. You cannot run the propellers at different speeds, thus limiting your maneuverability. There is considerable power loss through the hydraulic system and it is unbelievably noisy.

You still have the disadvantage of any drive system that has fixed propellers on shafts. The disadvantages are, catching trash, galvanic corrosion, vulnerability to damage, and the need to haul out for any maintenance.

5. Saildrive

Saildrive is a clever marriage between a fixed leg and an engine/transmission mounted directly on top of it, as a one piece unit. In addition to the advantages and disadvantages listed above for all diesel installations, saildrive has these additional ones.

Advantages

1. In some installations, it is possible to remove the entire unit without hauling the boat.
2. It is lighter, simpler, has fewer moving parts.
3. Eliminates the shaft log and stuffing box.
4. Drive is perpendicular to waterline, thus not having the propeller thrust loss associated with an angled shaft.
5. More compact, uses less space.
6. Some units have the water intake incorporated in leg, thus eliminating a thru-hull fitting.

Disadvantages

1. In too many cases it is not possible to locate the engine far enough forward to properly distribute the weight.
2. There may be more drag from this installation than a conventional shaft and folding prop (not true of all installations).
3. The vertical leg is a good candidate for catching trash and is in a difficult place to clear.

6. Miscellaneous

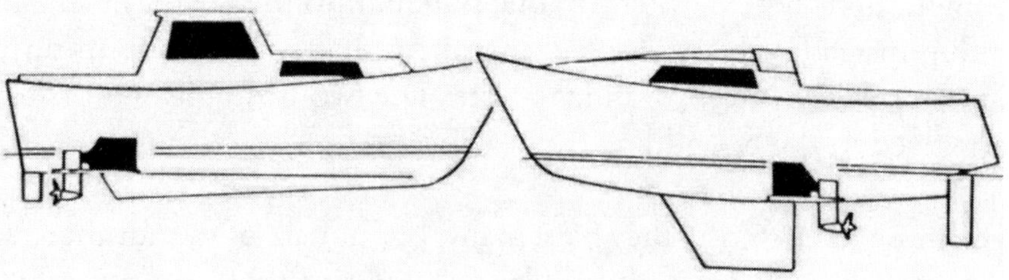

Examples of sail drive installations

Single inboard engine.

I owned a forty-foot catamaran (*Duet*) with a diesel in the starboard hull and a large fuel tank in the port hull. It was not apparent when under way, in which hull the engine was installed. After four years of continuous heavy usage, I could say that I was satisfied with the arrangement.

The vast majority of inboard equipped cats have the engines available from the cockpit. Unlike these catamarans, many monohulls and trimarans have the engines inside the boat. In those boats, you must carry used engine oil and all the replacement parts that go along with servicing, right down the companionway ladder, through the boat and across the furniture and carpeting.

The cockpit location is also excellent for allowing latent heat to escape. Just open the hatches (usually seat tops)and there it goes. Sure can't do that when the engine is right down in the bowels of the boat.

This makes an excellent sales point to those who do not care for grimy service people traipsing throughout their homes on a recurring basis. Remember, the usual procedure is to let the engines run long enough to get good and hot. Then you shut them down and open the access panels or doors. This procedure, of course, lets all the heat and odor into the boat (home). Next, proceed with the nasty job of pumping hot diesel oil into various receptacles .This releases yet more odors into the boat. (Your home). Oh, how many tales of disaster I could relate regarding hot, black, slimy oil spilling on carpeting, clothing, pets, ad infinitum!

Mixed engine types
There is precedent and good logic for having two different engines. The main propulsion engine can be quite large, thus giving maximum usable speed under power and the secondary engine smaller and sized to properly drive the accessories such as watermaker, SCUBA compressor, generator, refrigeration etc. It is common knowledge that running diesel engines lightly loaded is bad practice. Thus each engine is properly sized for the job required. Both engines are available for maneuvering, docking and emergencies. At this moment, one catamaran is being equipped with a diesel in one hull and an electric wheel in the other. Logic says its a workable arrangement. Time will tell.

Jet Drive
The advantage of jet drive is that there are no protuberances in the water to either slow you down or foul your prop. To this date, the unsolvable problems have been: corrosion and fouling. Someday, modern materials and their advantages will be usable to solve these problems. There are other yet unresolved disadvantages like: power in reverse and systemic power loss. Yet this technology has potential.

Electric

Trains run on diesel electric systems, why not boats? Multihulls have sufficient deck area to carry enough solar panels to generate enough electricity to power the boat. There is in fact, a fully electric, solar powered ferry operating in Sydney Harbor, Australia as this is written. Electric propulsion is an avenue that truly needs further exploration and exploitation. New technology abounds. Recent breakthroughs in electric wheel technology are being installed in production boats as I write. The implications are enormous. Maybe one of you out there will get a Government grant to study the problem? After all, this is our alternate energy, is it not?

Presently there are people exploring every avenue of this concept and I predict that within the next few years there will be diesel/electric (DE) systems commercially available for pleasure boats as they are for trains or ships.

Steam

There are interesting experiments in propulsion using steam as a jet propulsion medium. This concept has fascinating prospects of using sea water as a drive medium and eliminating all the underwater protuberances such as struts, shaft and props.

Helpful evaluation forms

The following three forms are designed to help you select the closest vessel to your own personal requirements. They have extra lines for you to add your own personal requirements and preferences.

They are for you to copy and use as you desire. I suggest making a set of copies for each vessel you are considering in order to compare in a relaxed atmosphere away from outside pressures.

VESSEL___________________________________Date:____

SAFETY rating form

1=Unsatisfactory 2=Poor but usable 3=average 4= very good/above average 5=Excellent

Feature	Grade	Type	Comment
Anchor handling			
Visibility from helm			
Deck slope			
Rub rails			
Toe rails			
Trampoline			
Cross beam			
Hatch opening			
Helm layout			
Interior Visibility			
Non-skid			
Companionway			
Life lines			
Hand holds on deck			

VESSEL______________________________Date:_________

ERGONOMICS rating form

1=Unsatisfactory 2=Poor but usable 3=average 4= very good/above average 5=Excellent

Feature	Grade	Type	Comment
Anchor handling			
Dingy handling			
Cockpit layout			
Winch access			
Galley layout			
Galley refrigerator			
Galley sink			
Galley stove			
Galley storage			
Head layout			
Seating comfort			
Cabin layout			
Bunk layout			
Cabin ventilation			

VESSEL______________________________**Date:**_________

UTILITY rating form

1=Unsatisfactory 2=Poor but usable 3=average 4= very good/above average 5=Excellent

Feature	Grade	Type	Comment
Rudder configuration			
Underdeck clearance			
Deck gear layout			
Traveler layout			
Winch layout			
Mast base access			
Running rigging layout			
Boarding ladder access			
Electronics layout			
Engine access			

How To Sail A Cruising Catamaran?

How does it feel to sail a catamaran or trimaran?

This is an interesting question because there are tremendous differences in boats of both types. I will try to describe them by type and class.

Trimarans

First, let me start with trimarans. The better trimarans I sail, sail about like the better monohulls I sail except for the heeling, rounding up when overpressed and being limited to a hull speed limitation factor. Few trimarans have bad habits, they sail pretty much like dinghies, tacking assuredly, responding to the helm nicely, turning on their daggerboards, etc. Unlike many monohulls, however, they will back-up straight as an arrow under power.

I extensively sailed the trimaran, *Moxie*, a Dick Newick design. On occasion, I sailed The F27, the F31 and a smattering of others including some extensive deliveries. (Book III: *Trimarans*). For better or worse, for whatever will come of the following statement from my fellow multihull enthusiasts, a good trimaran handles about like a good, ultra light displacement (ULDB) monohull. The difference in the feel of heel and feel of momentum. A trimaran has little of either.

If you really like the feel of sailing, as you experienced with monohulls and are willing to make the compromises in cruising amenities, a trimaran might be for you. I think there is one more item that directly relates to this, and that is the illusion, on an open winged trimaran, that the boat is not very wide, quite unlike a catamaran, which more often seems more immense than it actually is. Legendary sailor, Phil Weld, used to revel in watching the water sluice by on his several trimarans. The ability to watch the water from the helm was one of his design requirements. This is a little tough to design on a catamaran.

Catamarans

Catamarans are a different story. Even the Class 2, racing, catamarans that I sail aboard, are quite different. The very best handling Class 3 boats, like *Stiletto, Iroquois, Hirondelle, Edelcat, Shuttlecat*, do not feel like trimarans or monohulls. As you get further away from the light and racy cats, into Class 4 and 5, they begin to feel different yet. To quote one owner, talking about his forty-six foot *Wharram* catamaran "it's like sailing a dock!"

Let me relate a short vignette that might give you greater insight.

Several years back, I considered selling my Karmin Cobra MkII catamaran, *La Forza*. I had invited dear friends of mine, Todd and Maureen down to my winter residence in Marathon, Florida for a sailing session on the boat. Todd was very keen on buying it and going cruising. At the time, I was occupied charter-managing *Legal Mistress*, an Allied 39 monohull and had little time to devote to my own boat.

The day of the test sail was a typical Florida Keys winter day with blustery northeast winds averaging about 20 knots. Our combination tourist trip and demo sailing route would take us from downtown Marathon, about five miles offshore in the Atlantic to Sombrero Reef. Sombrero is a favorite destination of the local cruisers. It was a broad reach to Sombrero, and a beat back. I set my full main and working jib for the reach. We literally flew down to the reef, averaging well over ten knots. We were completely serene the entire way, doing what good catamarans are best at: reaching. *La Forza* seemingly flew along the top of water, apparently tip-toeing from wave top to wave top, never plunging, crashing, pounding or wetting anyone with spray.

With the easy part behind me, now came the crux of this entire sales effort. How would she go to windward? I tucked one reef in the main, changed jibs to my Yankee, which is about 80% of my fore-triangle, jibed and started back toward Marathon.

I swept around the lighthouse, hardened the sheets, lowered the leeward daggerboard and proceeded to drive dear ole *La Forza* as hard as she would go to windward. She got in the groove somewhere around 30 degrees making about eight knots. (*La Forza* is an exceptionally weatherly catamaran) After trimming and securing the sheets and making sure everyone and everything were copasetic, Corinne went below and made coffee.

Some 20 or thirty minutes later, I was in virtual euphoria from the outstanding performance of my boat and from thinking I was making all these sales points. After all, we were driving to weather at an outrageous speed, pointing as high or higher than many boats with thousands of dollars worth of go-fast gadgets, drinking our coffee which we could put down anyplace without fear of spilling. We were dry and comfortable. There was no spray coming over the bow. There was no pounding or falling off waves. The helmsman needed only the normal feather light touch to correct the degree or two we would lose from an occasional odd direction wave. Todd was at the tiller steering. He has considerable experience aboard *La Forza*, having crewed on many races with us.

Why shouldn't I be in euphoria? I could practically count the money, who could possibly resist such an impressive test sail?

Maureen, turning to face me, took a quick sip of coffee as if to lubricate her vocal cords. After just the right pause to set the stage, she said:

"Chuck, I could never buy this boat!"

I was nonplused. *Wh wh why?* I stammered, trying to imagine what I could possibly have done wrong?

"No, I could never buy this boat," Maureen continued. "Where is the heeling? Where is the sitting stiff legged in the cockpit straining against the tiller? Where is the green water running down the decks? Where is the sting of the salt spray? Where is the driving through the waves? This is not sailing the way I know it! No, I could never buy this boat."

I could not answer her. I had no answer. Cruising catamarans simply do not sail like that, ever!

You will hear from people that when sailing a catamaran, it does not have the crisp, responsive, feel that you get from a good monohull. You also may hear that it is much harder to know when you are in the groove going to windward with a catamaran. There is truth to these claims. As you leave Class 2 and 3 cats and get into Class 4 and 5 cats it is even more so and as cats get bigger cats, you do lose more sensation of sailing. Somehow, on a monohull, you never seem to lose that sensation, no matter how big they get because they still heel.

La Forza, Karmin Cobra MKII, 1974, 32 x 16 x 2'/5' draft 450 sq ft sail

On a monohull you often appear to be going faster than you really are; on a catamaran, you often seem to be going slower than you really are!

There are times on a catamaran when you have to look at your speedometer to understand your movement. I describe this incredible stability as feeling like a "magic carpet" straight out of the Arabian Nights. Whether this is to your liking is something only you can judge. On the other hand, seems to me this is precisely the desired condition for comfortable cruising!

It is for these reasons I believe the highest and best use of catamaran technology is for cruising.

Tacking

Some catamarans do not tack particularly well and need more attention and special technique. This is also true of some monohulls, however, a smaller percentage. There are a few inherent reasons why this is so and why it is seldom so with a trimaran.

First, just the fact that the catamaran has two equal hulls, means that one of them must drag through the water. The catamarans with nicely rounded bilges, some rocker and very shallow draft tack more easily than those with extremely straight thin hulls. A-symmetrical hulls compound the problem. My own boat, which has thin hulls with little rocker and a masthead rig is one of those boats that are tough to tack.

Any boat with a masthead 50/50 rig, (technically a cutter) that is, that has jibs as large or larger than the mainsail will be harder to tack than a boat with a true sloop rig. The reason is really simple. The proportion of drive in the main, compared to the jib, is less. As the boat tacks, there is a shift in the center of effort (CE) when the jib luffs.

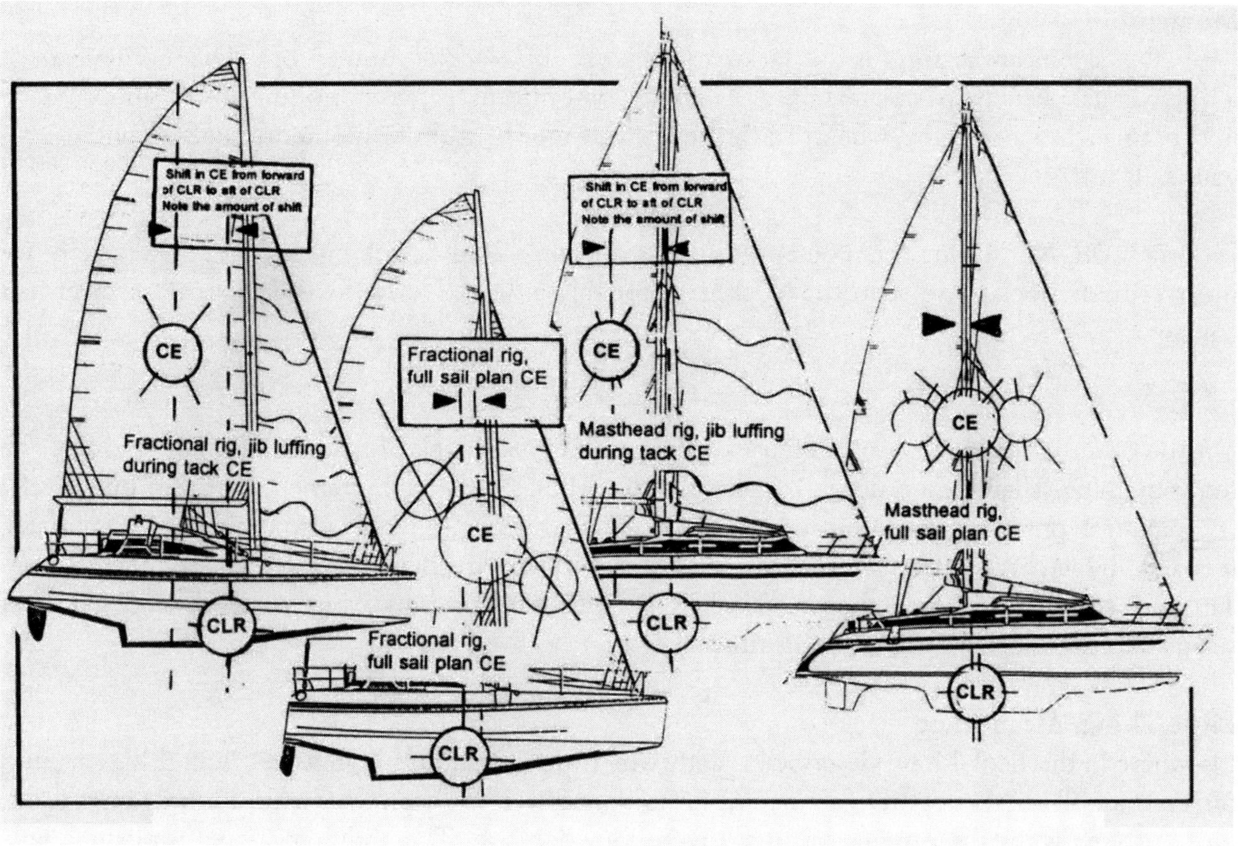

Without getting too technical, when you are going through your tack, your jib luffs before your main.

- It luffs first, because it is at a greater angle of attack to the apparent wind.
- As the jib luffs, the center of effort (CE) shifts aft. The operating center of effort is normally a theoretical spot between the center of the jib and the center of the main.
- Without the jib, the CE move aft to the approximate center of the main. The center of effort (CE) and the center of lateral resistance (CLR) are then no longer in balance.
- The main keeps driving after the jib luffs. It begins to drive before the jib fills on the new tack.
- If the center of effort of the main is far enough back, it will try to keep the boat in irons. You will have difficulty in tacking and you will need to backwind the jib to complete the tack.
- This is a similar phenomenon to the flogging flag, or luffing sail, where the pressure fluctuates from side to side.

Momentum

One other significant reason for this tacking problem, is the lack of momentum. The momentum of a heavy ballast keel helps considerably in tacking a monohull. On a monohull this is "shooting" as in shooting a mooring buoy, when you can fetch your mooring under momentum after you have let your sails luff.

It's a trade-off. Not having the momentum is wonderful around a dock but missing on a tack. Since most cruisers dock more than they tack, its probably a favorable trade-off (not true of racers, of course).

Jibing

On the other hand catamarans, because of their stability and lack of heeling eliminate most of the terror of jibing. That terror is deeply ingrained in monohull sailors. So ingrained, that they buy special equipment to prevent it, control it and modify it. While this equipment is also usable on catamarans, it is seldom seen because jibing is no big deal. There, you have the makings of a true user comparison. The monohulls tack better, the cats jibe better. Since most cruisers aspire to downwind sailing, it seems that the cats have the edge this time.

Little Things Mean a Lot

Elsewhere in this book I have described exactly why rough water and high waves fling things around on a monohull but do not on a catamaran. From the cruisers eye view, this is profoundly important, and perhaps is the deciding factor in a buy/no buy decision. The ability to carry on normal daily activities while cruising very often is the difference between having the time of your life and having a grueling, tedious trip. This is most emphatically so with spousal and offspring relationships. Here, less than totally committed wives rebel at the loss of tranquility and children often resent the lack of ability to carry on their routines.

Remember the Bing Crosby song with lyrics: "The weather outside is frightful, but in here it's so delightful!" That title is the essence of the cruising catamaran advantage. Often Corinne and I would hardly believe how terrible it was out there. Inside, Corinne might be baking a cake, using the computer, or anything else she chose. Our catamaran sports regular dinette chairs that we use for all purposes. So far, in 20 years and tens of thousands of miles of cruising and racing, we have never had them fall over. (and we only have a little 32 footer, as you get bigger, it gets even better)

Occasionally we will be motor-sailing down the Intracoastal Waterway and see monohulls trying to do the same thing. When you sail along side tall trees, the wind turns fluky and gusty, especially when you sail past a river or creek mouth. You can watch the monohulls suddenly heel as they go past an opening in the trees. They straighten up only to suddenly heel again just a little way further on. After a couple of hours of this punishment, someone will furl the sail, or sails. Subsequently, they will just motor. Corinne and I have watched that scenario hundreds of times and spoken with many participants involved, confirming our thoughts on the process. That process is the sudden disruption of routine. We, on the other hand, simply accelerate under those identical conditions, often reaching ten or eleven knots with no effort or trauma at all.

In June 1986, three cruising boats left England bound for the United States. They were to sail together for safety and security, a convoy being a quite usual thing for cruisers.

Two of the three boats were monohulls, the third was a Bill O'Brien designed, *Oceanic* Catamaran, owned and sailed by Chris and Anne Monkton and sailing with their three young children. Besides the adventure of crossing the Atlantic, their other agenda was to sell the boats at a considerable profit. At that time the monetary exchange rate was favorable for such a venture. All three married couples owned and sailed their own boats.

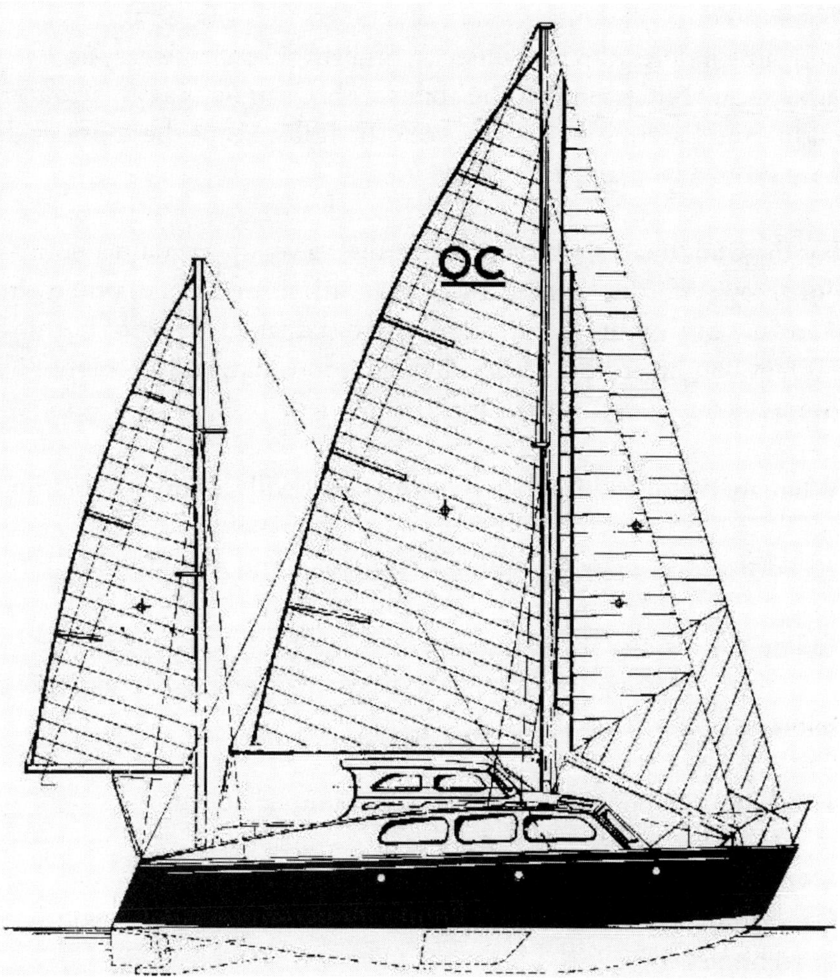

The famous Oceanic 30, made famous by the infamous Rosie Swale and her trip around Cape Horn

Upon arrival in the colonies, (sic) the two couples on the monohulls split, the wives returning to England in a huff. As Anne Monkton tells the story, they were just sick and tired of cooking on the floor having their meals flung around. Being tossed from their bunks when they tried to sleep exacerbated the situation and added insult added to injury. When asked by Anne about why they were cooking on the floor, they looked at her incredulously. They could not believe that she could have carried on domestic life in normal fashion in the same sea conditions they found so intolerable.

Comparison Summary
True cruising catamarans, Classes 4 and 5, though displacement vessels, sail level and more on top of the water than through it. They slam (pound) less, have less spray on deck and seldom if ever have green water on board. They have little or no tendencies to broach and best of all, have no surplus kinetic energy around to fling things (and people) about the boat. The results of these differentials are that the catamaran cruiser perceives conditions to better than his monohull counterpart. He sails much further in any given time frame, and arrives at his or her destination less exhausted.

There are further advantages at docks and anchorages. Cruising catamarans are cruiser friendly. They have little momentum to counteract. Straight sides create easy access to dock. Higher power to weight ratios create better control. The ability to back straight as an arrow and the tremendous advantage of raising and lowering anchors between the hulls, keeps the topsides dent free. Twin screw catamarans have more advantages still.

For these advantages, the catamaran cruiser gives up a certain amount of feel, or "rush" and often, but not always, effortless tacking. In some locations more than others, he gives up mobility due to beam. As a rule of thumb, you can carry up to sixteen (16) feet of beam with no penalty. Between 16 and 20 feet you begin to find penalties. Beyond that, you lose mobility rapidly. By the time you reach twenty-two feet, you are playing in a different ball game.

What are the differences between the way I will sail my catamaran and how I sail a my monohull?

The differences between sailing a catamaran and a monohull come in two distinct areas.

1. The first being: The catamaran does not heel.
2. The second being: The catamaran has a much wider sheeting base, therefore leaving you many more sail trim options.

There are a few myths we should dispel immediately.

Myth 1
Catamarans always have the wind forward of the beam, so you do not need downwind sailing sails or technology.

Reality
Smile when you hear this, as Porgy says to Bess, "it ain't necessarily so." Perhaps the Formula 40 class and other hot racers may experience this phenomenon. The average cruiser, which are the cruisers defined by sailing Class 4 and 5 boats, most definitely will not.

If you think you are going to tack downwind through most of the Intracoastal waterway, or over the Chesapeake Bay, or any location in which you must follow a buoyed channel, forget it. Often, your location will constrain you. Further constraint comes from the obstacles outside the buoyed channel. Obstacles like crab pots, fish traps, sunken wrecks, anchored fishing boats, trawlers, seiners, markers and a host of others. Many of your boats will not have the potential speed capability anyway. Even many of those that do, the problems created by high speed while cruising may nevertheless, overrule the possibility of tacking down wind. There are always exceptions and contradictions. This is no exception.

If you have been dickering on a cruising boat, as defined in the chapter: Classifying Multihulls, and the salesman keeps on talking about windward ability, maybe its time to start looking at a different boat. The reality of cruising existence is an attempt to always sail downwind. Of course, the perversity of sailing mitigates against this, nevertheless, that is the cruising ideal.

Myth 2
Catamarans don't go to windward.

Reality
Any well designed catamaran goes to windward at an acceptable angle. Perhaps not as close as a pedigree monohull race boat with thousands of dollars worth of go fast goodies and an outrageously deep high aspect ratio fin keel and a hydraulic backstay; but boat for boat, class for class they are just as weatherly and in some case, more weatherly than the competition. Anytime someone tells you: "catamarans don't go to windward," remind him about Dennis Conner.

> *Denis Conner, in his overwhelming win of the 1990 Americas Cup, pitted his 60-foot catamaran against the 120-foot New Zealand challenger. The challenger was the most sophisticated mega yacht built to that date. Dennis easily out pointed him. Once, he out pointed him by more than five degrees. The entire world watched this happen!*

In addition, the most weatherly boats in the World are the C Class catamarans. Boats like Cogito sail as close to the wind as ice boats, somewhere around 15-17 degrees.

Myth 3
Multihulls are hard to tack.

Reality
Some are, some aren't. Just like monohulls. I have tried to place a 1 thru 5 scale on the boats I personally have sailed extensively. This should give you a little insight into which are better at tacking than others. If I use our ex-charter boat, The Allied 39 monohull as my scratch boat, then all the others measure to that. The Allied was definitely a "2" or perhaps even a "1". I use this boat as my example because it is a good boat as described in popular literature. It has a mystique about it though I think, that mystique is undeserved.

On the other hand, many modern catamarans tack like dinghies. I spent five days sailing and cruising aboard the MANTA 38, and another series of trips aboard *Kestrel*, another *Manta 38*. Those boats sailed beautifully, tacked like a dream, handled impeccably and were a pleasure to sail. Completely the opposite of our old Allied 39 that we had not so affectionately named: "roll, wallow & broach!" Yet, the Allied has the good reputation.

Consider, however, the catamarans that tack the best are those with the balanced spade rudders under the hulls. This may be an aid for tacking, but is sure is a detriment for cruising. First, it creates deeper draft. Second, it leaves the rudders vulnerable to damage. Third, it is an invitation to snag crab pot lines. Forth, it creates holes in the hull that need some type of fitting. Fittings are a possible source of annoying chronic leakage, requiring haulout to remove the rudder(s) for repair. The ideal cruising rudder, according to the Kanter definition of the ideal cruising boat, is transom hung. It is not a perfect system, but at least it is accessible from outside and does not violate the integrity of the hull. It's the old proverb number one: "If you haven't got it, you can't break it!"

Keep in the forefront of your consciousness that balanced spade rudders are creatures of the racing rules. The reason transom hung rudders have disappeared from sailboats is they figure in the total waterline length of the vessel. There has always been controversy about this practice. One way to avoid the controversy is to put the rudder under the boat. It is true that the balanced spade rudder is more efficient. The trade off is clear, more efficient sailing, for higher vulnerability to damage and less accessibility for repair.

Understanding Leeway

1. What is Leeway?
2. How do you measure leeway?
3. How much leeway does the average boat make?
4. How do I prevent it? Control it? Take advantage of it?

1. What is leeway?

Leeway is described as slippage, or the amount of unwanted downwind progress you make through the water. In other words, progress to leeward. If not resisted by lateral plane devices, it would be so excessive you could not sail at all, except down wind.

2. How do you measure leeway?

Slippage, or perhaps, skidding, describes leeway. It is measured in angles. The angle between the actual heading, and compass course is the leeway angle. The course made good is the compass reading with the leeway angle added to it. Leeway angle is easily measured with GPS.

The closer to the wind, the more the leeway. The less efficient the lateral plane device, the greater the leeway. The slower the speed through the water the greater the leeway potential.

3. How much leeway does the average boat make?

My own boat has two highly sophisticated leeway preventing devices, daggerboards and hard chine aft. I consider it above average in the leeway prevention department. It makes up to ten degrees of leeway, depending, of course, on conditions. In moderate conditions I allow about 2 degrees for a broad reach, 5 degrees for beam reach and seven degrees for a beat. If conditions are bad, like light winds and rough seas, then I allow 10 degrees. This is just a quick rule of thumb, and adjusts to actual conditions.

People lavish much money on lateral plane devices, just as they do on ultra high tech sail rigs. As with sail rigs, you trade off a few degrees of windward ability for considerable expense, vulnerability to damage and deeper draft. Not necessarily a good trade off for a cruising boat.

4. How do I prevent it? Control it? Take advantage of it?

You can't prevent it entirely, but you can control it within parameters. The following is a descriptions of the various methods of control, with comments about how each method affects boats used for cruising.

Taking Some Advantage of Leeway

When sailing between dead downwind and beam reaching, occasionally, it is advantageous to deliberately cause the boat to have some leeway. You do this to better fill your sails. By pointing a little higher, you can get a better angle of attack to the apparent wind, therefore, more drive. By allowing leeway, your course made good will be downwind of your compass heading. This is especially handy when you are sailing downwind enough to blanket your jib, but not enough to pole out the jib on the opposite side. This tactic is best accomplished with vessels having retractable leeway devices.

This tactic is a favorite of the racers. Its application is a reason (although admittedly a minor one) for the switch from a-symmetrical to symmetrical hulls.

A Simple Leeway Gauge

Experiment by making yourself a simple leeway gauge. You can make this gauge out of any scrap plywood or similar flat material. It consists of a homemade large scale protractor and a piece of monofilament fishing line . The protractor has zero in the center, and one degree increments on either side. Attach the gauge to a clear spot on the transom and string out a piece of line long enough so that it trails back beyond your wake. The line should cross the zero mark when trailing straight back. You can then read the number of degrees of leeway directly from the protractor. Playing with your boards (and rudders) while measuring will allow you to create a leeway chart for yourself, just like a compass correction chart, and you will know just how and when to adjust your boards for maximum efficiency. Each boat is different enough to need to make its own chart.

Your GPS will show you how much leeway you are making, but not with the simple immediate accuracy of the gauge.

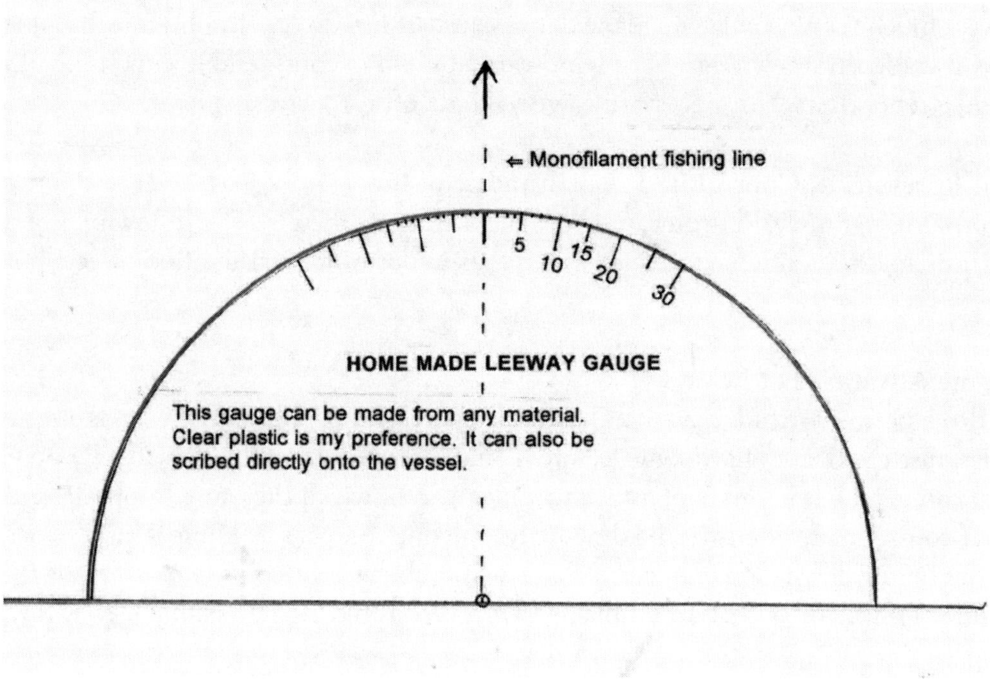

Preventing leeway (lateral plane devices)

3. Lateral plane devices

- Keels.
- Centerboards.
- Daggerboards.
- Vortex generators.

Keels

Many cruising catamarans have fixed, low aspect ratio keels. Some of the advantages are obvious, they protect the bottom, protect the rudders, are cheap to make and there are no moving parts or space lost from the accommodation. The disadvantages are also straight forward and obvious. There is greater drag due to more wetted surface and draft increases considerably. They are fixed in place. There is nothing you can do about their ability, or lack thereof. Downwind, they have the distinct disadvantage of not being retractable, thus creating excess drag. In addition, you cannot take full advantage of natural leeway by pointing just a little higher to better fill your sails, but still making good the desired course because of excessive leeway. Keels also increase your draft.

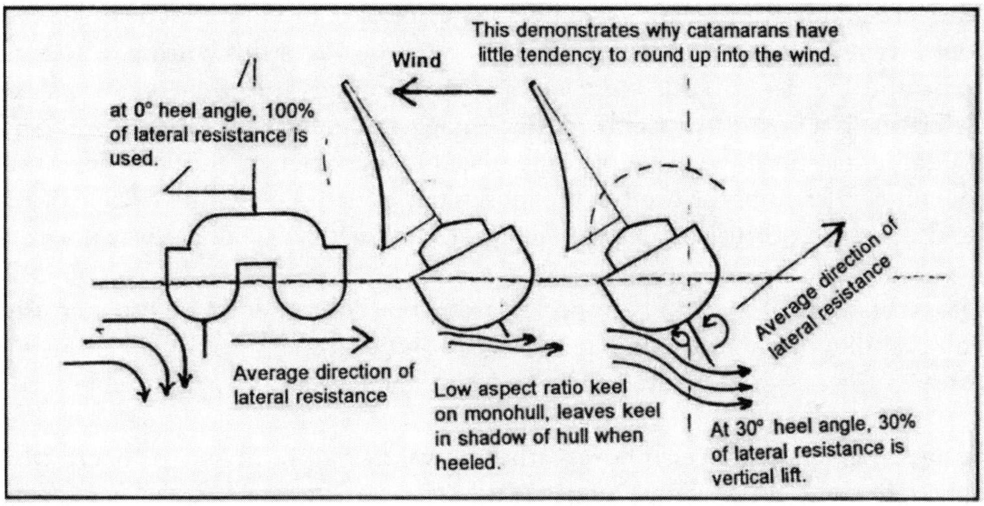

Monohulls round up because the center of effort is not over the center of lateral resistance as shown in this diagram. The more the boat heels, the more weather helm is produced and the more the boat wants to round up. Helping in the round up process is the lift being generated by the keel.

There is a great deal learned from monohull keel boats, both about the operation and shape of keels. Some applies to catamarans, but some does not. Catamarans can get far better leeway protection from a low aspect keel than a monohull, simply because the catamaran does not heel. Thus the side flow is blocked on a catamaran where it is disturbed on a monohull. The more modern and more International Offshore Rule (IOR) conforming the monohull, the fuller the bilge and flatter the bottom the worse this condition becomes. This is another reason for the trend to deep draft on the modern monohulls.

Centerboards

Centerboards have been very popular, until lately. They are not only popular with multihulls but monohulls and for the same reasons. Circa 2002 the overriding factor of production cost has virtually eliminated their availability. A centerboard is usually a pivoting board within a trunk that will kick up into its designed trunk if the bottom is struck. The biggest disadvantage to a centerboard is the inability to create an optimum foil section shape. Another disadvantage is the open trunk which creates some turbulence. A third disadvantage is a large trunk in the accommodation. A forth disadvantage is the often esoteric and complicated lifting or downhaul mechanism. Most catamarans with centerboards have boards that are buoyant and float up into their trunks. These boards are held down under way. This reverses the mechanism found on monohull centerboards. The advantages of centerboards are that they are retractable when not needed. This is especially important when taking evasive action under storm conditions at sea. Extremely shallow draft becomes possible by the retracted boards.

Daggerboards

Daggerboards, under many circumstances are the most efficient lateral plane device. They have incredible advantages but some safety concerns. Optimum foil shapes, optimum aspect ratios, optimum tip shapes are all possible with daggerboards. Most daggerboard trunks fit well and allow only a minimum amount of air to be drawn down to create ventilation. Minimum weight, ease of operation, ease of construction and installation and general all around utility describes most daggerboard design. However, designs like Catana managed to turn simple operating systems into hidden complexities.

They are unquestionably the lateral plane device of choice for closed course racing. However, recent sophisticated experimentation changes that premise. As of this writing, vortex generators, foil shaped bows and low aspect keels are making serious inroads into that premise. A daggerboard, as name describes, is a board in a scabbard, like a dagger. Centerboards, or daggerboards, should be used in the same manner. The principle difference between them is the way they work. The daggerboard is the more efficient of the two but the most likely to damage the hull or take damage itself if grounded. It, also, has an obtrusive top sticking out on deck when not all the way down.

Daggerboards fall short in two specific areas, both detrimental to a true cruising boat. First, is their vulnerability upon grounding or hitting a solid object at speed. Second, is the space they take up through the center of the hull, often splitting the best and most desirable accommodation space.

Some designers build crash boxes either forward or aft of the daggerboard case to absorb the shock of a serious collision. Some design the boards so they will just shear off at the hull line. Both techniques will save the boat from total disaster. But, will probably leave you unable to win the race. If you happen to be cruising, they will leave you in need of major repairs somewhere away from facilities.

Vortex Generators

An old idea that may be getting a rebirth in a new form. Essentially, it is an extremely low aspect keel that is quite straight and long, with squared edges. Some designers believe having an end plate increases the efficiency. However, I disagree with calling such an end-plate keel a vortex generator. The concept of a vortex generator is that water does not like to flow around a hard corner thus causing a vortex. A vortex is literally defined is a vacuum. This vacuum prevents the leeway. It is the strongest, least expensive leeway prevention method, unfortunately, apparently also the least effective. However, its effectiveness increases with higher vessel speed, meaning it gets better in heavy going, just when you need it most. End-plate keels have their own set of problems much like the wing keels on monohulls.

Not generally discussed is the fact that daggerboards need speed to be efficient. At low speeds, say five knots or less, the fixed low aspect keel is probably more efficient than the daggerboard.

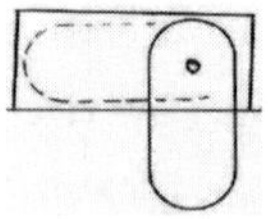

Pivoting centerboard. Diagram 1 shows a typical installation. A proper planform may be obtained with this type of board but when lowered, it leaves a big opening in the bottom of the trunk. This opening creates considerable turbulence which negates part of the efficiency of the board.

Hemispherical pivoting board that keeps the opening closed when lowered but has a very inefficient planform. Diagram 2

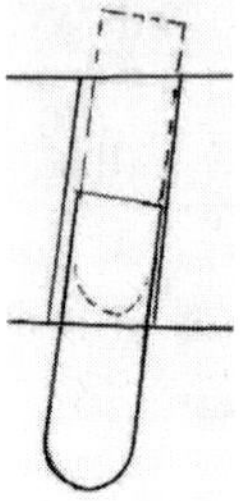

Daggerboard. The most efficient of all lateral plane formats. However, has several disadvantages. many people object to the top of the board sticking through the hull. Lastly, there is a certain safety concern in the event of a collision at speed. Diagram 3

Rudders

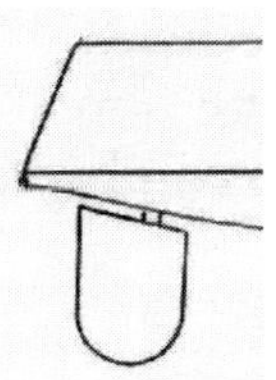

Balanced spade rudder

Most popular with modern production boats, both multihull and monohull. Favorite of racers because it is not measured in the LWL and gives best overall performance.

Advantages. Best overall performance because the underbody acts as an end plate. Lowest production cost.

Disadvantages. Snags pot lines and other refuse easily. Once snagged, it usually means a trip overboard to clear it. Most prone to damage from both floating debris and grounding. Contributes to quirky steering and oversensitive helm

Sample of vessels using this system: *Manta, Victory, Privilege. Fountaine Pajot, Seawind*, others. . .

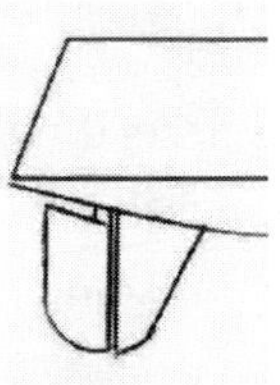

Skeg mounted rudder. Very popular.

Advantages. Assists in steering straight course. Protects rudder from damage. Allows vessel to sit on bottom.

Disadvantages. Must be well built to be effective.

Sample of vessels using this system: *PDQ 36, Edel 35, Prout 50*

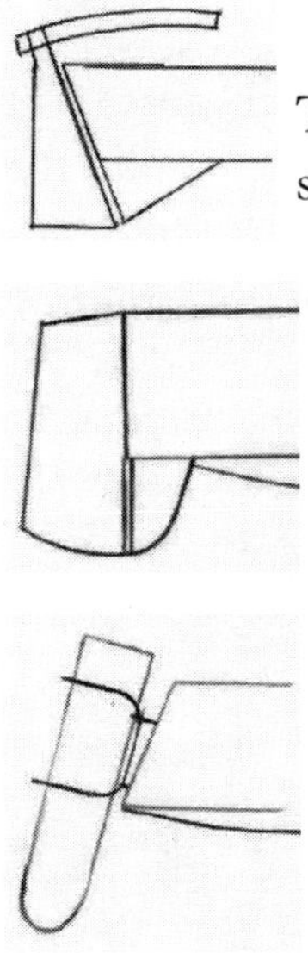

Transom mounted rudder. Very popular in earlier designs. *Wharram, Prout* and *Gemini* styles diagramed.

Advantages. Rudder is outside the boat, thus does not violate hull integrity. Adds to LWL thus helps that aspect of performance. Easiest to repair. *Wharram* style, (top) seen in many traditional monohulls, has forward sloping rudder post for balance and skeg protection.

Prout type (center) has skeg protection. They are normally hollow for additional buoyancy.

Gemini type, (bottom) has balanced slide out rudder blades in cages. Blades are balanced by tipping top back. Blades removed and stowed when not in use.

Disadvantages. Usually preclude other uses for stern area especially with tiller steering. Contribute to clutter. Exception, like *Catalac* 9 meter have a platform that covers over the kick-up transom mounted rudders. Rudders are exposed to damage.

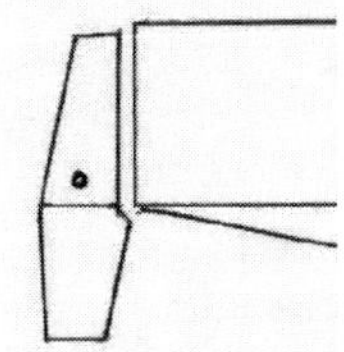

Kick-up transom mounted rudders. Typical style, *Louisiane* 37 diagramed. Once the standard fare for catamarans, now the trend toward shallow keels has eliminated the need for them.

Advantages. Has all the advantages of transom hung plus the advantage of kicking up for damage prevention and shallow draft.

Disadvantages. Lifting mechanism and pivot mechanism are weak spots and high maintenance items. Sample boats include *Catalac* 8 & 9 m, *Iroquois*, *Gulfstream 35*, *Louisiane 37*, all the sport trimarans and all the smaller day sailing boats. The *Gemini 105M* and many of the semi-custom boats use a kick-up system that is under the hull and has the advantages of kick-up plus the spade rudder advantage.

Windward Performance, Fact and Fiction

What is good windward performance? What is comparative or relative windward performance? Why is it necessary? Who seeks it? What design factors most influence it? Why the exceptional emphasis on it and what does it mean to the average sailor and especially the cruiser?

One very important reason for this chapter is to provide you, the readers, with reasonable, sensible, articulate and accurate answers for the inevitable question: "How does it go to windward?" First, let us understand some things about windward ability. Exactly what is it? Why we need it and most important of all, why does everyone who owns a monohull sailboat seem to always talk about it?

Windward Ability Defined

Windward ability is the ability of a sailing vessel to go upwind, into the wind or otherwise described as in the direction from which the wind is blowing. This characteristic often defined as weatherly. Anything, even a haystack can sail downwind. It is to reach destinations in the direction from which the wind is blowing that describes windward ability.

Thus far in the development of sailing vessels, none can sail directly into the wind, all must go at some angle to the wind. The amount of the angle is the arbiter of windward ability. The smaller the angle, the greater the windward ability of the vessel. A full description of all the tricks of the trade for sailing to windward is beyond the scope of this book, yet all the basics are need-to-know information. See the glossary for several good books on how to sail.

The polar diagram below shows the circle of windward ability.

- 35 to 45 degrees from the true wind: Good.
- 45 to 55 degrees: fair/poor.
- 55 degrees or more: Varies from mediocre through terrible.

This circle is constructed for class 3 through 5 boats only. There are boats that sail as close as 14 degrees and perhaps closer by the time this is printed.

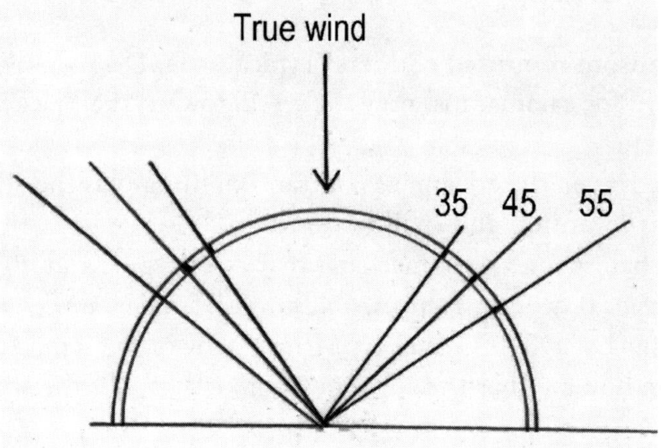

Often, you will hear the argument that a boat must go to windward well in order to claw off a lee shore. Other arguments are much more practical in that they say that the perversity of sailing finds you with the wind always on the nose. Either argument being the case, how close an angle does your boat need to attain to be considered weatherly?

Claw Off a Lee Shore

The next part of the Go to Windward syndrome is the claw off a lee shore group. As you get further into sailing, you will find and examine the people and personalities involved. The Claw Off a Lee shore group, usually sail the types of boats in which clawing off a lee shore really is a problem. Heavy displacement boats of less than forty feet, the "roommarans" of the monohull world, schooners, gaff rigs, antique Friendship sloops, double-ended boats under 40 feet and various other types of character boats; for those boats, clawing off a lee shore can be a real concern. On the other hand, if you have a modern production catamaran (there are exceptions) or popular trimaran, properly rigged and fitted, you will be far ahead in your ability to claw off a lee shore.

To the lee shore argument I would first suggest that not allowing yourself to be trapped in that circumstance is far more vital than any imagined clawing ability. However, if your vessel can maintain between 45 and 55 degrees to the true wind you will have as good an opportunity as anyone.

Decades ago this argument had more relevance. With the coming of the age of powerful, reliable auxiliary engines, universally installed in cruising boats, the lee shore argument gets less important.

To the perversity part, I would suggest any angle you can live with is the proper angle. By definition a modern boat is considered weatherly when it will tack between 90 degrees and hold course to windward in extreme conditions between 45 and 55 degrees. There are many boats that cannot do that, no matter how many hulls they have.

On the other hand, from the practical point of operation as a cruiser, you will find that even though your vessel might actually point higher, you will foot off a bit because it is dryer and more comfortable. Really driving to windward can be wet and rough.

Windward ability is the rational and the evaluating factor in monohull design. Sailors, believing themselves well versed and sophisticated in their sport will ask: "how does it go to windward?" or, "I heard they don't go to windward?" or some similar question. They will ask these questions thinking they are good, reasonable questions. They will ask, despite the fact and knowledge that most of the monohull so-called cruising boats are notoriously poor to windward. Many popular cruising boats are practically laughable to windward. Some boats need the assistance of the engine to tack. The thing you must get the questioners to understand is that the people who buy those boats don't care! Windward ability played no part in their, buy/don't buy, decision.

There is one more important aspect of windward ability and it applies almost exclusively to good monohulls and sport trimarans. That is the feeling , the rush, you get when your boat is in the groove, beating to windward, seemingly defying nature. That feeling is not obtainable on a class 4 or 5 cruising cat. They simply never sail that way. Sorry, but that type of sailing is one of the items you give in trade for the many other advantages of you cruising catamaran.

150 Years of Rule Making

The extreme emphasis on windward performance is a result of 150 years of various racing rules development. The universal uses of the Olympic racing course has dominated the design concept of monohull sailboats for the last 100 years, which is long before there was a popular fiberglass boating industry. That rule produced the trend toward deep fin keels, bendy masts, backstay adjusters and myriad other go-fast items. Race course results are the dominant hype in advertising. Due to the economics of production boats, it is only natural that the same molds produce both the racing and the so-called cruising boats.

Modern history buffs, will remember that automobiles were once marketed the same way. For decades, that was what auto racing was all about. Getting limelight, publicity, headlines, in other words, hype, were the reasons auto racing existed. The sailboat industry presently engages in the same routine.

What has racing around a track at 200 miles per hour have to do with you and me and the kids needing a sedan to visit Grandma? Not very much! What has a high tech speed machine hurtling across the ocean at thirty-five knots have to do with mom and pop enjoying the Intracoastal Waterway? Not much either!

Racers Emphasize Windward Ability

1.4 x the square root of the waterline length, is the formula expressing monohull sailboat speed limits: There is some variation in this formula. For ballasted displacement hulls, it is inconsequential to the scope of this book. From the practical perspective, it means that there is little gain from maximizing a keel boats downwind performance. You can only achieve a percentage of the wind speed based upon your waterline length. Most races are sailed in light air, further limiting any advantage. Battling the racing rules to eke out a fraction of a knot downwind is a classic study in the law of diminishing returns. Like using your old 180% Genoa, remember them? Only esoteric speed machines such as "C" Class catamarans and ice boats can sail faster than the true wind speed, downwind.

In addition, the design factors that create the downwind abilities are the very factors that are heavily penalized in the rating rules. Items like centerboards, which are retracted downwind thus relieving the boat of wetted surface, are the most heavily penalized. Therefore, it is reasonable to understand that considering a closed course, when you design a boat that will persevere off the wind, but gain advantage on the wind, you have a race winner.

Just contemplate for a moment. All other factors being equal, in a two-mile windward leg, the boat that can point one degree closer to the wind covers .33 miles less distance over the bottom. Having sailed a shorter course, by three tenths of a mile, it will be ahead of its competitors traveling the same speed, making the same leeway.

Carrying this further, the things not penalized under the rating rules are all the little skill and expense oriented go fast tricks like:

- The tension of your headstay (backstay adjuster etc.)
- The section of your mast (bend, spreaders, runners, etc.)
- The camber of your sails (cut, adjustment, cloth, age, etc.)
- The angle of your jib leads (more adjustment, bigger advantage.)
- The condition or coating of your bottom (clean, shiny, etc.)
- The size and layout of your winches (bigger, more convenient.)
- Your deck hardware, layout and design (travelers, clutches, etc.)
- The placement of your crew (rail meat.)
- The weight of your stores, equipment, fuel, water, etc.

These are but a few of the unregulated features that racing sailors and the marketeers that supply them love to play with. It is the most exhilarating part of monohull racing. In dealing with the various conundrums faced by monohull racers, tremendous amounts of energy, enthusiasm and experimentation are expended making their boats more weatherly. This is only right and proper. No criticism intended or implied. I enjoy it as well as everyone else.

My only thought concerning this wonderful and fun process is how little it relates to cruising.

Factors Influencing Windward Ability

Rigging and sails

- Cat/sloop/cutter/ketch.
- Cloth weight.
- Cut of jibs.
- Size of mast section.
- Diameter of wire & number of wires supporting mast.
- Number of protuberances such as radio antennas, radar, TV.
- How the mast is stepped, keel, deck, rotating.

From the perspective of a cruising boat, the rigs that are most useful are sloop and cutter. Cat rig, especially the unstayed cat rig, while being very efficient, is too cumbersome and not sufficiently versatile for cruisers (except in quite small boats.) Cat rigs also require taller masts, thus limiting cruising options because of fixed bridge restrictions. Ketch rigs have never worked well on any multihull with any speed potential at all. They are really not that terrific on monohulls either. But, they do break down the sail area and give you a mizzen mast from which you may hang all your toys such as various and sundry antennas, flags, mast steps, etc.

The greatest disadvantage of the cutter rig is in tacking. Since the mast is so far back, the main may put the boat into irons when tacking. There is also interference to the jib by the inner stay during a tack. This requires more skill and finesse. The same is true of the 50/50 masthead rig, the most popular monohull rig, when used on a catamaran. However, the inner staysail compensates for the disadvantage by its flexibility of use in heavy air. This sail is in just the right position to drive the boat nicely to windward. It moves the center of effort aft giving some weather helm. It allows you to have the sail cut specifically for the expected wind range and to allow you to dowse the forward headsail completely.

Most modern catamarans and trimarans use the fractional sloop rig with a full-batten mainsail and a roller furling jib because of the inherent efficiency of the rig, the simplicity of the rigging and the ease of handling. Even the monohulls are beginning to use this system. It is probably the rig of choice in the 35 to 50 foot vessel size range we are primarily addressing.

With a properly designed fractional rig, many go-fast goodies like hydraulic backstay tensioners are unnecessary. There are no fixed backstays to interfere with optimum mainsail shape and roach. Head stay tension is a function of main sheet tension, thus relieving the rig of the constant strain of over tight rigging.

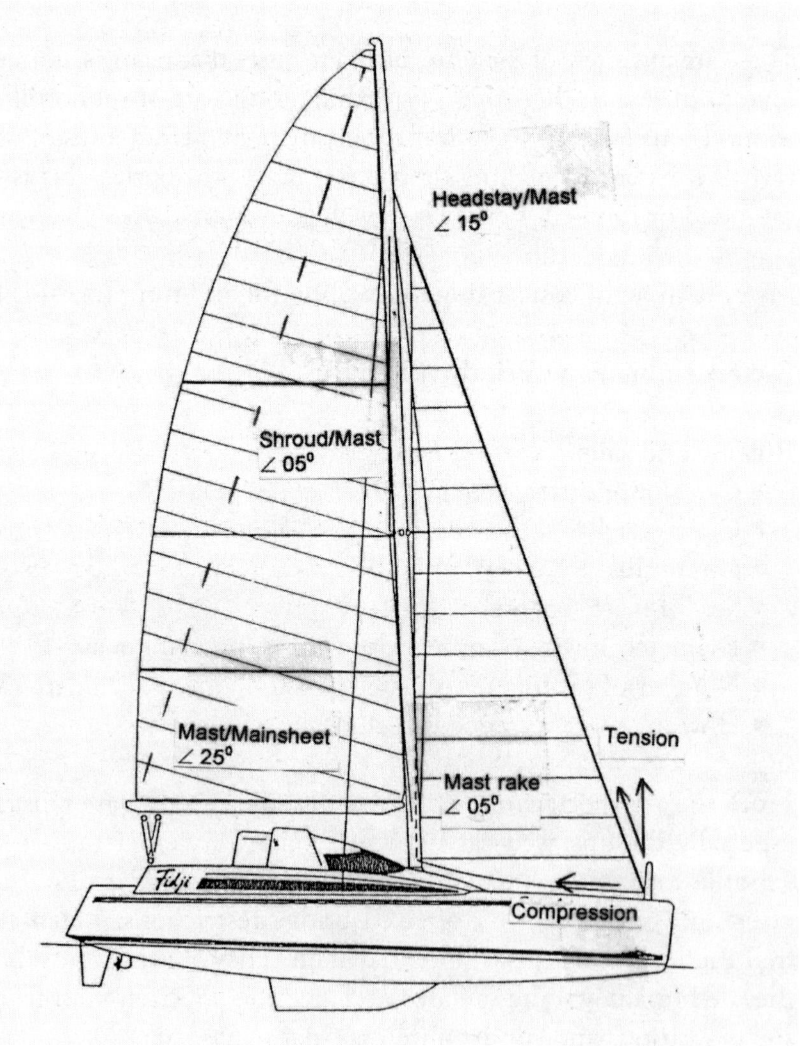

As you can ascertain from the diagram, it would be self-defeating to attempt a bar-tight headstay with shroud tension since the angle of incidence of the headstay to the mast is greater than that of the shrouds. This means that it is not mechanically possible to achieve it. On the other hand, note the angle from the mainsheet. That is your headstay tensioner. A further note distinctly shows the compression from the headstay towards the mast clearly outlining the need for a compression member as discussed previously.

Hull shapes

- Distribution of buoyancy.
- Hull entry shape.
- Waterplane shape.
- Underwater shapes.
- Stern shapes.
- Wetted surface.

The famous Cod's head and Mackerel's tail definition of hull shape from the days of yore is now supplanted in terms of NACA foil sections. Nevertheless, the ancient description is still relevant. Those general shapes, at the waterplane, are the shapes needed for downwind sailing. Something like the reverse of that is optimum for upwind sailing. There, you have another conundrum. You cannot have both in the same boat, yet! All the six facets of hull shape affect the windward ability of the vessel. You cannot have it all. Therefore, compromises to facilitate cruising, which subsumes downwind sailing and load carrying ability, affect the windward ability of the boat.

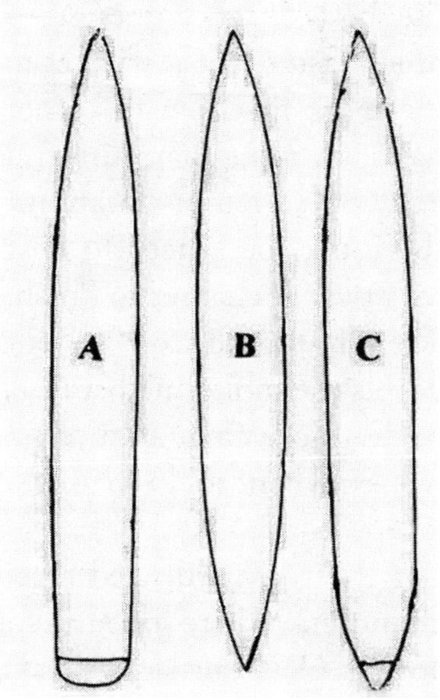

Waterplane

This figure shows three hulls with exactly the same LOA and HB but have quite different proportions and approaches to the distribution of buoyancy.

A. Carries buoyancy well aft: *Lagoon, Privilege, Norseman.*

B. Carries buoyancy centered. *Prout, Wharram, Karmin,* Proa designs.

C. Carries buoyancy 3/4 aft. The present compromise to accommodate the required loads.

The Polar Diagram

This diagram compares the VMG of a typical monohull racer with a typical multihull racer. In the case of the mono-hull, it arrives at close to its maximum speed close hauled. If it foots off a little to gain some more speed by reaching, it gains only a marginal amount of additional speed because of its waterline length hull-speed limitation. On the other hand, what is readily apparent on the diagram is that the multihull racer gains a dramatic amount of speed by footing.. Right there, you have the key to the mystery. Each type of boat is sailed emphasizing what the boat does best. This usage relates to the factors showing the greatest promise for winning the race. On the race course, comparing monohulls to multihulls and racing them together is always an eye opener to most racers.

The fact that the better multihull racers go to windward as they do, astonishes many racers. I constantly get the comment: "Gee, I didn't know that thing could go to windward?" There again, the reason is simple. In an around the buoys race, you may not get a close comparison. The multihull, though it may cover more ground to windward because footing off slightly is the better tactic, does it so much faster that it will get to the mark first. On the other hand, in races that start and end at different locations, there may be a need to beat up a river or channel to finish. The multihull often out points the monohull because, under those circumstances, it was expedient to do so.

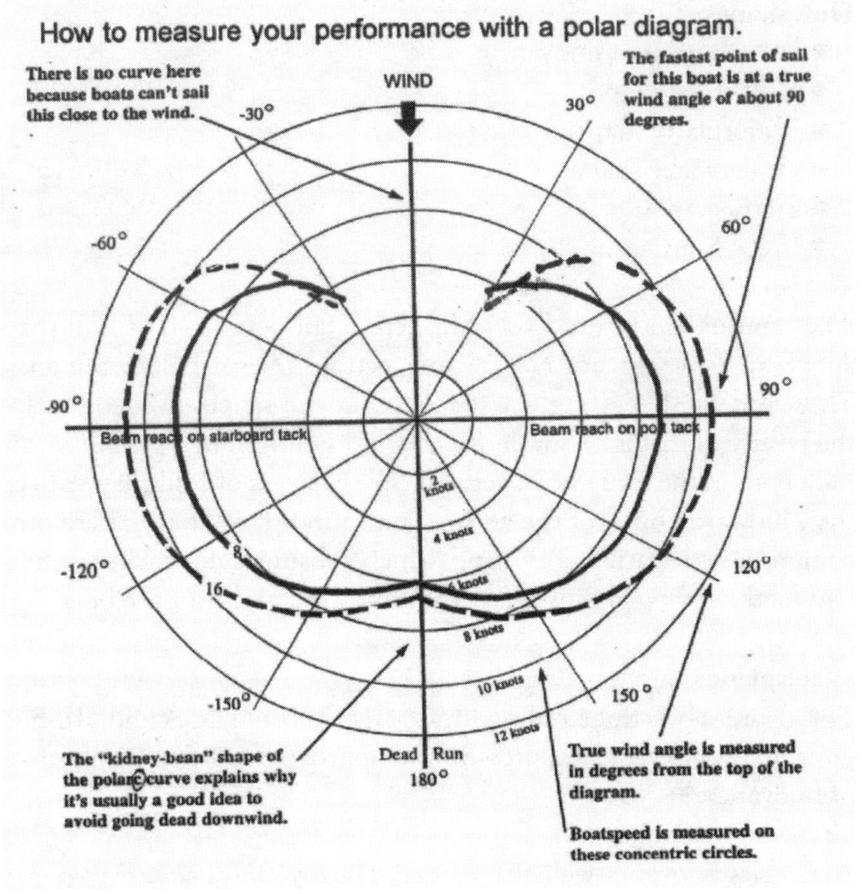

Monohull ---------------------- Multihull - - - - - - - - - -

For the truly dedicated racers, there are several types of vector analysis plotters available at most chandlers. There are also computer programs available for those with onboard PC's. The vector analysis plotter makes the distinction in velocity made good to windward (VMG). You factor in your pointing angle, your speed over the bottom, and compare the differences that you would achieve by falling off to gain more speed. Of course, the more modern way is not to fool with some celluloid discs but let your computer or GPS do it for you. (note: In the year I have been writing this book, all the above has been replaced by GPS)

Multihullers do this too. Mostly, falling off to gain more speed is the tactical preference because the differences between beating and reaching speeds are much more significant.

How to Use Centerboards and Daggerboards

Basic trim for both is similar. Differences are the raising and lowering arrangements, since the centerboard pivots and the daggerboard slides. From the theoretical performance point of view, the daggerboard is the most efficient device of all. They are constructed in an optimum foil shape and aspect ratio. Since they are retracted or extended vertically in their trunks, the foil sections do not change shape in relation to the angle with the water flow. Centerboards, because they pivot, do change relationships. An additional advantage is light weight and ease of removal for service. One disadvantage of a daggerboard is the ungainly top of the board sticking up through the deck.

How To Use Boards For Maximum Advantage

The preferred method is to use the leeward board and only the leeward board. In many boats, using both boards creates a lot of drag. It is interesting that the day sailing catamaran classes know this and routinely lift the windward board, but the larger boats do not. It is also notable that experiments with twin keel monohulls, have produced mediocre or worse, performance. Check your twin keelers, like Westerly, and you will see what I mean. The sales pitch and the paper descriptions belie the facts that the twin keel boats no matter how theoretically well designed, are less efficient than a comparable single keel. You will find your catamaran the same way.

An interesting aside to this is the fact that possibly some production boats that come with low aspect keels might perform better with boards. I have no comparative knowledge of this possibility. Low aspect keels represent the same production efficiency to catamaran builders as bolted on ballast keels represent to monohull builders. They are cheaper and easier to build.

In light air to windward, make sure the leeward board is all the way down. The heavier the air the less board you need except dead downwind, or when you are trying to take tactical advantage of your leeway. You always need some, depending upon your boat, of course. Boats of the same make and same configuration, Gemini, for instance, will vary greatly. This variation between boats in their basic windward ability is quite noticeable. How they react to treatment by various combinations of windward and leeward board trim depends on many factors. These extreme variations are usually the result of loading, bottom condition and total loaded weight. Weight distribution is very important.

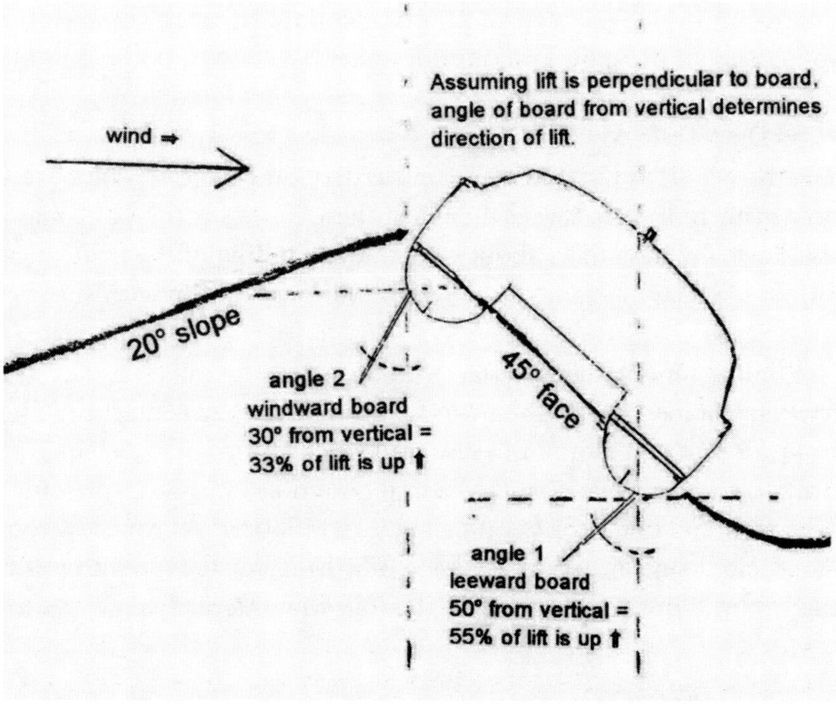

Study very carefully the diagram about using the leeward board. You will note that all boards have a certain amount of vertical lift because they all have some amount of angle beyond the perpendicular. They are foils, and act as such. Boats like *Iroquois, Sea Wind*, and many others, have boards angled from the outside, inward. These boards will develop lots of vertical lift. This vertical lift is not necessarily a bad thing if it only applies to the leeward hull. If you have both boards down, you will develop vertical lift on both hulls. Certainly, you do not want to develop vertical lift on the windward hull.

Use in extreme conditions

There are some that will tell you that in survival conditions you should keep the windward board down. Allegedly, to act like a person holding on to something above him whilst sliding down a hill. I submit that the person would be better off using his foot to push something ahead of him while sliding down the hill. The same principle applies for boards. The leeward board will lift the hull vertically, helping any side slipping. The windward board will also raise the hull vertically. Only it's the wrong hull! This will increase the angle of heel and raise the possibility of wind getting under the hull. Thus increasing rather than decreasing the possibility of capsizes.

In monohull circles, the knowledge of how the angle of the keel contributes to the rounding up motion of the boat is a known factor. Research details many attempts to thwart the vertical lift generated by the keel. You probably all remember the trim tabs on the trailing edges and the various full race boats, including the 12 meter boats. The most recent attempt to nullify this problem is the Harken experimental vessel, *Amoco Procyon,* which has a hinged keel that theoretically always stays vertical. Talk about esoteric complexity!

This conversation, of course, has little application to monohulls and trimarans as they only have one hull with a board, or keel in it, there are, as usual, exceptions. Many older trimaran designs have boards or fixed fins on the amas. Several classes of monohull have boards that slide into trunks on the leeward side. The attempt is to keep the board vertical while the boat heels. Otherwise, try to synchronize board changing with tacking. Keeping your boards clean and dent free is very important. Boards of the hinged type are not particularly efficient anyway, if the leading edge is all chewed up you create an incredible amount of drag.

How Do I Trim My Sails?

The three basic sail trim items, Camber, draft and angle of attack are the same for all boats, no matter how many hulls they have. You adjust these items for the same reasons you shift gears in your car. If you have any hesitation about your understanding of the principles involved, a good basic book on sailing like Gary Jobson's *How To Sail* will provide you with sufficient detail on those items.

Not unlike a monohull, you start with the after-most sail first, usually, the mainsail. You will find very few catamarans that like the main boom right on the centerline. There is a subtle reason for that. In reality, it is true of monohulls also, but since monohulls heel and are deliberately heeled in light air, the boom never really is on the vertical centerline anyway. Depending upon your boat, trim your sail's draft from full to flat. Flatter usually goes with faster. As you slow to monohull speeds, trim your sails like them, using greater draft. Use your three, (or four) basic controls to shape the sail. They are all important, but think of them in the following hierarchy:

1. Mainsheet.
2. Halyard (and/or downhaul or Cunningham.)
3. Outhaul.
4 Traveler.

The halyard, downhaul and outhaul control the location of the draft. Draft (camber) is centered for average speed ranges and moved forward for higher speeds or stronger winds. Tightening the halyard will move the camber forward. Tightening the outhaul will flatten the sail by reducing total draft. The mainsheet determines the sail twist and the position of the leach. After that, the often neglected but all important, traveler takes over.

A Cunningham is simply a downhaul that hooks into the sail above the boom and is strictly the province of racers. The purpose is to be able to stretch the luff without changing the total luff lengths of the sail. Note, officially measured race boats have black bands around their masts that delineate the maximum hoist permitted within the boats measured handicap. If you use a regular downhaul, or the halyard to stretch the luff and, therefore, move the camber forward, you might exceed these limits. The Cunningham allows you to stretch without changing the position of either the head or tack of the mainsail. Obviously, this sail control has no relevant use on a cruising boat.

Traveler

The mainsail traveler is very important, yet least understood sail control on your boat. Ask any Hobie or Tornado sailor, he can show you exactly the preferred usage. Simply put, you adjust your mainsail for the wind conditions with your three (or four) mainsail controls: halyard, (or downhaul, if so equipped) outhaul and mainsheet. Next, position your main to its optimum angle of attack with the traveler. As the air picks up, you can ease off your traveler to reduce drag and depower. The closer your angle of attack, the less the drag, therefore, the less heeling and the feeling of being overpowered. As the wind picks up you flatten your sail and slack off on your traveler. Slack the main sheet for more twist and you have the right combination of mainsail adjustment as the wind picks up.

Use Your Telltales

It has been my experience that most people tend to over trim their sails. The simplest way to learn to trim your main is to attach tell-tales about 18 to 20 inches from the sail luff. Three on each side is really sufficient and it is possible to get by with only two. One pair in the top third, one in the center and one pair in the lower third. A properly trimmed main will have all six tell-tales streaming straight back. Get it trimmed, then use your traveler to adjust for course change, wind shift, wind speed change etc. Keep easing off the traveler until they are all streaming straight back. Use your mainsheet to get them all to stream back evenly, by providing the proper twist. If the top tell-tales are not streaming, slack off the mainsheet, if the bottom ones are not streaming, harden up the mainsheet. If the inside tell-tales are not streaming, adjust your halyard and/or outhaul (or down haul) tensions or let out on the traveler.

You need to flatten the sail and move the camber forward for heavier air. Conversely, more camber and the boom closer to the centerline as the air lightens. My experience has been that most people tend to over trim. Don't over trim. When in doubt, let it out.

Using a vang

It is possible to do the same trimming that you do with your traveler using a vang. Depending upon which piece of gear you already own, it may not be necessary to purchase the other. It is difficult to sail a cruising boat without a vang for a host of other reasons. There are different types of vangs, but most are simply a four-part purchase tackle using fiddle blocks, with snaps shackles at the ends.

Improvements to the traveler are a worthwhile investment. The secret to getting maximum lift and minimum drag to windward is to not try to force the boom to centerline. View the main as an extension of the jib or genoa. Try to get the sail combination to act like a single sail. If you have experience with a cat rigged sailing dinghy, you will remember that the boat won't sail with the boom on the centerline. You must have some angle off center, usually around 15 degrees.

Boom centerline and sail combinations

Treat your sail combination the same way. Remember, a monohull heels when it sails. If you have its boom on the centerline on deck, it really isn't on centerline in relation to the sails. The real waterplane is the heeled immersed cross section and the boom is actually at an angle to that. Usually, that angle of heel is somewhere between 15 and 25 degrees. This represents a major difference between sailing monohulls and catamarans. This is especially true with a modern fractional rig.

Light air sailing in cruising catamarans

Emphasis always seems to be on the heavy weather, but most sailing is done in light air. Light air being described as true wind under ten knots. Production cruising multihulls have certain differences in approach for satisfactory performance in light air.

Tricks of the trade

The wide beam of a cruising catamaran provides opportunities to use freestanding sails more efficiently. Moving the jib tack to weather will increase the usable range of any free standing sail. Many people install a moveable jib tack across their bows so they can haul the jib tack upwind to get the sail out from behind the main. This tackle was standard equipment on the Stiletto. On some points of sail, moving the jib tack to leeward results in remarkable drive. It requires a little experimentation on each point of sail to determine which is the most advantageous jib tack position.

Gemini bow traveler as a way to move a freestanding sail's tack point to windward or leeward depending upon objective.

It is not necessary to have a sophisticated set up as pictured above. A tackle set between the bows will also work. That is the way the original Stiletto was set up.

Sail selection

Popular nowadays are the single luff, a-symmetrical spinnakers. They are simple, effective and easily handled by a crew of two. Yet a conventional spinnaker is also easily handled from the wide decks of a multihull with little effort if it is contained in a *Chute Scoop*© or comparable sock. There are also freestanding drifters which are very similar in application to a-symmetrical spinnakers, but far more versatile. Becoming ever more popular is the Berig Camberspar®. This sail also has advantages for light air not usually considered.

Single luff a-symmetrical spinnaker

The advantages of the single luff a-symmetrical spinnaker are that it tacks to the center of the foredeck like a conventional jib and is handled with a pair of sheets to existing or auxiliary turning blocks and winches. In essence, it is a free standing jib with a balloon-like shape. It's greatest disadvantage is its difficulty in sailing dead downwind because the main blankets the sail as it would with any jib. There are "dead spots" usually about 15 degrees either side of dead downwind where the a-symmetrical spinnaker will not fill because it is blanketed. Therefore, the sails lends itself best to tacking downwind. Jibing (or tacking) is just like any jib. However, you must be prepared to manhandle the sail past the headstay. Combining it with the moveable tack point, as described above, greatly enhances the sail's versatility.

Conventional spinnaker

Since a spinnaker is really a downwind sail, not blanketed but augmented by air spilled into it from the main, it does not have the problem of a "dead spot" on downwind courses. In the majority of cases, the cruising sailor would prefer to sail a rhumb-line course rather than tack downwind. In many cases, tacking downwind is not an option and having the flexibility of being able to jibe easily and perhaps even sail by-the-lee for periods of time is a distinct advantage. The trade-off for this is a more complex system, but still not as complex as a conventional monohull.

To fly a spinnaker as a cruising sail, consider the following. As a given, I assume that your spinnaker is housed in a Chutescoop® or similar spinnaker sock device. You can very successfully fly a spinnaker without a pole. You need a spinnaker halyard, spinnaker sheets port and starboard, that go outside all rigging to a convenient turning block near the stern and back up to your primary winches and port and starboard guys that lead to turning blocks on the bows then back to bulls-eye jam-cleats in the cockpit.

Therefore, the chute is hoisted in the conventional manner. The clews are controlled by the sheets and guys (downhauls) the beam of the boat at the bows acts as a spinnaker pole. The guys will need additional purchase power once your chute gets larger that about 1000 square feet. Most users in the larger size range have auxiliary winches. You trim the spinnaker just as you would any spinnaker. The leeward sheet controls the draft and the windward guy controls the tack position.

Berig Camberspar®

While not really considered a "light air" sail, it has several advantages over a conventional jib. The fact that it will retain its foil shape despite the shortage of wind is one advantage but the ability to boom it out, wing on wing without any additional poles, guys, downhauls, etc. is of exceptional merit. It has its own pole built right into the sail. For people who do not want to fiddle with any additional sails, this is definitely an option.

How to maximize light air

A cruising catamaran or trimaran has certain advantages and a few disadvantages over its cruising monohull counterpart in light air.

The disadvantages lie in the fact that a monohull will have less wetted surface in relation to sail area (not always true, but mostly true.)

The advantages lie in two areas, the cat is lighter, therefore, can take better advantage of puffs. The second is the big, stable, wide platform which allows the use of spinnakers without any real pain, problem or exotic handling equipment.

A spinnaker, normally thought of strictly in racing terms, is a routine sail on a catamaran cruising boat. Corinne and I fly a one thousand square foot spinnaker by ourselves with little more than ordinary handling.

First, our spinnaker stores in a *Chute Scoop®* This eliminates the problems of hour-glassing, headstay wrap and panic take down.

Next, the chute has four lines, two forward guys to blocks on the bow cleats which come back to the cockpit through cam cleats and two sheets to the sheet winches. With those four lines you can do anything you want to the sail. A spinnaker pole, the bane of spinnaker handling, is not required.

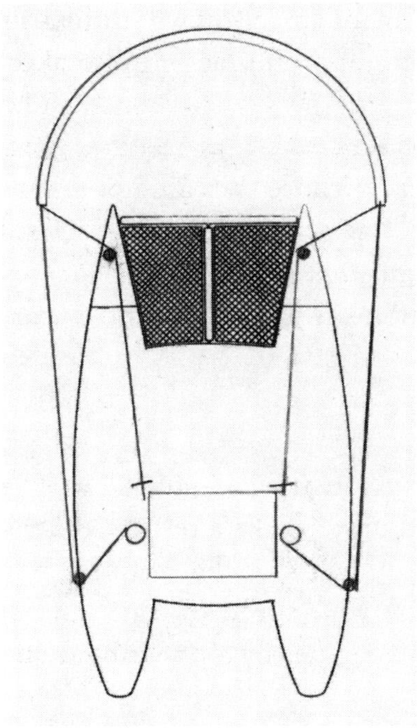

Pole-less trim on a run

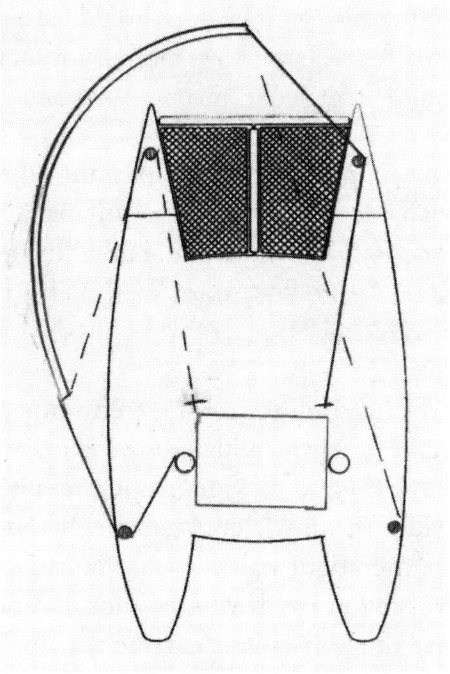

Pole-less chute trim on a reach

Our chute, purchased second hand from Bacon Associates, in Annapolis, was designed for a Tartan 37. I can carry it with the wind forward of the beam up to 50 degrees apparent. I believe if I played around with movable tack points I could get it closer than that. However, at five knots true wind I can get *La Forza* to sail six knots at 50 degrees to the apparent wind. If I need to go in that direction, it really is terrific.

If conditions are appropriate (very seldom) you might try tacking down wind. Monohulls can do this too, of course, but if the air is very light and you really are sick and tired of listening to that engine, try it! Who knows, you may like it!

CAUTION!

In any catamaran it is possible to get yourself into difficulty by being overpowered sailing downwind. As the wind builds you simply sail faster and faster until you suddenly recognize that the true wind is stronger than you realize. You might be sailing dead downwind around 10 to 12 knots. This is easy with almost any of the production cruising catamarans. You note that your wind speed indicator reads 8 to 10 knots. The low wind speed on the indicator reading has lulled you into a false sense of the true conditions. A comparable length monohull would either never be able to achieve that speed anyway or be wild and almost uncontrollable, needing a full crew to handle it. Your cruising cat has calmly, innocently, gotten you into a situation where you have too much sail up for the conditions. The true wind, in that scenario, is most likely around 25 knots! Downwind, add the two speeds together to get a reasonable approximation of the true wind speed.

Those of you have been reading *MULTIHULLS Magazine* for years will recognize this scenario. A classic case of pitchpole of a 50-foot racer. It was sailing downwind, in big seas, at over 20 knots, in about 30 knots of breeze. While the boat was sailing at more than three quarters of the wind speed, thus showing only about ten knots apparent wind, it was perfectly under control. However, when the boat sailed down into a wave trough, its speed dropped to about 10 knots. When the boat came up out of the trough, it again got the full force of the wind. But, now it showed over 20 knots apparent. It meant the boat was seriously over canvassed. The boat drove its pointy bows right into the seas and pitchpoled.

Your cruising catamaran should have far less exposure to such a predicament since you will ordinarily carry less sail and have more buoyancy forward. Nevertheless, this is an important factor in how you handle your boat in strong winds.

I have outlined a few important differences and safety procedures here because they are not readily available elsewhere. For a full explanation and serious discussion of How To Sail, there are several excellent books recommended in the glossary. I suggest reading *Catamaran Racing For The 90's*, by Rick White and Mary Wells. This book will give you clear instruction on just how to take every possible advantage of whatever gear you have on your boat.

Advanced motor sailing technique

Many times you will be in the situation where you are sailing nicely but you are going too slow. The wind is light, you are making reasonable time, but you need to speed up to make a restricted draw bridge or dinner engagement. If you have a twin-engine catamaran, you will find that running one engine about half its usual cruising rpm will be sufficient to increase your speed to your normal cruising range. If you do a lot of motor-sailing, spread the load on the engines by alternating them. Use the port engine on one trip and the starboard the next.

For single engine vessels, the technique is about the same.

Sail trim will change because you will move the apparent wind forward. When the wind lightens sufficiently that you cannot keep the jib from flogging, hand the jib and sail on mainsail alone. Centerline the boom for best results. You will find that you can retain drive from your main with as little as seventeen degrees apparent wind angle if it is trimmed in tight.

Heavy weather sailing

First, let us make the distinction between heavy weather sailing and survival sailing. Heavy weather sailing is done in full control of your circumstances and your ability to manage the vessel. Survival sailing is when you pass that point and must take action to preserve life and property.

Historically, many situations that would have been survival situations in other craft, were just heavy weather sailing for catamarans. There is the classic story by Robin Knox Johnson of sailing his catamaran through force 12 winds. His tactic was to press on rather than attempt to heave-too or lie a-hull. There are other similar experiences that support the notion that keeping the vessel underway is perhaps the best survival tactic.

Literature is sparse dealing with survival situations in multihulls. The basic reason for this lack of literature is the astonishing safety record compiled by ordinary sailors in off-the-shelf catamarans. For instance, at the time of the famous Fastnet disaster, of which volumes have been written, there were two Prout catamarans in the vicinity. They were shadowing the fleet as unofficial entries. The racing fleet sailed into a serious storm. The carnage caused by the storm was so great that a *Committee On*

Safety From Capsizing was formed. It consisted of The Society of Naval Architects and Marine Engineers, (SNAME) The United States Yacht Racing Union, (USYRU) and The United States Naval Academy (USNA) and others. The purpose of this committee was to design a minimum stability formula for ocean racing monohull yachts. The weather that was disaster to the Fastnet fleet was considered merely "beastly weather for sailing" by the two adjacent catamarans.

Then there is the tragic "Queens Birthday" typhoon off New Zealand in which there was loss of life and property. The two catamarans that were part of the fleet caught in that storm, showed that even with aging and somewhat handicapped short crew, the vessels provided predictable and adequate safety. There are numerous stories of people simply furling all sails, lashing the helm amidships and going below to wait out the storm whilst their catamaran simply bobbed around looking after herself.

Heavy weather sailing basic tactics

Knowing when to reef is the most important skill to develop for heavy weather sailing. Then comes sail shape and sheeting angle. There are a few generalities that will help you get started learning about heavy weather sailing tactics.

- As the wind increases, move the sheeting point to leeward. This is one of the best features of multihull sailing. Multihulls have a wide sheeting base which allows a greater angular choice for sail trim than narrow boats. Ease off the traveler to move the main to leeward and use an outside rail attachment point, stanchion base or toe track and car to move the jib to leeward.

- Allow more twist in the sails as the wind strengthens. This is achieved in the main by slacking the main-sheet and in the jib by moving the sheeting point aft.

- Reduce camber (flatten) the mainsail. This is done with the outhaul.

- Reduce sail. (Reef) You will find most multihulls just as happy with reduced sail and you will be surprised how little speed you really lose.

- Learn to use a barber-hauler to control your jib. A barber-hauler does not need to be a fancy multi-part tackle, but can be made from most anything, including just a length of line from the clew of your jib, an appropriate turning point or a winch.

- Slow down to a more moderate speed. Like slowing down your car when you come to a rough road.

- Choose sail combination according to direction. Downwind, reduce main area first then the jib. Upwind, reduce jib area first then the main. This is just a rule-of-thumb. It must be tailored to the individual vessel.

- Expect to hand-steer downwind in big waves. Your auto pilot most likely will not be able to react fast enough, nor can it anticipate waves. The trick is never to allow yourself to be perpendicular to the wave, but always to be at an angle to them. That way you are always sailing downhill and you eliminate the chance of pooping or pitch-poling. Pooping is almost unheard of in cruising catamarans and pitch-poling comes from excessive speed perpendicular to the waves and to my knowledge, has never happened in a production cruising catamaran.

Getting to windward.

If you seriously feel the need to maximize your heavy-weather windward ability under sail, then you must be prepared to change headsails for the purpose. Roller reefing sails are good to a certain point. It must be remembered that a sail is sewn to pull against the head, the tack and the clew with the luff supported by a stay. When a sail is partially rolled, it is no longer pulling against the designed strong points but is pulling against the luff and the foot where they roll around the stay. While it is possible to do reasonably well with a partially rolled jib, it is without doubt much better to have a sail specifically cut for that purpose. In other words a good storm jib.

On the other hand, if you do what most cruisers do and simply run your engine about one third your normal cruising speed you will go to windward just fine. This motor-sailing concept is almost the universal choice of cruisers for going to windward in heavy weather. In a catamaran with twin engines, it is only necessary to run one of the engines to gain the desired effect, usually the leeward engine.

Heavy weather in restricted sailing areas

Occasionally you will be faced with the difficulty of going upwind through a narrow channel in a waterway. Those that traverse the Intracoastal Waterway know exactly what I mean. You need to get to the head of this river or bay but force five or better winds are right on your nose. If you drop your sails and motor, your speed will be seriously reduced and you will slam on the bridge deck. The tactic of choice is to tack up wind with only a tightly sheeted mainsail and your engines running about half or three quarters your normal rpm.

Make short tacks. Fall off just enough to keep the main full and driving and cutting the waves at about a twenty degree angle. This tactic will allow you to reach your destination in relative comfort without a lot of pitching and pounding.

How do I know when to reef?

This is the most often asked question concerning multihull safety. The answer is deceptively simple:

It is time to reef when you first think about it.

This is not meant to in any way belittle the importance of knowing that time. As you get more experience with your particular boat, the more feel you will get for the process. I will divide this into two categories: Subjective and objective.

However, first I would like to suggest a format or concept for learning a new and different subjective skill. Let us compare learning how to feel your boat with how you learned to feel your car.

When you first learned to drive, how did you know when to start slowing for a stop? If you started slowing too soon, you created a traffic hazard. If you started too late, you wound up with a panic stop or a rear end collision. How did you learn this subjective judgmental skill? How did this judgmental process become habituated?

If you drive your car too fast around a curve, you may lose control, skid out, or depending on conditions, roll-over (capsize) your car. Yet, you drive every day without giving it a thought. How did you learn the parameters of safe cornering? At what point did the few words of an instructor metamorphose into a routine, subjective, habituated skill?

I suggest that sailing your catamaran is quite like driving your car. The results of lack of skill are quite similar as are the consequences. You could conceivably skid out or even roll-over (capsize) your car any day, any time, any where, buy pushing the limits or bad judgement. You would have to work much harder at doing that same thing in your cruising catamaran. Most of the catamarans available to you (production catamarans) are of the fool proof type!

Let me attempt to be the voice of your driving instructor, as you take your fledgling jaunts in your car, only in this case, your catamaran.

Let us split the descriptions two ways: Subjective and Objective

Subjective

From the subjective point of view, when you begin to feel uneasy, apprehensive, concerned, it is time to reef. When the boat no longer has its feather light touch at the helm, it is time to reef. When the boat's motion changes from its normal light, resilient feeling to one of petulant obedience, It is time to reef. When the lee bow seems to want to plunge and bury, it is time to reef. Perhaps I can explain it this way. How do you know when you are driving your car too fast? How do you know it's time to slow around the curves? The answer is a subjective value and the experience of the operator. Therefore, I suggest you might take these same gut feelings that you have habituated from your driving and apply them directly to your catamaran sailing. You can feel when you are over-pressed. It is the same as you feel when you screech around a curve in your auto and you are over-pressed.

Objective

From the objective point of view. For most boats, when the apparent wind speed goes over 15 knots, it is time to reef. When you are heeled 2 degrees more than normal, it is time to reef. When you are no longer strong enough to crank in the sails, its time to reef.

Reefing, as referred to in this section, includes both headsail and mainsail. As a general rule for masthead boats, upwind, reef the jib first, downwind reef the main first. It is hard to generalize about fractional rigs. When I was sailing a Stiletto, I would always hand the jib first on either point of sail. Sailing under main alone is far more controllable. Then, when necessary, I would reef the main. I think, the easiest sail to control is the full batten fractional mainsail. It has the most sail controls, is held on two sides by spars, and can be given optimum size and shape. Jibs, only supported on one side by a flexible stay, flog like crazy, ball up the sheets or tangle them around protuberances.

Reducing sail downwind

If you believe you need to reef and for any number of reasons cannot come up into the wind to do it, there are a couple of techniques you can try.

Reefing the mainsail downwind

This a basic study for reefing when wind or waves have come up suddenly and you dare not turn into the wind. Or, you cannot turn upwind due to other circumstances, like being in a confined channel or among crab traps.

The basic idea is to slide the entire sail downward while keeping the head of the sail and the leach of the sail pointed into the wind. Do not let the sail press against the rigging. The friction of pressing against the rigging will prevent you from lowering it. You can also damage the sail, or worse, get it snagged in the rigging. If you have a sliding boom with a downhaul, you can pull in a reef from the bottom up, then haul the sail down with the downhaul and mainsheet. In other words, you can raise the boom up to meet the reef points instead of pulling the head down. You may raise the end of the boom by hauling on the reefing lines, then, slowly and deliberately hauling down into position by using the downhaul and mainsheet. This procedure is pretty much a two-person operation. One must slowly release the halyard and haul in on the mainsheet while the other tightens the downhaul.

A Procedure For A Fixed Boom

1. Center the traveler and feather the sail into the wind, leach first. If it is not dangerous, tie the boom in the center so it cannot flog. On some vessels this is a cinch, on others, impossible. On *La Forza*, I do it by clipping on a vang on one side and moving the traveler over on the other. The topping lift holds the boom up, thus forming a perfect triangle. In many cases you can tie the boom on center between the backstays or stanchion bases.
2. While slowly releasing the main halyard haul down on the first reefing line as tight as you dare trying to keep the leach feathered into the wind. Haul down on the luff the same way. A single line reefing system, makes this procedure easier. Fully battened mainsails are much more responsive to this procedure because they do not flog.
3. When you have the boom tied on center, you may want to slack the topping lift so that it cannot interfere with keeping the leach taught.
4. Slowly lower the sail with the halyard. When the twist becomes excessive and it looks like the sail might hit the rigging, stop!
5. Go back to the reefing lines and pull down some more. If you do not have sufficient power in your reefing tackle, slack off the main sheet and raise the end of the boom with the topping lift. Now pull the reef down to the main boom. Sort of raising the boom to the sail instead of pulling the sail to the boom.

6. Continue working the sail down in this manner, going from reef point to reef point as necessary. You can continue this process until the sail is all the way down or at least through the third reef.
7. If the sail does get pinned against the mast, slack the mainsheet and use your reefing line to lift the boom, then tighten the mainsheet to tighten the leach. If you still cannot get the sail away from the mast rigging, put as much tension on the leach as you can. Then very slowly and carefully begin a jibe. As the tension begins to lessen on the sail, pull down on the luff end of the reefing lines or downhaul, depending upon which you have. Then tighten the leach, working your way down a little as a time.
8. A good 4:1 downhaul is a valuable asset under these circumstances. Rig it from the most convenient luff cringle, to a padeye on deck. Preferably, this padeye is just below the boom behind the mast. On most boats, this downhaul can be led right back to the cockpit, if necessary. Rig a downhaul by using a line led through a snatch block to a cockpit winch. If a reefing hook is unavailable, poke the end of the line through the cringle and use an ordinary figure eight stopper knot in the end of the line. The purchase on this line doubles by feeding the line through the cringle and securing it to a fitting at the bottom of the mast. This gives you 2:1 purchase.

Experience Counts!

There is no substitute for experience, and I suggest taking your catamaran out in a controlled environment in strong winds, with some capable crew. Sail it as hard as you can and try to lift a hull. You may or may not be able to do it. You sure will learn a lot about how your catamaran feels when you begin to reach the realistic usable limits. Never sail your boat that hard again. I have little doubt that the vast majority of us learned about our cars in the same manner when we were teenagers.

Trimarans:

A trimaran is far easier to judge when over-pressed than a catamaran since the incredible initial stability of the cat reduces most sensory and visual clues. A trimaran heels almost like a monohull and you can visually see the lee ama being depressed. Since there is more heel, there is more familiarity. However, like any multihull, your wide beam allows you trimming angles not available to monohulls, thus you can have far more control and keep your power up longer.

On the other hand, a good multihull requires less power to move at speed. This is true not only because it is lighter, but because it does not have the same hull speed limitation factor as described in another section of this book. As a vessel approaches its hull speed limitation factor, it increases the need for larger amounts of power to drive it into its own bow wave.

Cruising boats should never need to face these problems. Just reef. You will probably find your boat goes just as fast anyway. Besides, most times you will want to slow down a bit to remain comfortable. A bit like slowing your car down when you come to a rough stretch of road.

There is one other basic concept associated with reefing: Monohulls reef to the gusts, Multihulls reef to the lulls. This is a reasonable rule of thumb for cruisers.

Possible Sheeting Angles

The older, narrow beam catamarans, BMAX 14' has a maximum sheeting angle of 40^0.

The modern, wide-beam mono-hull cruising boat has a maximum sheeting angle of 15^0.

Sail shape degrades rapidly when sheets are eased, turning from lift to drag if sheets are eased too far attempting to create optimum sail shape in relation to the apparent wind.

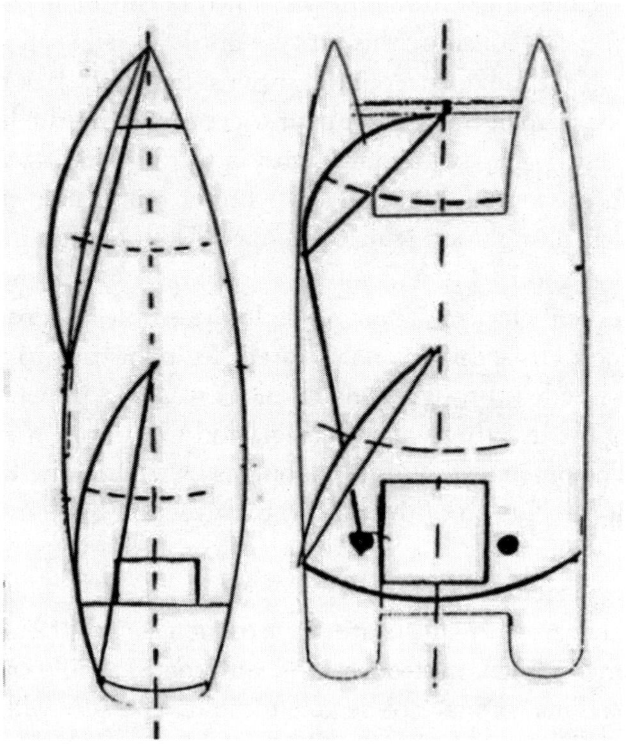

How to tack a catamaran that is hard to tack
(or monohull)

1. Get the boat going as fast as possible for the given conditions.
2. Get the boat pointing as high as possible, without pinching.
3. Put the helm down slowly and evenly. Sail the boat through the turn. Be careful of pushing the helm too far over, or the rudders might stall the boat. Watch the rudders for excessive turbulence which will indicate this condition. 35 degrees is usually about maximum rudder angle for tacking.
4. Do not release the jib (backwind the jib) until the point during the tack is reached where the mainsail fills on the new tack. Often about fifty or sixty degrees on the new tack
5. Release the main sheet or move the traveler well to leeward, which ever is more appropriate for your particular boat. Moving the traveler is better if possible, because you may retain more forward drive.
6. Trim in the jib on the new tack as quickly as possible, keep the helm down to gain way.
7. Do not re-trim the main until you are sailing well on the new tack.

(Don't let any of your buddies give you any BS about this paragraph, some of the most popular monohulls ever built need these tactics also.)

How to get sailing again from a stalled position (in irons)

1. Release the mainsheet.
2. Turn the helm so the boat backs in the opposite direction from the one you wish to go.
3. Let the boat sail backwards. If you have to, because of light air, back the jib. Back it by holding the clew forward on the opposite side of the boat from the way you wish to go.
4. When the boat gets far enough around, approximately 45 degrees from the wind, trim the jib.
5. Turn the helm the opposite way, that is, turn for the new tack.
6. After the boat is moving well, trim the main.
7. With fractional sloops it's different, you can trim the main when the bow has past the eye of the wind. The jib has much less effect Fractional sloops seldom have this problem.

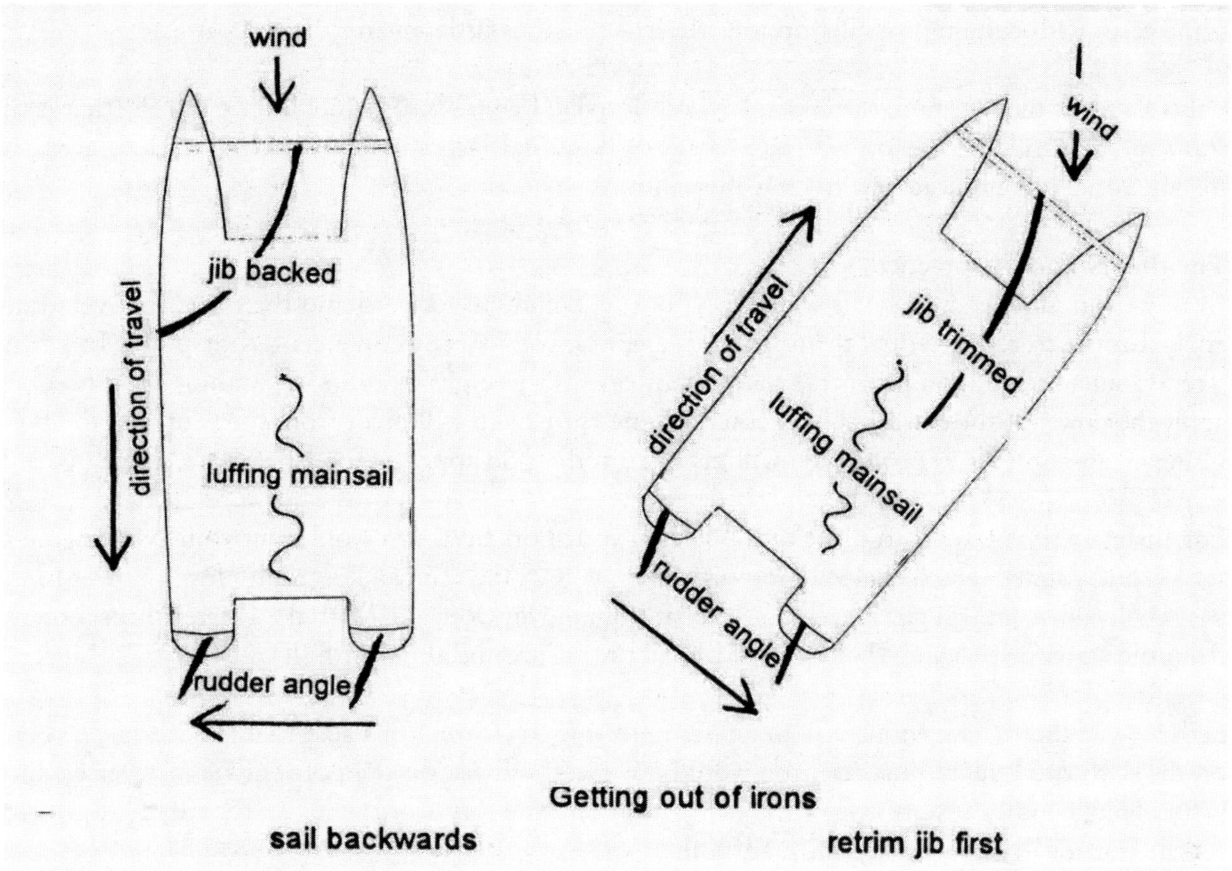

Maneuvering twin-screw catamarans

At the dock

Many of you are familiar with maneuvering twin-screw power boats. As good as they are to maneuver, twin screw catamarans are even better! Actually, you have every potential maneuver available to a track vehicle. The reason they are better is because their wide beam creates options not available in a narrow boat. Twin screws provide confidence in docking. Knowing the procedures and the theory behind those procedures allows you to quickly master the techniques and build the skills to handle most situations. If you can handle docking, you can handle any other maneuvering situation including tying to pilings, approaching and leaving slips and anything else you may encounter.

The helm

The helm is not normally used in these maneuvers. All maneuvering is done with the throttles. Simply leave your wheel with the rudders centered and perform all steering operations with the two throttles. There are some caveats, however. This article subsumes an "average" production twin screw catamaran with spade rudders and three-blade fixed propellers. Two-blade props, folding props and feathering props will change the amount of thrust available. A single engine with twin hydraulic drives has different operating characteristics since you cannot create greater and lesser thrusts at the props. In some cases with certain propeller options there is precious little reverse thrust to use.

I also assume single-lever engine controls. Occasionally, I come across dual lever controls, that is the transmission and throttle on separate levers. While dual levers may be routine on single engine installations, it complicates twin-screw installations

The theory and the practice

In most twin-screw power boats when you put one engine in forward and the other in reverse and apply throttle to receive equal thrust to both engines, you will spin on a centerline point. In a twin-screw catamaran, you can pivot the boat on any one of three pivot points depending upon how you apply the power to the engines. Using equal engine thrust you will pivot from the center of the boat. Using greater and lesser power, you will pivot from the side applying the lesser power. [see figure 1]

For instance, if you want to pivot to the left on your port hull, you would put your port engine in reverse and your starboard engine in forward. Then you would apply a lesser amount of power to the port engine, just enough power to keep the boat from moving forward. With the greater thrust coming from the starboard engine, the boat will literally swing around the port hull.

Let's say you didn't have enough room on the right side to clear an obstacle so it is necessary to pivot on the starboard hull. In that case, you would put greater thrust on the port engine in reverse which would literally pull the port side of the boat away from the forward moving starboard, thus pivoting on the outside or in this case, starboard hull.

Not every case is identical, not every situation calls for the use of thrust exactly the way it is portrayed in the following diagrams. A little bit of practice and the exercise of prudent judgement and you will be an expert in no time. The same technique applies whether you are underway, at rest or docking. This is a related technique to that used in operating a track vehicle such as a tank or a bulldozer. The concept of greater and lesser thrust is similar, except that in a track vehicle, the inside, or pivoting side may be held in place by a brake. Your propellor substitutes for the dozer's brake.

Terms

When using engines for maneuvering in the terms of this article, the terms are defined as:

1. Lesser Thrust = enough thrust to keep that hull from going forward or aft or relatively stationary. Your dozer brake equivalent.

2. Greater Thrust = enough thrust to make that hull go forward or aft.

Thus the hull with greater thrust pivots around the center of lateral resistance of the stationary hull. Equal and opposite thrust pivots the boat around its center point, directly under the mast. [Fig. 1]

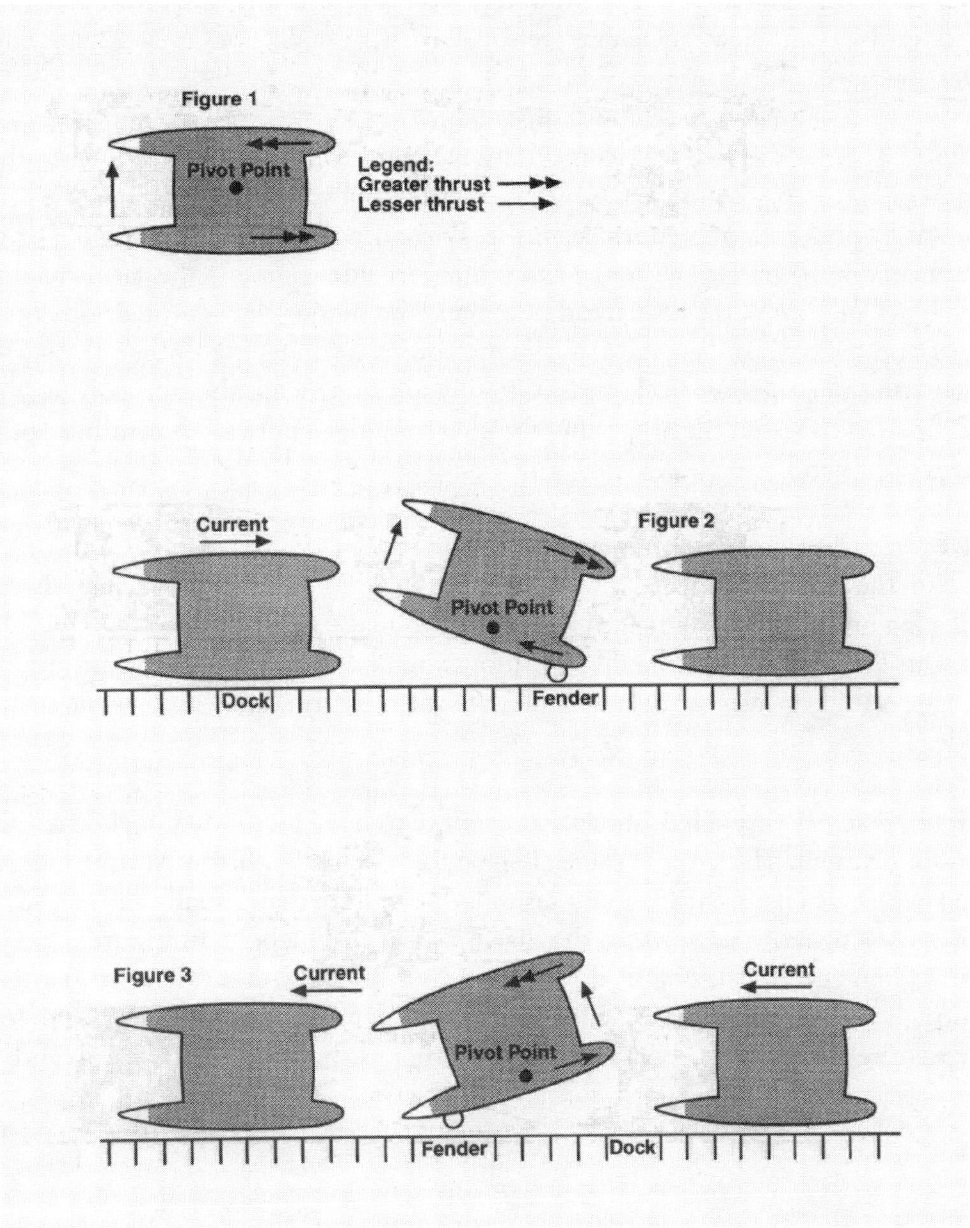

Figure 1 Shows a twin-screw catamaran applying equal thrust forward and aft and spinning at the center point of the vessel center.

Figure 2 shows a catamaran in a tight docking situation with strong current coming from forward. To prevent abrasion with the dock, a fender is placed in the stern quarter. During your learning curve, it is advisable to also use a spring line as a safety feature.

To prevent rolling the fender out, the technique is to literally pull the starboard hull around until the bows are clear of the forward boat. The port engine maintains just enough power to keep the boat from going backwards into the dock. At the point that the bows clear the boat in front, the starboard engine is taken out of reverse and put in forward, more power is applied to the port engine and the boat moves quickly out of the slip, the current keeping the vessel clear of the docked boat.

Additional help may come from a crew member who stands on the transom and holds the fender in the appropriate location. That person will whip the line around a stern rail or stanchion or whatever is handy to increase holding power. No one should hold it directly to reduce the possibility of rope burns.

Figure 3 shows the identical technique except the current is reversed making it expedient to go out backwards. Twin screw catamarans are equally maneuverable forwards or backwards (exception to that is if you have folding two-bladed props.)

Figure 4 shows the most difficult situation for any sailboat, being pinned to the dock by current or wind or both. The technique is similar to figure 2, but you must be prepared to either walk the fender forward or have several fenders available. Preventing fenders from tangling or jamming in pilings is the major problem. Use more thrust on the port side engine as the cat turns into the current. Once the bows are clear of the boat ahead of you and the stern is clear of the dock, continue enough reverse thrust on the starboard engine to assure the bows continuing to turn into the current but not so much that you force the boat backwards. Often, you can just slip the starboard engine out of gear at this point so as to lessen the aft-pulling force. It's a judgement call depending upon the strength of the current or wind.

As soon as the port bow passes clear of the boat ahead of it, give more power on the port engine and less on the starboard to speed the pivoting action until the boat into facing directly up current. Keep pivoting until you are facing into the current to make sure you do not get swept into the boat ahead of you. The figure shows the final position when both engines are in forward but the port (leeward) engine is using greater thrust to compensate for the current and to bias the boat towards the favored side.

Figure 5 step 1 through 3, shows a procedure to leave a crowded dock when there is no significant wind or current. This is the skill area to practice at every opportunity. It is possible to do this without any fenders against the dock and it a similar procedure that actually allows you to "walk" a boat into or out of a tight slip sideways. In the diagram, first tip the bows in towards the dock to pull the stern out. Then reverse the engine directions to pull the bows away. If your first try pulled the stern far enough away so you can clear the dock, just keep turning until clear. Otherwise, repeat the first procedure to "walk" the boat out a little further as shown in figure 5a, 5b and 5c.

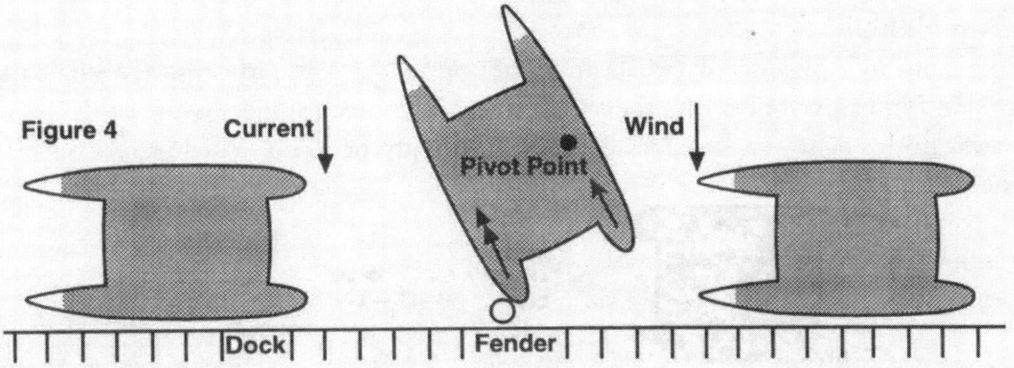
Figure 4
Current
Wind
Pivot Point
Dock
Fender

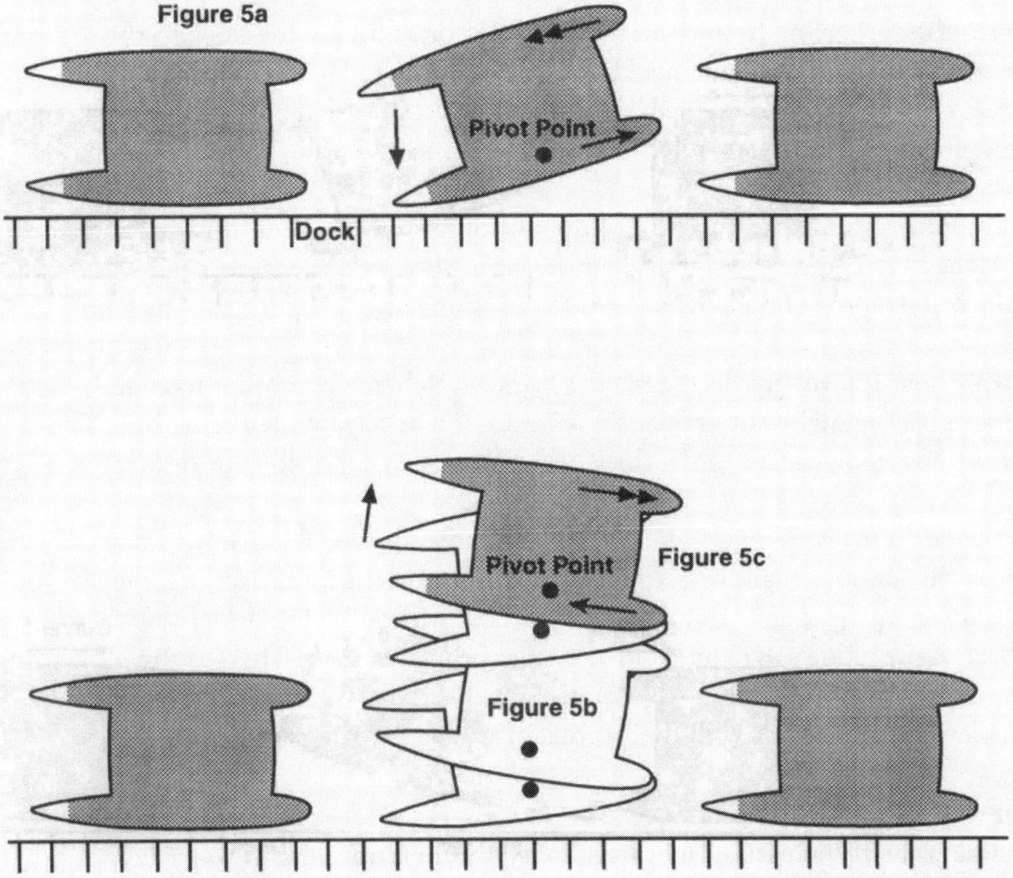
Figure 5a
Pivot Point
Dock
Pivot Point
Figure 5c
Figure 5b

Maneuvering Under Power

Single screw multihulls

Most cruising multihull single-screw vessels are either outboard powered or have a single diesel with a sonic drive leg. Both are handled with the classic single screw technique. Assume this situation, you must back away from a tightly packed dock, backwards, up-current. With a single-screw vessel you need a spring-line and a fender.

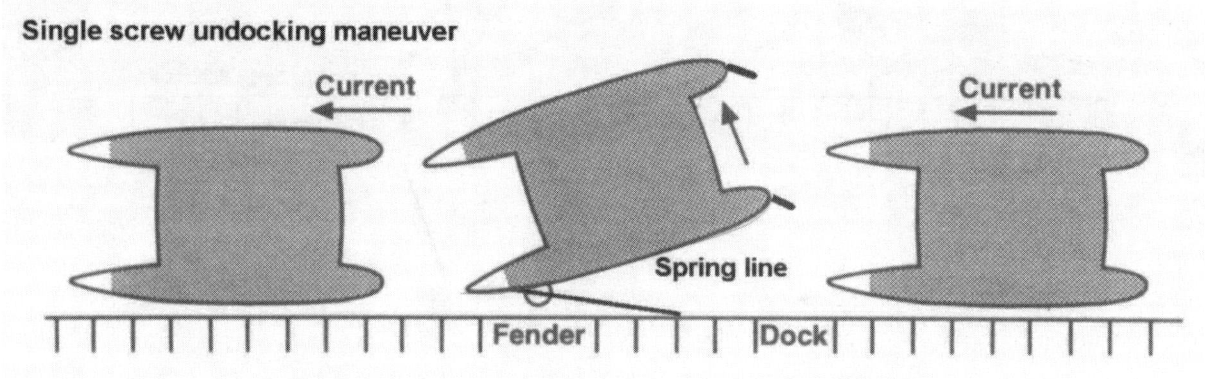

1. You attach the spring-line from the a suitable cleat or piling on the dock to the forward cleat on the boat. Use a method that you can quickly detach. In most cases that means running a line around the dock cleat or piling and back to the boat so you can just release it at the appropriate moment and haul the loose end in. Place the fender forward where the bow will contact the dock.

2. Turn your rudder hard in the direction of the dock. Put the engine in forward and slowly increase power until the boat comes up tight against the spring-line and the stern begins to move out, away from the dock. The force of the engine against the spring-line will push the stern out.

3. When your boat is forty-five degrees to the dock, straighten the rudder to center, put the engine in reverse and back straight out against the current as you release the spring line. The current will keep you from sweeping back into the dock. Once out, steer off onto your desired course.

In general, catamarans with single-screw systems handle almost identically to any other outboard-powered boat. Since you actually turn the propulsion unit, be it outboard or sonic leg, you have the distinct advantage of being able to maneuver the boat without being underway, that is, not having water flow across the rudders as required in any fixed propeller drive. This feature allows single screw cruising catamarans to be among the most maneuverable of all cruising boats and far more maneuverable than monohulls with their props in an aperture or with folding props.

Powerboat wakes

Powerboat wakes are the nemesis of many sailboaters. When traveling in waterways or other marked channels, your vessel will encounter irritating wakes thrown by the parade of powerboats. Powerboat wakes represent one of the major challenges of inland waterway cruising. They are quite different from the storm generated waves or the shallow chop and wind against the current waves of some notorious bodies of water. There are two types of wakes to contend with, those from vessels passing you and those from vessels going in the opposite direction. The antidote for these waves is very similar in both cases. Theoretically, try to take the waves at approximately a forty-five degree angle to your centerline. This will vary somewhat depending upon your type of boat.

Wakes from astern

Wakes coming from astern represent the lesser problem. You can often surf on them under complete control if you catch them right. Different powerboats throw wakes at different angles. Some come almost parallel to your boat, while others come perpendicular or directly abeam. The beam-on wakes are the most difficult to handle.

If the wakes come up from behind you at a forty-five degree angle or less, hold your course. If the waves come anywhere from forty-five degrees to beam on, hold your course to the last possible minute, when the wave is only inches from your boat, then fall off and head down until your stern is at forty-five degrees to the wave. Hold that course until the last wake-wave, then return to your original course, compensating to regain any distance lost.

In the event you do not have sufficient sea-room for this maneuver, you can steer each individual wave or you can use the standard tactic plus throttling back on the engine. By throttling back you allow the waves to pass under you without giving you any forward motion. If you are under sail, make sure to keep enough room to leeward to complete these tactics.

Most keel boats do not do this maneuver because of the possibility of pooping, broaching, or having water splash up into the cockpit. You will sometimes see them spin around and head into the wakes bow-on. This creates problems on a crowded waterway. On a multihull, if you allow the waves to hit beam-on your boat will roll, but not nearly so badly as a keel boat in similar circumstances. Multihulls tend to have a quick snapping roll, as explained elsewhere in this book. A snapping roll is one that does not leave excess momentum to throw things around.

Wakes from ahead

Wakes from boats passing in the opposite direction are a little easier to judge. Again, try for a forty-five degree angle. Experiment with closing that angle to less than forty-five degrees. The closer you can get that angle, the less under-deck slapping you will get. A little practice and you will be able to arrive at a good angle that will neither roll the boat nor slap underneath.

Hobby-horsing

After crossing a wake you will often wind up hobby-horsing. This is especially so in narrow channels where you get reverberation from the sea-walls. All boats may do this to some degree, but some catamarans are worse than others. The quick and easy way to stop this motion is to again turn forty-five degrees into the wave train. Turn quickly to break the pattern, then return to your course quickly in order not to pick the harmonics back up again.

Safety is a Day to Day Concept

SAFETY:

One day Corinne and I were in Annapolis harbor when a 40-foot yawl docked in the end slip at the dock often called "ego alley." It was obviously a cruising boat on its way south. Long distance cruising people are usually easy to spot by the nature of the equipment and accessories lashed on deck.

This couple, backed into the slip and tied up with the traditional six line tie: two stern lines, two bow lines and two spring lines. Then they went below to prepare lunch.

Moments later we heard an explosion. As with most sailors, we were ready, willing and able to help. En masse we all grabbed our fire extinguishers and headed for the boat. Before any of us in the adjoining slips could get to this boat, it had sunk!

Fortunately for this couple, they had done a proper job of tying off. They avoided drowning because their boat only sunk to the extent of the slack in their dock lines.

Upon investigation of the accident, we learned that they had their dinghy engine gasoline supply in a one gallon Clorox bottle in the lazarette. This bottle had either chafed through or received damage from another piece of gear. The gasoline leaked out and ran through a weep hole in the separating bulkhead, and into the bilge. When the couple lighted their alcohol stove, somehow the spark ignited the fumes in the bilge causing the explosion. The explosion knocked out a section of the hull next to the keel. This section had been weakened by the installation of several thru-hull fittings.

This incident really raised my consciousness about safety aboard boats. It graphically pointed out that general safety was far more the realm of the little daily things than it was worrying about crossing the ocean. The blue water considerations of safety are a tangential subject to the real issues of safety. Separate them accordingly.

Through the intervening years, both my profession and my curiosity have prompted me to catalogue a list of safety considerations. For every deep sea sailor, there are a thousand weekend sailors and two thousand dock sailors! Thus, this little epistle on recognizing just what safety really is.

SAFETY 101

What is safety? What is different about safety as it relates to multihull sailboats? What safety gear that is common aboard monohull sailboats is seldom seen or used differently aboard multihulls?

Safety is an elusive concept. No one is ever 100 percent safe in any location under any real circumstance. Safety is always a matter of degree. You are safer in a low crime area than a high crime area. You are safer at home than on the highway. You are safer driving a car than a motorcycle. I expect to prove to you in this chapter that you are safer cruising on a catamaran than any other type of vessel.

Defining terms
First, let us define our terms and set our parameters. There is everyday, routine safety that affects all of us all the time, then there is special consideration safety such as:
- Routine safety.
- Transoceanic or offshore procedures.
- Survival procedures including life rafts, flares, etc.

I will cover each aspect in a different section since they are truly different subjects. Most universal, however, is our routine safety.

Sailing, cruising, boating, are risk-taking sports. The magnitude of the risks are directly related to several factors including, but not limited to vessel, climate (including weather) and location. For instance, the risk from hypothermia is greater in Narragansett Bay, Rhode Island than in Biscayne Bay, Florida. Since we can agree on those basic givens and variables as parameters, let us look at the other side of the equation, the variable over which you have the greatest control, your choice of vessel specifically your cruising sailboat.

Routine safety
Heeling, rolling and pitching related falls and spills are the most ubiquitous safety considerations. Therefore, the less your cruising boat does that, the greater your margin of safety. Heeling in most sailboats is related to initial stability which is directly related to hull shape. Any of you reading this who are familiar with the monohull racing rules fully comprehend that the characteristics that create an initially stable, seakindly hull are the very same characteristics that are most penalized by the racing rules. Following that to its conclusion, we get all the strange compromises that give us narrow beam at the static waterplane, long overhangs (lots of wave slapping at anchor as well as low initial stability) and what I sardonically describe as "hidden agenda" stability which comes into play when the boat heels over far enough to sail on its true designed waterline.

I am not criticizing hidden agendas. They are what makes sailing and especially sailboat racing so intriguing. I am only pointing out that there is serious compromise to your comfort, convenience and routine safety when you use a boat designed like that for cruising. It's a little like buying a sports car, bolting a luggage carrier on the trunk, and expecting it to perform as a station wagon! There are a multitude of reasons why the sailboat mass-producers market these racing rule clones as cruising boats, not the least of which is cost; but that is a story for another day.

Companionways and stairs
There are often falls related to companionway stairs. Most monohulls have a threshold and a two-part opening consisting of a sliding hatch on the horizontal portion and wash boards on the vertical portion. In the Navy, we were trained to go down ladders and steep stairs backwards. That way you can grip the hand rails properly. Going down forward or swinging down from the edge of the hatch leads to falls.

Galley safety

Refrigerators

Back at routine safety, let us consider our galley, galley range and refrigerator. While you may muse about why a refrigerator falls into a safety article, let me say that with the majority of cruising boat refrigerators being top loading, a host of accidents occur related to slamming tops catching fingers, sliding tops hitting heads, eyes and related body-parts whilst said body is inverted and partially immersed in that undignified search-fridge position. Ah, for a front loader.

On my cat and a majority of others, we *do have* front-loading refrigerators! Since we don't heel, and stuff does not take off flying on its own, front loading makes sense and is practical. The monohull guys try to tell you that top-loaders are more efficient because when you open them, the cold air does not escape. Read the refrigeration section in chapter titled, "Catamaran Galley Features," for the details.

Stoves

The stove, on the other hand, is pretty obvious. Spills of hot food being the first danger. Sudden heeling can put that steaming pasta pot right down your front, that is why on many serious offshore yachts you will find a full length leather apron, a harness with appropriate attachments and a gimbaled stove. A gimbaled stove sets up its own environmental impact. For instance, sudden heeling may send the base of the stove crashing into your shins or knees, or higher parts depending upon your height and the stove's depth. It also may crash against its compartment limit because most manufacturers only allow for twenty degrees of heel. Liquids that may be protected from spilling while on the gimbaled stove top, may spill when taken off the stove and placed on a non-gimbaled counter top. If a stove is improperly gimbaled, it becomes dangerous when a large sized tall pot it placed on it because it could become over balanced.

All things considered, the location of the galley stove on a cruising monohull sailboat is the most import single layout item. It must be mounted athwart ship so it can gimbal on the fore and aft axis to accommodate heeling. It must be at the center of vessel rotation to dampen vessel pitching motion. Therefore, on a serious cruising boat, the interior layout of the boat is designed around the location of the stove. Think about all the compromises that entails! [See section on galley]

In my career, I sailed, delivered and surveyed scores of offshore cruising catamarans and have yet to see any with leather aprons or gimbaled stoves. Nor do catamaran cruisers ever mention which orientation their stove has or its location as a function of safety. On catamarans installation varies from the far forward galley down configuration of the *Endeavour* 30, and *Maine Cat* 30 to the aft-facing galley-up location of the *Maldives* 30, *Manta* 42 and every possible location and configuration in between.

Fuel

Then there is stove fuel. While most of us are using the eminently satisfactory propane (LPG.) there are still a few holdouts using alcohol and even more rare is compressed natural gas (CNG.) While alcohol is probably the most dangerous of all stove fuels it also has the least BTUs per volume, is the most expensive and the most difficult to obtain. Tragically, alcohol users believe it is safer because the fire can be put out with water. Small consolation for the large number of explosions, spills and burns caused by that antiquated technology. Most people have a problem with it because it is hard to see the flame and they believe the stove is not lit. When they try to add fuel to the primers, it explodes.

CNG

Compressed natural gas, CNG, on the other hand, claims to be super-safe because unlike either LPG or alcohol, the gas is lighter than air and will dissipate harmlessly into the atmosphere if it leaks. While the merits of this claim are debatable, one must remember that you have a steel tank of gas sporting four-thousand-two hundred pounds of pressure lurking somewhere in your hull.

LPG

The greatest objections to using propane, LPG, have been overcome. Even the lethargic Coast Guard acknowledges that and approves it for use on vessels. The main danger was always blowing out an unattended burner thus allowing gas to escape. Happily, that problem is solved with safety push-to-turn burners and thermocouples. Piezo lighters eliminate the need for matches, thus eliminating the cause of much burned arm hair. Propane safety has advanced to the point where a large percentage of catamaran refrigerators are propane fueled.

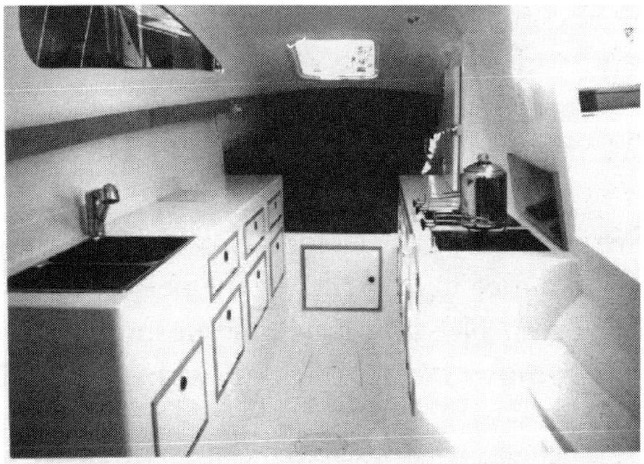

Maine Cat 30, galley, stb. Hull, photo facing forward

Doubting Thomases always say: "what if?" Good point, on a monohull or trimaran you might have an actual bilge in which gas could lurk awaiting a spark to set it off. In my own catamaran I not only have no true bilges but no machinery that could cause a spark down there and to make matters even better, my propane installations are all on the bridgedeck, way above the waterline. Most catamarans are happily in this same circumstance. Those that have galley-down usually have watertight bulkheads separating engine compartments from the possibility of LPG intrusion. Besides, well before the LPG could get to an explosive mixture, the smell would drive you out of the boat. Accidents involving LPG in sailboats are few and far between. I have not heard of an actual insured loss due to LPG in years.

Safety equipment, some differences in focus

Jacklines

Every once in a while I come across a catamaran that has jacklines installed along the side decks. This type of installation is a throwback to monohull offshore racing where dire warnings are given about not using the lifelines for attaching harnesses. What I get from these regulations cum warnings are that lifelines are unsatisfactory on three counts. First, that you must un-clip and re-clip every time you come to a stanchion. Second, that the lifelines are not strong enough and third, that they are too close to the edge of the boat, thus do not give real protection from falling overboard.

The catamaran side-deck jackline installations I see, use the same wire and the same fittings as used in the lifelines, vinyl coated wire and swage end fittings. I suggest that if that material is not sufficiently strong for life lines with many attachment points and stanchions to lend both support and shock absorbing quality then how on earth could a single wire of the same material stretched bow to stern be sufficiently strong as a jackline? Especially when considering the low angle of pull which multiplies the force against the wire many times?

To be useful, jacklines should be run inboard and when not in use, certainly not lie on the deck to get underfoot and become an ankle twisting menace or a dirt catching nuisance. Jacklines should also be of a material that is strong, resilient, shock absorbing and easy on your hands and feet. I suggest to my catamaran readers the following concept for easy to use, sufficiently strong, properly placed jacklines without resorting to expensive tape systems.

First, they should be temporary. When not in use or needed, they constitute another annoyance and maintenance item on deck. The perfect materials are right at hand on your boat and those are your docklines. Needing jacklines subsumes not needing docklines. Therefore, you have the perfect opportunity to do double duty with your equipment.

Next, is placement. On a monohull, you have the distinct possibility of being tossed overboard as the boat rolls, corkscrews or plunges through green water. Because of the nature of the motion on a catamaran, those actions are much diminished and the likelihood of a momentum generated overboard launching curtailed. Nevertheless, jacklines are a good safety item to use. Fasten one end of the dock line cum jackline to a convenient strong point close enough to your cockpit that you can hook up before you leave its security. The other end of the line should be securely tied using the appropriate knots to a convenient strong point at the center of the bow crossbeam or stem head fitting.

Safety harnesses

Catamaran harnesses are also a little different. Learned from the day racers is the use of the safety line with three ends and three clips. The base clip and clips to the harness can be released by the wearer in the event he or she is being held under water. The other two are at the ends of a longer line and a shorter line so that you can clip-unclip from one to the other without risking being unclipped even momentarily. It also means you can tether yourself between two strong points if need be.

It is easy and routine to see published pictures of monohulls with their bows buried in water and/or their decks awash, but it is pretty tough to find one showing that condition with a cruising catamaran and when you do, at least it is always level.

Safety from sinking

The greatest fear of a monohull sailor is sinking. With a ballasted boat with between thirty and fifty percent of the vessel weight in ballast (lead, Iron) it is practically impossible to keep it from sinking if its watertight integrity is violated. Without the thirty to fifty percent dead weight in a vessel, it is both practical, feasible and routine to build vessels that are virtually unsinkable.

The history of sinking catamarans is a remarkably brief tale. There are the few that would have been sunk no matter what kind of boat they were because they were run over by barges being towed, hit by submarines, destroyed by category 4 and 5 hurricanes and other life risks that bear little relevance to the type of vessel. Interestingly, the history books are replete with stories of multihulls surviving incredible catastrophe that would not have been possible with sinkable vessels.

I leave it to you as the cognitive elite reader. In defining risk, which has the greater risk, a vessel designed with watertight compartments capable of floating it's displacement and built from foam sandwich materials with enough buoyancy to float it in any event, or a vessel having thirty to fifty percent of its displacement in lead or iron that could not possibly float if holed?

Safety Is a Day to Day Concept!
I divide safety aboard a vessel into two classifications:

1. Perils of the sea.
2. The ergonomic considerations of practicality, usage, layout, equipage and function.

Perils Of The Sea

Cruising Catamarans	Other Vessels
1. Essentially unsinkable	1. Sinkable
2. Low or no bilge volume	2. Bilge has high volume and machinery
3. Watertight bulkheads routine	3. Few if any watertight bulkheads
4. Basically rides on top of water	4. Rides through water
5. No rolling	5. Severe rolling
6. Quick motion, little momentum	6. Slow motion, much momentum
7. Little heeling	7. Considerable heeling
8. Little tendency to broach	8. Broaching a major problem

Ergonomics

Cruising Catamarans	Other Vessels
1. Safer usage of dangerous fuels, propane, gasoline	1. Explosive fumes in bilges a possibility
2. Total perimeter lifelines	2. Some have no protection
3. Cabin entry doorway	3. Steep companionway ladder
4. Large deck lockers	4. Often scant deck storage
5. Easy anchor handling between hulls	5. Banged and scratched topsides
6. Sail handling on huge foredeck	6. Sail handling on skinny bows
7. Most have reasonable visibility	7. Fly bridge powerboats are better
8. Most have good visibility from inside the cabin	8. Few have good visibility from inside the cabin
9. Interior stowage mostly orderly	9. Interior stowage often haphazard
10. Docking from almost square shape	10. Docking from pointy ends
11. Dinghy stowage routine	11. Dinghy handling problematical
12. Mostly good visibility	12. Good visibility an exception

Anchoring
Safety really includes anchoring. You need to anchor for all sorts of reasons, but safety is always the primary concern. While this book does not go into all the aspects of anchoring, there are some modern marketing problems generated by folks who really do not understand the requirements.

Anchor location

Look carefully at photo and you will note that the anchor is centrally located *under* the trampoline. That's right, *under!*

Underdeck anchor location. Note large opening in critical area of bridgedeck, inability to clean or clear debris or hooked anchor lines of other boats.

I can see it now. Its a beautifully Sunday afternoon. The factory marketing staff is out on a demo ride with some prospects. They are waxing eloquent about the features of their catamaran when of them notices that the dirty old anchor chain is pulled across the nets. "That's awful," exclaims the first marketeer, "how dirty, and disturbing to the guests."

"Yes," chortles the second marketeer, a sales-staff-in-training. "I believe I have the answer" the first marketeer continues. "We will drop the anchor from the center of the boat from the dedicated anchor locker. After all, the windlass is right there, there is the ideal way to hook on the bridle and everyone knows you must keep the weight out of the ends of a multihull."

"Besides that," says the sales-staff in training, "it will save the factory a lot of time and money."

Topside view of anchor UNDER trampoline. Also note there is no central walkway-compression member. How in the world are you going to clean the anchor? Will you take all the mud and grass into and around the windlass? Do you think it prudent to handle dangerous machinery in that below deck position?

The Captain, having a little experience in these things, thinks to himself: "That's the craziest idea I ever heard of! It probably will work in the most benign conditions, but the minute some neophyte tries to drop the anchor in a cross wind or while the boat is still moving in any direction except straight backwards, the anchor chain will be tearing up yards of gel-coat in between the hulls. In addition, this system leaves a gaping hole in the center of the bridgedeck to accommodate the anchor which leaves a potentially dangerous situation in head seas when waves have a direct route inside the main structure and will inundate the windlass at the least."

Then, of course, come the operational problem: what happens when inevitably, someone hooks something on the bottom and must clear the anchor? How do you clear an anchor you cannot reach? Do you take out your trusty knife and cut away the trampoline?"

Further, how do you simply clean a muddy anchor chain? How do you untwist a twisted chain? What happens if someone inadvertently drops the bridle hook before hooking to the chain? How do you get it back up if you do drop it? How many charter people will not even bother with the bridle and let the anchored vessel hang right off the windlass just waiting for a squall to send the boat anchor-sailing and tear up yards of gel-coat at the least?

The Captain, in fear of losing his job, says nothing and marketing designs yet another gimmick to dazzle customers.

Other aberrations of the above scenario are popping up.

Anchor handling

First rule of sailing: your anchor is the single most important piece of safety and navigation equipment on your boat! Thus it is only common sense that anything that interferes with proper or appropriate usage is detrimental to safety. Handling heavy, dirty, clumsy items that have the potential to do grave bodily damage must be arranged in the safest and most user friendly possible manor. Anything that interferes with that is suspect.

The following check list is a good start

- The ship's main anchors should be launched from the bow, not the middle of the boat.

- Equipment, especially items like windlasses, should be easily accessed. Two people are often required to operate them.

- No piece of heavy duty machinery that is vital to safety and security that might need to be operated in emergency conditions should be installed in locations that require your fingers to work in tight little corners.

- Chains or rodes should not pass through long inaccessible pipes, especially on the tension side between the windlass and the anchor.

Note: Certain boats have taken to hauling the anchor through the tubes that support the trampoline. This has the potential for disaster by jamming virtually anything in between the chain and the tube or worse, in the event of an accident or other occurrence that puts a crimp in the tube you will no longer have the ability to lower or weigh your anchor. Shackles and other connectors must fit through the tube giving the possibility of all sorts of mischief. Here you have the potential for the irresistible force against the immovable object scenario.

- In the regions where anchoring with two anchors from the bow, Bahamian style, is the norm, you will encounter additional difficulties and dangers to your fingers with any enclosed system.

Helms

One of the primary safety features of a cruising vessel is helm location and its ergonomic layout. This is especially true when the crew consists of a cruising couple. If you are a cruising couple or in a situation where you will be the primary responsible party with only minor assistance from guest, children or visitors, then it is more important than ever to have a safe, secure helm that is comfortable and out of the weather.

The following may appear overly critical, but I think not. My experience has shown me that it is exactly this type of problem that has crushed the cruising dreams of so many people and it is the escape from precisely this type of problem that has generated a great part of the popularity of catamarans as cruising yachts.

A helm on the top step of the starboard transom.

Is this any place to locate a helm?

Look at the photo of the starboard helm (right) and port helm photo pg 154. Study them. They show the most dangerous helm I have ever seen designed into a boat, either monohull or multihull or power! In my 20 + years as a surveyor I have seen some pretty awful designs but nothing even coming close to this one.

What is crystal clear in the photo is that the helm station is actually the top step of the transom stair. It is not in the cockpit. It even requires climbing over a coaming to get into the cockpit. There is a little plastic protrusion bolted to the stern rail that is supposed to be a "rest" for the helms-person. There is a single piece of life-line with a pelican hook that is all that exists between the helms-person and eternity. Perhaps that MOB gear is there so the helms-person can hook it to his belt whilst steering, just in case.

In a full crewed racing boat, such an aberration of common sense may be defensible; but this vessel compares to a racing boat in the same manner a Maserati compares to a Winnebago. Would you buy a car in which you had to stand on the rear bumper to steer? Picture yourself in that situation, would you want to forever consigned to be "away from it all" in the broiling sun, freezing rain, dark of night guiding a vessel from which you can only see straight ahead on your side?

Study the picture some more. Note the lines underfoot on that step. Note the placement of the winch. Explain to me how you are supposed to get back and forth from one helm station to another without climbing over all that stuff? Does all that climbing and all that stuff represent yet another intense danger? There is not only no place to put lines safely out of the way, but when they trickle down the transom steps they more than likely will wind up in the props or jammed between the hull and the rudder Murphy's law, you know!

Visibility from either helm station on that boat is miserable. Even viewing the sales brochure showing a six-foot man at the helm, it is obvious visibility is limited to a sector of the side of the boat you happen to be on. If you leave the helm to place the boat on auto pilot, your visibility plummets from miserable to nil. Sorry, nil is not an exaggeration. Once inside the cockpit, the main bulkhead is so high you cannot see over it and even if you were a basketball player it would not matter because the continuous Bimini prevents that. If you go inside and try to look out the windows, you will find you have no windows facing forward only two diminutive opening ports, one on either side of the mast. The radically curved roof-line precludes any other arrangement.

Imagine yourself standing behind this helm. Your head is about level with the Bimini top, you cannot see over it even if you are quite tall. Below the Bimini forward vision is impaired by a solid bulkhead. You cannot see through the cabin and out the front windows because there are none.

Further examination of the photos detail that there is precious little room to install all of your electronic toys at the helm. There is hardly even room for a radio let alone RADAR, digital charting, depth sounder etc. If you do push the issue, you need repeaters for the other side and remember, you will need high quality covers for both sides when not in use.

The "Rambo" helm stations as pictured above, are completely exposed to the weather with no practical covering solutions in sight. Perhaps great for a little fantasizing by a Caribbean charter customer sailing an hour or two a day, but absurd for any private owner wishing not only protection and visibility but community.

Ah yes, I can see it now, stand straight before the onslaught of the trade winds, wind whistling through your hair, the sting of the salt spray, your bare torso exposed to the elements, Rambo would be proud. A couple of hours later Rambo is at anchor in a serene anchorage preparing to dinghy ashore for sundowners. Private owners on a trip someplace are still out there at the helm, especially if they are in a confined waterway someplace.

Picture this scenario. You are out cruising on the ICW. A rain storm comes through. The channel is too narrow to anchor in, you must press on to an appropriate destination. The helm station of your cruising catamaran is way back in the stern quarter. Visibility from the helm is terrible to start with because your line of sight from the helm is interrupted by the enormous mass of a cabin. In the best of times, you cannot really see forward. Steering from this position is manageable at sea in good conditions because the helmsperson can toggle back and forth between helm stations thus somewhat mitigating the poor visibility. Besides, you are on auto pilot most of the time at sea anyway. Piloting in close quarters is a horror story needing all eyes on deck, with crew (if any) shouting back instructions to the far removed helm station.

Now you are forced to steer because of the circumstances. It is cold. It is wet. You have neither a nice Bimini top over you nor a windshield in front of you because in that remote helm location, none can be designed. You would love a hot cup of coffee. You do not ask for it because you hate to have your wife climb through the cockpit and over the coamings to reach you, thus getting soaked herself. Besides, there is not even a place to put the cup down out of the cold biting rain!

It is two days later. You are heading for a typical cruise, the Dry Tortugas or Bimini if you are in South Florida, Catalina if you are in Southern California. It is light breeze with full sun. For some reason your auto pilot refuses to work so you must hand-steer. Now you are faced with standing behind a wheel which gets so blistering hot you can hardly grasp it, standing on decks that get so hot you must put on shoes and, of course, getting the full sun yourself. You try various clothing combinations to protect yourself from that brutal sun but the best you can do is roast instead of fry. Your wife cannot relieve you very much, she is fair skinned and can only take a limited amount of sun.

If you are cruising your way out, you can turn back and just lose cruising time. If you are on your way in, you just tough it out. Either way, you will really suffer tomorrow. To add insult to injury, some boats built like this do not even have a place to sit at the helm! Does that sound like "pleasure sailing" to you?

In my first book, *Cruising On More Than One Hull,* I pointed out the importance of having good visibility from the helm. "The helmsperson must be able to see all four corners of the boat. In addition, he or she must be protected from the elements and must have a comfortable seating arrangement." The best of cruising boats also have sufficient accommodations at the helm for a small chart or cruising guide. Most people who buy the larger catamarans insist on full instrumentation. None of that opportunity exists with the helms in the stern quarters.

In fact, one of the most expensive catamarans with stern helms not only has no provision for the tails of sheets and it is awkward to crank the winches but you are consigned to sitting on the traveler, if you dare sit! When you sit, your visibility goes from very poor to nil. One boat brand has a steering wheel so large there is no clearance between the deck and the wheel. A perfect place to jam the tail of a sheet because there is no place to put the sheets' tails and they are free to wander around on their own.

The latest onslaught of boats built like this to hit the United States not only have all the above problems, but are manifestly unsafe as well. Unsafe? Picture this. You operate the helm from outside the cockpit perched on the very end of the afterdeck! There is only a single strand of wire to prevent falling overboard and that wire may often be left un-clipped by people using the transom steps. There is a little sitting rest for the helms person, I hesitate to call anything without a backrest, a seat. One misstep, one errant powerboat wake and its Davie Jones Locker or at a minimum a serious fall down the transom steps.

Then there is the operating procedures problem. Mainsheet, traveler and jib sheets are operated from the helm position yet there is not even the slightest thought about where to put the tails of the lines. Since the helm is already outside the cockpit, it would be expected that Murphy's Law would be in full operation and the sheet-tails would migrate down the transom to be snared in the props or jammed between the rudder and the hull.

Even under the best of circumstances, visibility is bad, exposure to the elements is total, the helmsperson is isolated from the social activity, it is awkward to operate the sails, there is very limited room for modern electronics and the location is dangerous as well as uncomfortable.

The next question, of course, is "how come there are so many boats built like that?" The answer is not at all complex. First, there is the mechanical equipment advantage. Just as in a monohull where the helm is in close proximity to the rudder, there is less linkage, less gear and an easier less expensive installation. But that is not the crux of the matter. The main objective is to keep the hired crew as far as possible from the paying guests. You must remember these boats were designed and built in continental Europe to European aesthetic and social standards. They were built to "entertain" wealthy charter customers, not to be family cruising vessels.

Note possibility of lines in close proximity jamming under wheel or even jamming your hand between the wheel and coaming. You have reach through the wheel to operate the throttles and if you want to sit, you do it on the traveler.

There are some uninformed individuals that believe the helms were placed like that in order to give the helms person a good view of the sails, similar to an America's Cup boat or a radical "A" boat. Not so. First, those boats heel and have entirely different performance parameters plus a highly trained youthful full crew. Second, even on those boats the wheels are still in the cockpit, not out on the transom. People who purchase enormous home-like cruising catamarans do not buy them to experience the "rush" of sailing. No matter what you do, you will never ever get that rush on one of those charter boats. The reason is simple. They do not heel, are so seakindly and so comfortable and so level, that rush hardly develops.

A sailor taking a spin on a Catana, a high-end catamaran featuring daggerboards and many carbon fiber parts. It has an aft-facing power winch that handles all lines. This subsumes at least three in crew. One directing the winch operator, the operator and the third crew steering. All the lines are lead through mostly inaccessible concealed channels. Failure of any one line will easily spoil your whole day. Note the mess of spaghetti in the cockpit. Note the path required of the helmsman should he or she need to toggle between helm stations. Does all that massive seat construction create sufficient windage to negate the perceived advantages of the high tech ultra-light construction?

If you want to "feel the road" you buy a sports car. If you want maximum comfort, you buy a sedan. If you want maximum utility you buy a van. Thus it is with boats. Putting helms on cruising catamaran sterns to ape the "A" boats is like putting wire wheels on your van.

Capsize ! How Real, How Imaginary?

I agree, multihulls can capsize. What does that mean? What is the extent of the risk involved? Is the risk greater than any other reasonable risk? Is the risk as great as comparable risks? For instance:

- Is the risk greater than monohull sinking?
- Is the risk greater than monohull capsizing?
- Is the risk greater than capsizing a sportfisherman?
- Is the risk greater than capsizing a trawler?
- Is the risk greater than capsizing your automobile?

Always, the answer is resoundingly no! The risk of capsizing a class 4 or 5 production cruising catamaran is considerably less than any of the above. All multihull sailboats, including the featherweight flyers and exotic racing boats are statistically safer than any of the listed vessels or vehicles.

In 2001, *The Race*, which pitted the fastest, most exotic racing catamarans yet built in a no-holds-barred around the world race with dizzying speeds of up to 680 miles per day, had a better safety record (gear failure but no capsize) than similar races where sailors were lost in capsizes because of keel failures. My friend and long time idol, Tony Bullimore almost lost his life in a capsized monohull. The study of that epic makes one realize how much better off he would have been had that accident happened in his previous trimaran racer, *Apricot*.

This chapter could be titled: *How To Solve A Non-Problem*. It is difficult for me to keep explaining the creation of the capsize straw man. People created it with either vested interests, hidden agendas or a high threshold of ignorance about a subject that should be completely self evident.

Well designed cruising catamarans do not capsize any more than your family car does. Operator error can destroy anything. Mother nature can destroy anything. Catamaran sailors generally have no overriding concern about capsize. Except in the most unusual of circumstances, it does not happen. These circumstances would compromise any boat, no matter what its stability system. This is only a small part of the facts, that should be obvious and blatant to all sailing participants: sailing is a risk taking sport! That is part of sailing's appeal! As it says in the United States Declaration of Independence: *We hold these truths to be self-evident.*

Sailing, like mountain climbing, flying, surfing, auto racing, biking, fishing, SCUBA and a host of other outdoor sports, has a risk taking component. There are degrees in all those sports that pass beyond the threshold of completely safe. Sailing a small vessel offshore, any vessel, passes beyond that threshold. However, some vessels are safer than others. If you want maximum security, stay home and play croquette. Let me relate just a small vignette that might be appropriate.

Some years back when I was chartering my 40-foot catamaran out of Tarpon Lodge, Marathon,

Florida, I used to have local people as crew. One such crew was a young woman, pregnant with her first child. Her husband was a local air conditioning technician and they lived aboard a converted steel life boat, meticulously refurbished as a sailing vessel and fitted out by her husband. One day she confided in me, that the boat terrified her. She had constant nightmares about waking up and finding the boat sinking!
It was an interesting insight into another's perception of the cruising life. What that revelation did, however, was to raise my own consciousness about this sinking phobia. Over the years, I have found it to be more common than one might think. It is interesting, that neither I, nor anyone else I have asked, have a similar phobia concerning capsize. I think about capsize aboard my boat in the same way I think about overturning my car. When you press the limits, you press your luck. Drive prudently, sail prudently and in your lifetime you will have only normal, rational, concerns about capsize. This obviously excludes forces beyond your control: hurricanes, tornadoes, earthquakes, etc. just as it would in your car, or for that matter, in your home. Ask the victims of hurricane Hugo or hurricane Andrew about that.

Design the preparation for natural disasters using appropriate technology. Multihulls are very lightweight boats as opposed to ballasted monohulls. Since the circumstances are different, therefore, the approach to property preservation will be different. Compare this to the differing approaches used to preserve shore-side property versus mountain property. This is a subject for another book.

A Recurring Theme

Themes of multihull capsize buffet newcomers to both racing and cruising catamarans. How much of the purported risk, is true? How great is the risk? Before we can begin any serious discussion, we must set some parameters.

First, we must that there are differences between racing and cruising boats and vast differences between catamarans and trimarans. So much so, it is not technically accurate to use the generic term "multihull" to cover the subject. Therefore, I will refer to the different types of multihulls under their proper classification.

A Search for Numbers

Figures don't lie, but liars figure! *Mark Twain*

Beginning this discussion, The Seven Seas Cruising Association (SSCA) is a world-wide organization of cruisers. I am a twenty-odd year member of SSCA and faithfully read the monthly newsletter. A distressingly large percentage of the monthly bulletins contain a terrible tale of disaster, in which someone has sunk their boat. From a five thousand member (approximate) organization, several boats sunk per year is a frightening statistic. The actual statistic from 1989 was 15 boats lost at sea, and one person lost overboard. That is more than one boat per month! (All this reported from one organization, yet customers at boat shows only occasionally ask the salesmen about sinking!)

As a Marine Surveyor, I carefully scan the journals for information in my specialty. Each month there are several sinkings of monohull sailboats. The primary causes vary, from just plain neglect, to dead batteries, to frozen pipes, to galvanic corrosion in thru-hull fittings through back siphoning heads, but the secondary cause: the ballast keel, is all too obvious. In most cases, a production cruising catamaran *would not sink* under identical conditions. Most boats that sink at the dock, sink from fresh water. The source of that water is from a hose to the dock, either being used for domestic water or air conditioning. In some situations, snow and ice.

The *Exchange For Marine Professionals*, (Boat/US), publishes a quarterly report, titled *Seaworthy*, concerning these topics and invariably there are reports of monohull and powerboat sinkings. This service of BOAT/US to which I subscribe, has to be of exceptional value. I highly recommend it to anyone in the marine business.

On the other hand, there are only four documented cases of capsizes of cruising catamaran while being cruised by owners or charterers. Three of the four cases were operator error. On one 32 foot *Maldives* in the notorious Bay of Biscay, a squall hit and the charter operators had no foul weather gear. Nobody on board wanted to go out and lower the sails! It is quite hard to fault the boat on that one. To the date of this publication, there have been only two reported capsizes of a production cruising catamarans while it was taking evasive action at sea. Considering this spans over a 50 year period, it is truly a remarkable and impressive record. This record is fact, not hearsay or rumor. Based on this record, why are people continually talking about capsize?

I specifically address myself to cruising catamarans. There are only a few custom built class 4 or class 5 cruising trimarans over 30 feet in length. This nullifies any statistical relevance. However, the U.S. Coast Guard does not recognize the trimaran classification. All statistics relating to any multihulled vessel, whether catamaran, trimaran or proa, are reported as catamarans. This reporting system even includes tunnel hull boats. Therefore, it is especially important that you examine USCG data and reports to verify the type of boat involved.

As a matter of some interest, if you check the definition of catamaran in Webster's unabridged dictionary, you will find that: *catamarans are vessels noted for their safety!*

Years ago, we used to blame racing as the cause of multihull bad press and publicity about capsize. It was true, when a racing multihull capsized, it made world wide headlines. However, a careful review of the major ocean racing events shows that there are proportionately many more monohulls lost than multihulls. It is the realization of that fact that has brought multihull racing into predominance.

Prior to the modern era of production catamarans, there were more capsizes and vessel failures. I, publicly, thank all those brave and hardy souls that put their money where their mouth is, took a chance and looked toward the future. These were our nautical test pilots. We routinely fly airplanes today based on what we learned at the crashes of yester-year and we sail vessels based on the experiences of our nautical ancestors.

The turtle gets no place unless he sticks his neck out. We, the beneficiaries of the early experimenters, are reaping the rewards of their innovation and daring. The modern age of production catamarans began in Hawaii, circa 1945 and England circa 1950. In the fifty odd years since the introduction of production cruising catamarans, those vessels have amassed a safety record *second to none!*

That record is so good, that a vignette comes to my mind. Upon receiving the news that the Schooner America had won the race, Queen Victoria asked: ". . . and who is second? The answer came back: "I am sorry your majesty, but there is no second!"

Throughout the World class grand prix racing, multihulls are now the majority. This is a difficult fact for some so called traditionalists to grasp, nevertheless it is true. Of course there have been accidents. Isaac Newton said: "The price of motion is accidents."

That is not to say you can buy a catamaran and be free of accidents, which is obviously not true. It is reasonable and prudent to say, that the risk of a capsize in a production cruising catamaran is a lesser risk. The risk of capsizes or sinking in a ballasted sailboat, power cruiser, sportfisherman or trawler, is a greater risk. This is a statistical fact, borne out by Lloyd's Of London insurance premiums, which often are lower for production catamarans than for comparable cruising monohull sailboats.

I might suggest a scenario for how the capsize myth started. Most of the world's vessels carry no external ballast. Bolting ballast to the outside of a vessel is historically, a new phenomenon. A skinny, round bilge monohull, is an unstable form that is ballasted for self-righting. It is elementary to grasp the idea. Just like a punching bag, or a child's ballasted toy, no matter how often you hit it, it returns to the same position. A multihull is not like that. If you push it too far, it falls over and does not self-right. (Like a motor boat or a freighter or an automobile)

To understand the concept of why it doesn't fall over in the first place, you need a bit more technical knowledge. There lie the seeds of the problem. Salesmen and others discussing their particular self righting concept, can use any number of training aids to demonstrate the concept. On the other hand, a catamaran obviously cannot do that. It requires sophisticated technical discussion to understand basic vessel stability. Since it is not in the salesman's interest to do so, the innuendo remains. That innuendo is clear. Because the monohull is self righting, it is safer. Yet that is definitely not so. Would you accept a similar argument to prove that a motorcycle (two wheels) is safer than a car (four wheels?)

Of course, the salesman also neglects to inform his prospect that most of the worlds' boats are not externally ballasted, and not self righting. The overwhelming majority, probably more than 90% of the extant boats world wide, power and sail, rely upon form stability rather than ballast to stay upright. A catamaran simply put, has the highest form stability of any non ballasted boat.

It is also critical to keep in mind that the more likely a boat is to capsize, the more energy and time are spent in discussing its prevention. That is why you have the combined SNAME, USYRU, USNA *Committee On Safety From Capsizing* (monohull, not multihull!) This committee labored for several years and came up with a stability formula.

All this energy was as a direct result of many capsizes. This list includes losing Angus Primrose, a designer of heavy displacement monohulls, in a vessel of his own design. It includes the Fastnet disaster and other examples. Of course, you have the expected screaming by certain boat manufacturers who dispute the numbers because their boats don't measure up thus creating certain product liability problems.

For a more complete look at the report of the *Committee On Safety From Capsizing* and the stability formula they have derived, see the bibliography.

In his book: *The Cruising Multihull*, Chris White put in proper perspective not only the statistics of capsize but also the basic mathematical aspects involved as they relate to why monohull people are forever talking about capsize. [See bibliography]

Put it this way. Suppose you only had one leg and had to carry a bowling ball to use as balance to keep yourself upright? Capsize, or falling over, would be a primary aspect of your concentration. On the other hand, having two legs, you have natural balance and except in unusual circumstances, seldom think about falling.

Monohulls relieve excess pressure from wind gusts by heeling and or rounding up. Multihulls do it by accelerating. It is easy to grasp the idea at the occurrence. A gust comes along, the monohull vessel heels smartly, recovering when the gust passes. If the gust is too severe for the sail carried, the vessel rounds up into the wind and the sails luff, often, with the vessel wildly out of control. The multihull, on the other hand, not heeling, leads someone unfamiliar with its energy relieving techniques to question the possibility of capsize. The basic energy relieving technique of a multihull is acceleration. In actual mathematical measurement, it takes only a little acceleration to equal to power absorbing ability of much heeling. Since the multihull does not have the same hull-speed limitation factor as a keelboat, theoretically, you can increase acceleration indefinitely. However, in the real world of sailing, as we have seen in previous chapters, there are, of course, limits. Even a little increase in speed absorbs an enormous amount of energy. It is just not quite as spectacular a process as heeling way over.

To reduce the possibility of capsizing, there is the standard technique of reducing sail. In addition, there are several additional techniques used while sailing. Two of these are moving the mainsail traveler to leeward and moving the jib tack point to leeward. These two techniques change the angle of attack to the apparent wind, thus changing the lift/drag ratio to one that is much more favorable. Having considerably less drag, therefore less heeling moment, dramatically reduces any possibility of capsize. The wide beam of a multihull allows a broad selection of tack points and traveler widths. Therefore, many trimming techniques routinely used on a multihull are impossible on a monohull. On a monohull, a headsail oriented rig exacerbates the problem. On the other hand, the engineering in modern production catamarans precludes most problems. Most simply do not have enough sail area to let you get into trouble.

USNA, USYRU, SNAME

Committee On Safety From Capsizing
I watched the films of the tank testing done by the *Committee On Safety From Capsizing* with great interest. It was a considerable revelation that losing the mast made the monohulls more prone to capsize. Time after time the tank test waves would capsize the various model boats. What was strikingly evident was that those same waves would not have capsized a catamaran!

An interesting personal insight into this perceptual problem comes from Tony Smith, producer of the Gemini cruising catamaran. Most of the Gemini purchasers are novice sailors, and/or cruising couples who want to adventure under sail. For year's Tony worried about people making foolish mistakes and wrecking one of his boats. Now, in 2002, with 750 boats sailing, the record is clear and unequivocal; he had little to worry about.

(In May, 2001, Tony with his son Neil, delivered a stock *Gemini 105* from Annapolis to England via the northern route)

The compelling advantages of most catamarans and some trimarans are: shallow draft, seakindliness, ease of handling, safety, comfort, room and for some, perhaps performance. They are sufficiently different from ballasted sailboats that they require additional study of sailing technique, and basic concept. Yet they are not sufficiently different to preclude beginners from purchasing them. Most purchasers are first time yacht owners.

There are many reasons for choosing a specific yacht. Whatever your choice yacht, you should know the truth about them. You should know their good points and their bad points and their advantages for your particular sailing requirements. The issues of capsize, however, hopefully, being finally put into proper perspective.

> *I have tried to add my own subjective feelings, my personal research and observations to the vast body of assembled data. It took me thirty years of introspection to arrive at these conclusions. I can find no other precedent within the considerable scope of my background for such a hardcore resistance to the plain facts, provable truths and undisputed scientific principles. These truth show up in practical ways, like lower insurance premiums and Webster's dictionary definition, yet this hardcore resistance remains. For those of you who cannot, or will not understand this, you have my utmost sympathy.*

For additional information and the most recent compilation of actual statistics, read: *The Cruising Multihull*, Chris White. For the original historical database, read *The Capsize Bugaboo*, Charles Chiodi.

Book 3, Catamarans in production

In this section, I will critique 26 of the 151 catamarans of different manufacture with which I have significant experience. They will be subdivided into two categories; boats from current manufacturers and boats no longer in production. Each section will be listed alphabetically with a cross reference by length.

It is not feasible to list every catamaran I have surveyed thus I have culled through my list and critiqued those most likely to be of interest to those looking for new or previously owned boats. In any case, reading the descriptions of sailing and handling may open up some ideas or avenues of exploration for you. It will also help broaden your understanding will of the nuances of cruising catamarans.

Each review has its own special viewpoint. That viewpoint may provide input from actual users and owners rather than simply the opinion of an old salt. Often more information flows from unwitting sources than from professional reviews.

Catamarans alphabetically.

Catamarans by length

Aquilon 265 Class 2.75, Racing/Cruising

Designer: Paul Stanek

For those following the internet generated hoopla concerning the alleged performance of the Aquilon versus the Farrier trimarans, from my observations, those guys ought to stick to the fabulous features of the Aquilon and quit trying to be something they are not.

In the light air of the Chesapeake during Annapolis Demo Days, the F boats blew right past us. The following day when the winds picked up a bit, the difference was much less noticeable yet it was still there, especially to windward. Which would be expected between a cat with diminutive keels compared to a trimaran with a maximized daggerboard.

Interestingly, three decades ago I was faced with exactly the same dilemma. Should I buy a Hirondelle 24 or a Telstar 26? Corinne and I agonized, gnashed our teeth and created much angst over that decision. We finally settled on the Hirondelle. I must say that in those days the choice was really much closer because the Telstar 26 was much more of a cruising boat than today's comparable F boats and the Hirondelle had daggerboards. Why do I say that? Because the Telstar 26 has hinged solid wing decks, life lines and considerably more interior room than the comparable F 27 or F 28.

Of course, the Aquilon has streamlined a bit from the comparison with the Hirondelle and it is demountable and trailerable, again narrowing the parameters. Rather than try to compare one to the other, I think looking at the positive features that meet the most needs of the buyer is the right way to look at it.

Going on the basis that the Aquilon is a "performance oriented cruising boat" (let's split hairs at class 2.75) where a comparable F boat is a " cruising oriented performance boat") (lets split hairs again at class 2.5) I list the following table.

Positive aspects of F boats for racing:

1. Exceptional performance
2. Highly mobile due to exceptional folding system
3. Large racing class, hundreds of boats

Negative aspects of F boats for cruising

1. Trampolines as only decks
2. Wet sailing
3. See-saw syndrome
4. Cramped and limited interior
5. Difficult to install Bimini top, dangerous low boom

Positive aspects of Aquilon for cruising

1. Layout and room
2. Solid deck in cockpit
3. Visibility and sail handling
4. Ease of installation of Bimini & dodger

Negative aspects of Aquilon for racing

1. Average windward ability
2. Lengthy disassembly time
3. Few boats available
4. Little race history

Aquilon

Sailing summary

Aquilon 265

1=Poor 2=Fair 3=Average 4=Good 5=Exceptional

	1	2	3	4	5
Motoring			3		
Docking			3		
Backing			3		
Sailing				4	
Weather helm/lee helm				4	
Windward ability				4	
Ease of tacking				4	
Visibility from the helm				4	
Tracking				4	
Seakindly/ride				4	
Underdeck Slamming				4	
Convenient Deck Layout			3		
Interior layout			3		
Adequate storage areas			3		

Vital Statistics

Aquilon 265

LOA (length overall)	26' 3" ft	7.99 m
LWL (waterline length)	25' ft	7.65 m
BMAX (maximum beam)	14' 9"	04.5 m
SA (sail area)	371 ft^2	37 m^2
Draft: 2' 0"		Keels
Designed displacement:	1,875 lbs	850 kg.
Mast hgt above WL:		41' (est)
Tankage:		
Water:	gal.	liters
Fuel:	6 gal.	22 liters
Auxiliary power:	one	9.9 outboard

Type .... Open bridgedeck
Foredeck .... Trampoline
Cabins ...... three
Galley up/down down
Rudders ..... Balanced spade
Rig .... Fractional sloop

Paul Stanek touts his double headsail rig. That rig changes the model of the *Aquilon* from 26 to 265. My experience with double headsails on boats in that size range has always been negative unless the apparent wind is midships or aft of midships, which it so seldom is in a swift sport-multihull. (There are exceptions not applicable to the Aquilon) It seemed no different with my experience on the Aquilon. During our sail you could get the vessel trimmed perfectly with all three sails apparently drawing well, then walk up to the foredeck and touch the jib and it would collapse having no pressure on it at all. It was obvious that the flow through the slot was keeping the jib full but it was creating drag rather than lift. The Screacher or forward-most sail was obviously under pressure and drawing well.

I understand that many of you are going to accuse me of hairsplitting, but I don't think I am at all. I think that a prudent sales person ought to recognize these subtle differences and try to fit his clients into the best boat for their purposes. Perhaps that is a little too idealistic for this business but I will keep trying.

There is an optional hardtop for the cockpit, but I prefer the regular canvas Bimini as it is more flexible. The vessel as shown also shows a boomless mainsail. My experience with them has been entirely negative. In fact, if you follow modern mainsail development, variations of the traditional gaff rig or leg O mutton rig are appearing, mostly as the fat head or square top. A mainsail is generally easier to control and to shape than a jib because two of its three sides are controlled by spars. Lacking a boom is lacking a certain amount of control, especially off the wind.

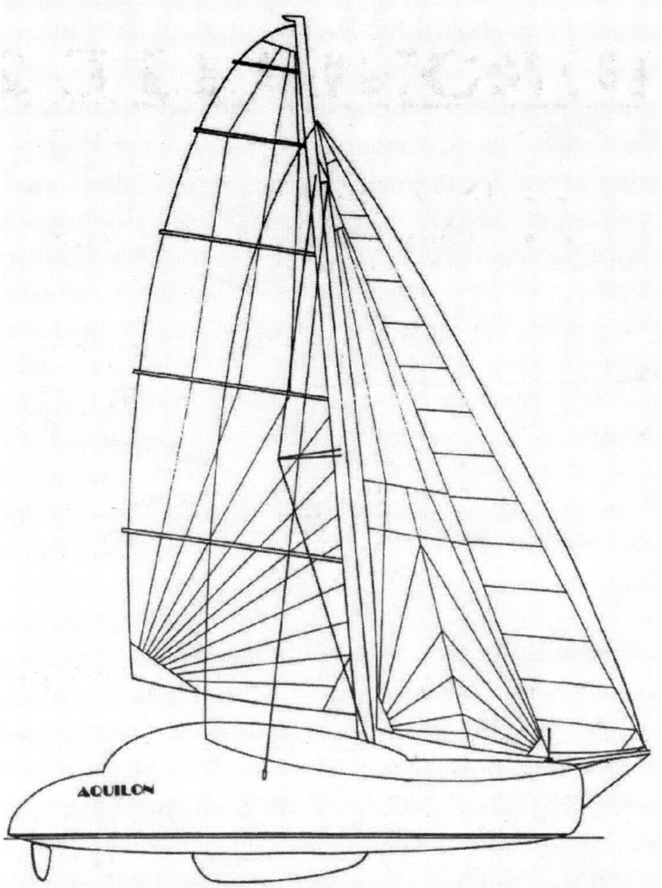

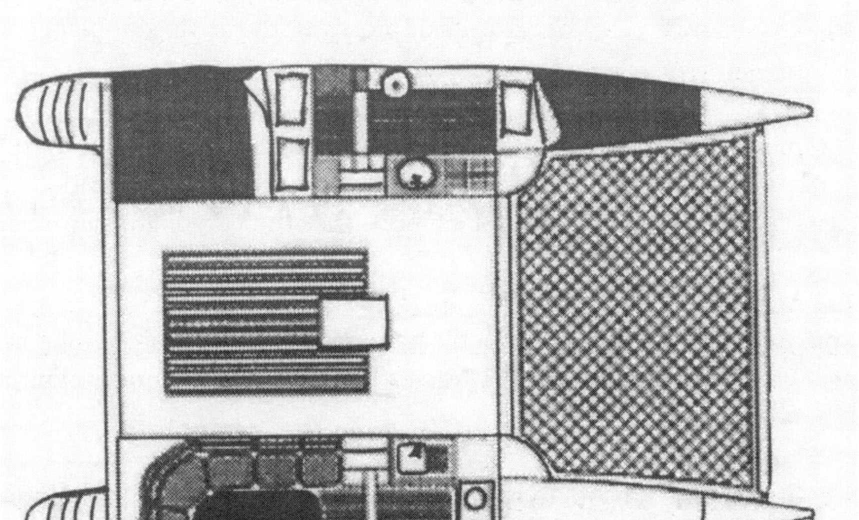

Endeavourcat 30

The Endeavourcat 30 Class 5, Cruising

Some ten years ago, I sailed the new Endeavourcat 30 in Tampa Bay off the St. Petersburg Pier. The weather was perfect. A steady easterly wind at 10 to 15 knots, puffy little fair weather clouds, and moderate seas. The new owners were aboard for orientation to their spanking new craft.

As we were motoring out of the harbor, the Yanmar diesel outboard auxiliary coughed, sputtered, and quit. Air had entered the injectors. This first emergency happened right in the center of the harbor basin. No problem, we lowered the Delta anchor which was neatly stowed on a bow roller and stopped. This is a sailboat. No minor shakedown problem would stop us from sailing. Besides, the photographer was waiting for us at the end of the pier.

Instead of fussing with the engine, we raised the sails and tacked out of there. I must call the unusual rig a mini-sloop or a maxi-cat. This is a tongue in cheek label because the original rig, a wishbone-boom cat rig, was modified to include a small, non-overlapping, self tacking jib and a standard boom with boom brake. The rig designer says it increases the efficiency and helps in tacking. It worked well while I sailed it. On the other hand, I participated in the original test sail several years earlier with Bill Symons, the designer at New Symrna Beach and could not fault the original rig as far as sailing was concerned.

We tacked, jibed, beat, reached, and ran all over Tampa Bay. At one point, a photographer engaged to do a sales video of the boat demanded we sail full speed directly at the St. Petersburg Pier and tack away at the last possible moment. This proved to be a thrilling exercise (for us, that is...he was firmly on shore). After several very close encounters with the seawall, I could not help but be impressed with the ease of handling of this vessel. To be completely objective, I must also add that, while tacking is a breeze, jibing requires the assist of specialized equipment: a boom brake. This rather low aspect-ratio mainsail and micro- jib is really a nice sail plan. Because the center of effort is never behind the center of lateral resistance (the way it is when you tack most masthead rig sloops or cutters), there is no tendency to go into irons.

We cured the Yanmar air lock and motored back in, but not before we tacked up the narrow channel to the marina. Frankly, the pointing ability was better than I expected. Considering this is an honest to goodness Class 5 cruising boat, I did not expect either the very high pointing angle or the crisp tacking.

Docking was a piece of cake. Just like any good catamaran, it backed into its slip arrow-straight, with no tendency to yaw with the application of reverse. The well thought out cockpit and deck layout is intriguing.

Details include built-in spaghetti boxes to manage the tails of the lines, secure lifelines, and handholds and a comfortable helm seat with an excellent panorama and view of all four corners of the boat. This is one of the very few catamarans that has visibility forward while seated in the cockpit.

My first impression of the current fetish of leading all lines back to the cockpit, is one that simply turns a one-person job into a two-person job. However, the Endeavorcat 30 layout is such that it reduces friction, lines disappear from underfoot to their own environment, and with minor exception, the system works (you still have to leave the cockpit to attach the halyard to the mainsail.)

Inside, this is a luxurious vessel. It's amazing what you can get into a thirty-foot boat. The appointments are tasteful and the quality is obvious. The head, excuse me, bathroom, has a separate stall shower. The galley has more counter space than many condo apartments. Best of all, there is a pair of real chairs in the saloon.

I am not enamored with the bunk layout. Two double berths in the aft end of the hulls is not my preferred arrangement for comfort, ease of access, or ventilation.

Since that 1992 initial sail, I have delivered two 30s and surveyed four others. On one delivery, I had noted marine engineer Russell Bartell with me and we put the boat through its paces. We spent a good part of the day in light to medium airs off the Florida Keys doing figure 8s, tacks, jibes, beats, reaches, and runs. The conclusion was the Endeavour 30 was one fine handling vessel.

The low bridgedeck has always been a concern, not only of mine, but of certain multihull designers. Therefore, I spent what amounts to an inordinate amount of time monitoring it, the owners, and the results of observation and survey. First, there is no question that this bridgedeck is noisier than many others. However, that must be tempered by saying it is really only noisy more often, but the maximum noise is less than getting a good belt from a wave on a boat with more clearance. I think it works like this: being closer to the water, it gets hit more often, but not as hard. None of the owners I spoke with complained about the noise.

The next thing I examined with a magnifying glass is the structure. Does being low to the water cause failure? My results. Apparently not. Of the boats I have surveyed and delivered, I have not found any structural failure of any kind that could have been caused by the design. It also must be considered that this is not an extra-wide beam boat, which means the ratio of exposed underdeck to the hull immersion surface is far less, meaning less force will be applied per square meter.

Always in the forefront of my concern is inappropriate use of the vessel. Because the vessel...any vessel...seems to be a well found vessel, does not mean you can indiscriminately use it in any circumstances. Like any other man-made contrivance, it has its design limitations. It was not designed, nor intended, to be a blue water circumnavigator. You should not expect it to be so.

Since the time of the Endeavour 30, three new models have been introduced: a 32-, 34-, and 36- footer and most recently, a 40-footer. Elsewhere in this book is information on those vessels.

Endeavourcat 30

Sailing summary

Endeavourcat 30

1=Poor 2=Fair 3=Average 4=Good 5=Exceptional

	1	2	3	4	5
Motoring			3		
Docking				4	
Backing				4	
Sailing			3		
Weather helm/lee helm			3		
Windward ability			3		
Ease of tacking				4	
Visibility from the helm					5*
Tracking			3		
Seakindly/ride			3		
Underdeck Slamming		2			
Convenient Deck Layout				4	
Interior layout				4	
Adequate storage areas			3		

Vital Statistics

Endeavourcat 30

LOA (length overall)	30' ft	7.99 m
LWL (waterline length)	27' 6"	7.65 m
BMAX (maximum beam)	14' 6"	04.5 m
SA (sail area)	482 ft^2	137 m^2
Draft: 2'1 0" Keels		.86 m
Designed displacement:	1,875 lbs	850 kg.
Mast hgt above WL: 41' (est)		12.5 m

Tankage:

Water:	75	gal.	284 liters
Fuel:	6	gal.	22 liters

Auxiliary power: one outboard

Type	Full cabin
Foredeck	Full bridgedeck
Cabins	two
Galley up/down	down
Rudders	Balanced spade
Rig	Cat original, sloop later

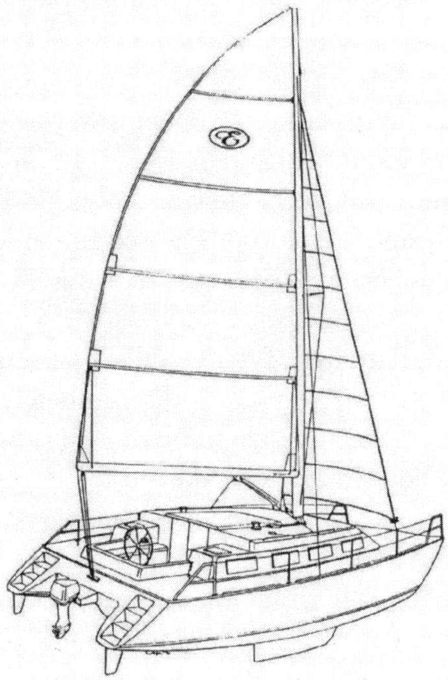

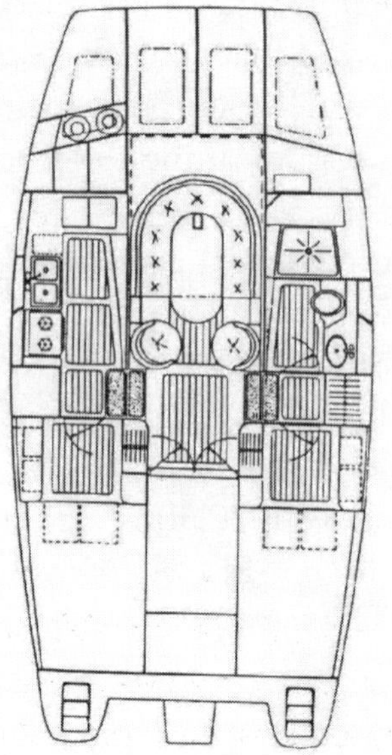

Gemini 30, 31, 32 Class 3/4 Cruising/racing

Depending upon how the boat is fitted out, it could be either class.

A 32-foot Gemini has a full queen-size stateroom, two double-sized cabins, an apartment-sized galley, and a convertible dinette capable of seating six adults. It has a large chartroom, huge sail locker and enough lazarette storage to satisfy anyone. How large of a monohull do you need to equal that space?

It is important that you compare like boats when judging performance. You simply cannot take some multihull that is like a Winnebago and race it against a monohull that is like a Corvette. If you race the multihull against monohulls with similar design parameters, you will be satisfied with the performance, especially in heavy air.

Assuming that everyone understands the difference between a Chevrolet and a Cadillac, let us approach the Gemini from that perspective. Few sailboats live up to their advertising hoopla, but I believe the Gemini is unique in that regard. Advertised as "the most boat you can buy for the money," it achieves that claim in my opinion.

It is not the fanciest nor the fastest (but quite possibly the fastest and most weatherly in its size class.) It is a reasonable boat that will meet the real sailing needs of many potential cruisers. My files show that I surveyed 22 Geminis, delivered three, given sailing lessons on three, and informally helped owners do repairs and have raced against and cruised with several others. Having followed the progress of the company since hull number one, I believe I have a good grasp of the vessels strengths and weaknesses.

The various Gemini models, starting with the original Gemini 30, and advancing up through the 31, 3200, 3400, and now the 105, always sailed better than their monohull production counterparts when raced. The reason is pretty clear: pivoting centerboards and the transom-hung rudders. If you want a cruising catamaran that will stay with the better monohulls to windward, sail past them on a reach, and better them on a run (and you are willing to keep the boat light), Gemini will be a good choice.

The boat sails nicely, tacks securely, carries her canvas well, and has no bad habits. It is a handy boat around docks, moorings etc. Those are the best points. Other points are less than terrific. The dog house and one-sided helm hinder visibility (the side varies by model year) If you are in cold or rainy weather and sailing with the windows closed, you are looking through two layers of glass – one potentially fogged from the inside and the other dripping with rain on the outside. It is difficult to see forward from around the cabin sides and a deck-sweeping Genoa makes it just that much more difficult.

Other chronic problems with earlier models include steering system failure, leaky windows, and a baffling water course between the sail locker on the port side and the hanging locker, starboard side. The window problems eventually were corrected with a combination of better installation techniques and better adhesive. The steering system problems require scrupulous maintenance on the part of the owner. The steering is a pull-pull Teleflex cable system. Because of the reverse slope of the transom, there is a bight formed in the cable that collects both fresh (rain, washing) and saltwater. If water is allowed to accumulate in this bight, it eventually destroys the cable both through common corrosion and galvanic action.

Several approaches have been used by owners to avoid the problem. The first is using accordion rubber boots covering the outside shafts of the cable attaching ends. The second step is the installation of grease fittings with the concomitant addition of grease. (Gemini website for vendors)

The water course was repaired by sealing it off from the inside of the sail locker. A factory advisement was sent out about this problem. This only applies to very early models.

I just love the steering and rudder system, even with its serious shortcomings. The reasons being that you can slip the blades out of the rudder housings and stow them out of the sun and water. The housings are easily repairable in the event of damage. They allow you to reduce your draft and thus avoid damage. Transom-hung rudders in general are easier to maintain, do not violate the integrity of the hull, are quick to rid of crab-pot lines and other fouling, are the easiest to repair, replace, service, or jury-rig, and they extend the sailing waterline of the boat. Of course, there is always a negative side. They clutter the transom, are easily damaged by backing into things, and are not as efficient as spade rudders. The 3400 and 105 have spade rudders that kick up – the only ones in the industry.

While the transom hung rudders have been an occasional irritant for some during the early model years, nevertheless, the basic design concept and execution are just the ticket for a gunkholing cruiser. Draft is reducible to a trifling 18 inches. A unique shock cord device protects the dagger type rudders from collision damage. Rudder blades easily remove for storage or service. This is an above average sailing boat.

Perhaps the following little vignette might say it all.

Tony Smith, the designer/builder was demonstrating a Gemini to a potential customer, I was also along as crew. We had a lively conversation about catamarans and one of the things that was a topic of discussion, was Roger MacGregor backing his MacGregor '36 catamaran into its slip in Marina Del Rey. Roger raced all the time and had no engine on the boat.

Now backing any boat into its slip under sail is no feat for the faint hearted, but it does require a vessel that will respond to the helm at very low speed.

Two dramatic advantages a catamaran has over a monohull at this game are that the catamaran has little momentum to stop and the almost square shape of the catamaran allows line handlers both forward and aft to wrestle the boat into the pilings without damaging a-midships.

Tony rose to the challenge. When we approached his dock under sail, the wind was blowing across the slip at 10 knots. From sailing downwind, we rounded up just in front of the slip, stalled, got the boat sailing backwards on a reach and gently sailed directly into the slip backwards, hardening the sheets to create some forward drive as a brake. Once in the slip, we simply released the sheets, grabbed the dock lines from the pilings, made her fast and stowed the sails. The customer bought the boat.

From the structural point of view, the vessel, as you would expect, is on the minimal side. That does not mean it is not strong — it is. It does mean that you have less leeway in fixing or repairing or installing stress-inducing equipment. To my knowledge there has not been any failure that can be traced to being under-built.

From the perspective of volume, value, performance, and shallow draft, the Gemini is a tough act to follow. From the perspective of finish, detail, and visibility from the helm, you can understand how the price has been kept in line.

I must note that an overwhelming number of the persons that buy Gemini cruising catamarans are retirees or life sabbatical types whose basic interest are going someplace, doing something and living aboard, they are not sailing aficionados per se.

Sailing summary

Gemini 30, 31, 32

1=Poor 2=Fair 3=Average 4=Good 5=Exceptional

	1	2	3	4	5
Motoring			3		
Docking			3		
Backing			3		
Sailing				4	
Weather helm/lee helm				4	
Windward ability					5
Ease of tacking			3		
Visibility from the helm			3		
Tracking			3		
Seakindly/ride		2			
Underdeck Slamming		2			
Convenient Deck Layout			3		
Interior layout					5
Adequate storage areas				4	

Vital Statistics

Gemini 30, 31, 32

LOA (length overall)	30'6"	9.30 m
LWL (waterline length)	27'6"	8.38 m
BMAX (maximum beam)	14'	4.27 m
SA (sail area)	425 ft^2	130 m^2

Draft: 18"/46 cm bd. Up 4'6"/1.37 m down
Designed displacement: 7,000 lbs 3175 kg.
Mast hgt above WL: 41 ('est) 12.8 m
Tankage:
Water: 40 gal. 151 liters
Fuel: 16 gal. 61 liters
Auxiliary power: one outboard

Type .... Full cabin
Foredeck .... Full foredeck
Cabins ...... three
Galley up/down down
Rudders ..... Transom hung
Rig .... Masthead sloop

Gemini

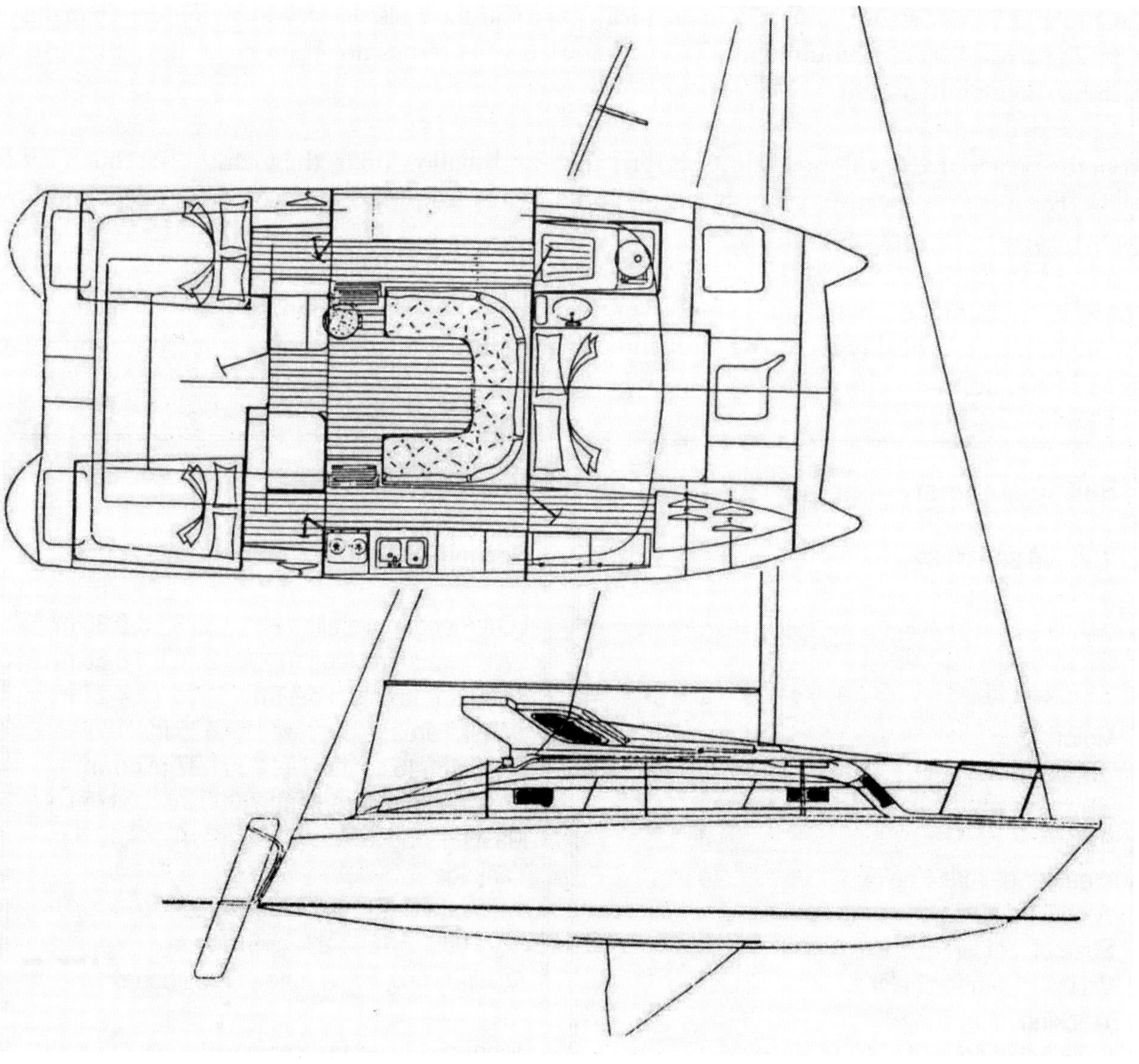

Gemini 105M Class 4, Cruising/racing

Designer: Tony Smith

I surveyed several of the new 105M models and found that there are significant improvements over the earlier models incorporated in them. In the summer of 2001, Tony Smith and his son Neil delivered one across the Atlantic from Maryland to England via the northern route. You have to have faith in your product to do that. The Gemini 105M is CE certified as blue water.

Sailing summary

Gemini 105 M

1=Poor 2=Fair 3=Average 4=Good 5=Exceptional

	1	2	3	4	5
Motoring			3		
Docking			3		
Backing			3		
Sailing				4	
Weather helm/lee helm				4	
Windward ability				4	
Ease of tacking				4	
Visibility from the helm		2			
Tracking				4	
Seakindly/ride				4	
Underdeck Slamming		2			
Convenient Deck Layout			3		
Interior layout				4	
Adequate storage areas				4	

Vital Statistics

Gemini 105 M

LOA (length overall)	33'6"	10.2 m
LWL (waterline length)	31'9"	9.7 m
BMAX (maximum beam)	14'"	04.3 m
SA (sail area)	510 ft^2	47.4 m^2
Draft: 1'8"/ 5'		.12 / .46 m
Designed displacement:	7,300 lbs	3311 kg.
Mast hgt above WL:	40' 6" (est)	12.4 m
Tankage:		
Water: 60 gal.		227 liters
Fuel: 36 gal.	136 liters	
Auxiliary power: one		30 hp Diesel

Type .... Full cabin
Foredeck .... Full foredeck
Cabins ...... three
Galley up/down down
Rudders ..... kick-up spade
Rig .... Masthead sloop

The latest model Gemini is the 105Mc Convertible. It has several options for full cockpit enclosures.

There is also a roller furling single luff spinnaker option that is controlled on a curved track out on and across the bows.

Gemini 105M

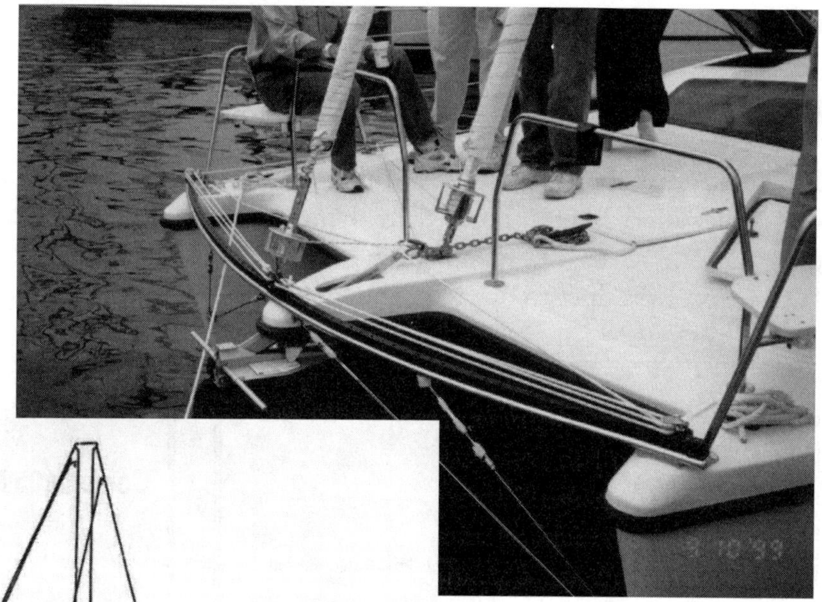

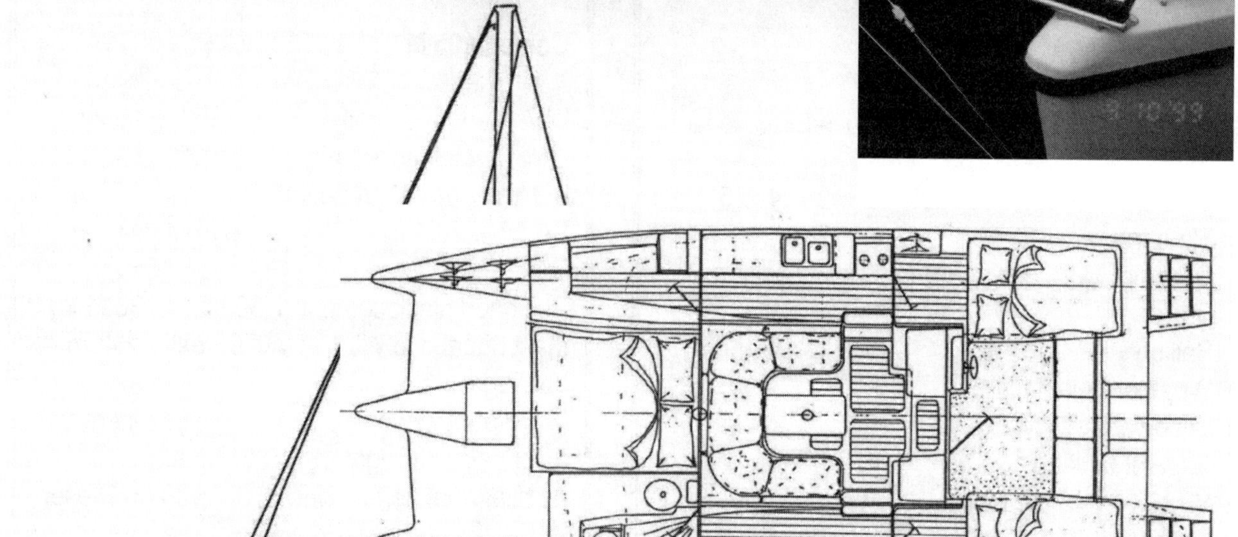

Gemini 105M

Hirondelle Class 4, Cruising/Racing 

Designer: Chris Hammond

My very first cruising catamaran was a Hirondelle, a 1971 Mark I, hull #51. It had box-type dagger rudders, daggerboards in each hull, tiller steering, a seven horsepower Chrysler outboard and a full complement of sails. I bought it from Bill Symons at the Annapolis Boat Show in the fall of 1971. The Hirondelle met my cruising needs at that time. First, and absolutely most important, was the shallow draft—13 inches.

I was living on the Little Magothy River in suburban Annapolis, Maryland. There was a bar across the river entrance—about two feet was maximum draft. We were sailing a swing-keel Venture 24, draft eighteen inches, at the time and upgrading to a larger boat with shallow draft seemed impossible until we came across the Hirondelle. We considered several trimaran offerings including the TelStar but the entrance bulkheads made it impossibly narrow for a shallow draft trimaran.

The following year, my relocation from Annapolis, Maryland, to Freeport, Long Island, New York, required an ocean sail from Cape May, New Jersey to Jones Inlet on the Long Island south shore. That sail really whet my appetite for the blue water. My Hirondelle handled the conditions beautifully. Despite heavy seas, strong winds, and all the usual open ocean roughness, she did not pound, did not hobbyhorse, had no tendency to bury her bows or display any of the other alleged cruising boat bad habits.

Having become fascinated with these amazing craft, I took a side job with Bill, assembling, delivering and trouble shooting the boats he sold. At that period in history, Bill Symons was Mr. Multihulls. He imported all the overseas produced models and had his own models built in Japan.

During this period, I delivered three Hirondelle from Long Island to New Jersey, including one delivery with the new owner and his wife aboard. On our way to the delivery destination, at the Atlantic Highlands Yacht Club, we sailed around the entrance sea buoy to Raritan Bay at Sandy Hook, New Jersey, the southern end of greater New York Harbor. As luck would have it, we ran right smack into the racing fleet from the yacht club. The racing fleet was using that same sea buoy as their leeward mark. We sailed right into the middle of the fleet and it only took a few minutes to realize that we and the racing fleet were headed for the same place, the entrance buoy to Atlantic Highlands at the south end of Raritan Bay.

Hardening up the sheets, adjusting the daggerboards, it gradually became apparent that we could hold the windward course better than the rest of the fleet, which consisted of boats between 25 and 30 feet. Not only were we pointing a degree or two higher than leading boats, we were also making about a half knot better speed than the leaders.

The owner practically went berserk with glee! He stood on the foredeck, his hands cupped over his mouth as he shouted toward the rapidly disappearing fleet. "Multihulls can't go to windward! Multihulls can't go to windward!" We were amply demonstrating that this one obviously could.

Over the intervening years, my experiences with this exceptional little vessel grew. If you really want to see what a good cruising catamaran can do, compete against Ernie Linke. Race against him. Unless the winds are really heavy, most folks just see his transoms. His Hirondelle Nip & Tuck during the CCMA races on the Chesapeake Bay is usually out in front.

In the years of ocean sailing, I have had my Hirondelle in Fire Island Inlet and Jones Inlet on the south shore of Long Island, New York; the Shark River Inlet, Manasquan Inlet and Barnegat Inlet, on the New Jersey shore and some other notoriously boisterous inlets in Florida. Only in Barnegat did we take any water across the decks. This boat handled all these rough inlets in grand style. I sold my lovely Hirondelle three years later to buy a larger boat, a forty-footer with which I planned to circumnavigate the World.

So much for history. At the Miami International Boat Show in 1995, a new, updated Hirondelle with a new rig was displayed. What had they done to my glorious Hirondelle? I had to find out.

The new model had no daggerboards nor dagger rudders. It had fixed keels and spade rudders. Personally, I have never been a fan of that configuration for several reasons. First, you lose your true shallow draft. The Hirondelle draft increased from thirteen inches to 2' 6". Perhaps this does not seem significant to you, but when you live in a region where four feet is considered deep water, you would understand.

There is no question that a vessel performs better with daggerboards than keels, and spade rudders are always a maintenance headache. The beam on the new design was widened from 10 feet to 12 feet and those stylish reverse transoms were added. Adding two foot of beam gives you forty square feet more boat area. This means the boat not only is potentially considerably heavier but gives the owner the ability to carry aboard much more "stuff."

Overloading a catamaran destroys its performance. If you widen the beam without proportionally raising the bridgedeck, the boat will pound in heavy seas. With this negative mind set, I climbed aboard the boat at the Miami show to see for myself. First thing I noted was that the unladen boat floated sufficiently above her lines so that a normal cruising load would bring her to proper trim. Joe Norwood, the importer, and David Trotter, the builder assured me that there was sufficient additional buoyancy in the keels and the transoms, plus the use of lighter materials, to provide a proper load carrying capacity for the vessel. They claimed the new version was actually 250 pounds lighter because of engineering and materials advances.

I could not help but admire the new Aero rig. It is very similar to the speed-demon catamaran Elf Aquataine where it was called a cruciform rig. Essentially, it is an unstayed rotating mast with a fixed boom that projects forward as well as aft. It has a three-quarter hoist self-tacking jib that clews to a fixed horse at the intersection of the mast and forward boom. The jib needs no adjustment after it is hoisted since it is self-tacking. The full-batten mainsail is contained by lazy-jacks. Most of the paraphernalia associated with sailboat rigging is missing. There are no stays and shrouds, no topping lifts, no hydraulic backstays, no vangs or preventers, no travelers, no banks of shiny winches, if fact, not much of anything by way of sail trimming devices that you are accustomed to. Merely an ordinary mainsheet.
You raise the sails, which you trim to optimum for wind conditions, then you rotate the entire rig to trim to the apparent wind. Not only is it simplicity itself, but it makes sailing infinitely easier. No matter which direction the wind is coming from, you can weathercock the sails into it. Therefore, there is never any flogging or luffing or panicky takedown. You can sail backwards as well as forwards, and I suspect—properly handled—you could even sail sideways into a dock. With no winches, no flailing jib sheets, and only one string to pull while sailing, it sure is a terrific cruiser's rig.

I was enthusiastic and needed to go for a demo ride. Our test sail was in a rather confined area behind MacArthur Causeway near Miami Beach. There were dozens of anchored boats and boats in motion at every speed and direction, making a nimble vessel a necessity. The flexibility of the Aero rig, with its ability to accelerate and de-power, was matched by the quick helm response of the boat. It was almost like using the accelerator pedal in your car. Are you familiar with the kind of boat that seems like slow motion to get response when you put the helm down? Well, the Hirondelle wasn't like that. While not a 505 or a Lightning, it's better compared to a dinghy with a light touch and tacks well.

Tacking a catamaran, any catamaran, even the very best, will never be like a keelboat. There is a difference in feel based upon beam and momentum. I was impressed with the agility of the vessel and its ability to maneuver in tight quarters. Our Hirondelle with Aero rig deftly wended through the waterborne turmoil of the varied boat show traffic without a problem. I have feedback from others that the Hirondelle does not pound in ocean waves and handles powerboat wakes satisfactorily, but I want to see for myself how it handles in the big seas of the ocean. It goes without saying that the interior room and exterior areas are huge relative to comparable vessels. The interior layout has some good features: a convertible dinette, two large quarter-berths and a galley down in the starboard hull with the head in the port hull.

Basic specifications for the new Hirondelle are:

Hirondelle 2
Length Overall . . . . 23'
Beam . . . . 12'
Draft . . . . 2' 6"
Sail Area: . . . Aero rig 212 sq. ft. Standard rig 245 sq. ft.

Hirondelle

Sailing summary

Hirondelle 24

1=Poor 2=Fair 3=Average 4=Good 5=Exceptional

	1	**2**	**3**	**4**	**5**
Motoring			3		
Docking				4	
Backing				4	
Sailing					5
Weather helm/lee helm					5
Windward ability				4	
Ease of tacking				4	
Visibility from the helm				4	
Tracking				4	
Seakindly/ride					5
Underdeck Slamming				4	
Convenient Deck Layout				4	
Interior layout					5
Adequate storage areas				4	

Vital Statistics

Hirondelle 24

LOA (length overall) 24' 0" 6.9 m
LWL (waterline length) 20' 6.1 m
BMAX (maximum beam) 10'0" 04.5 m
SA (sail area) 250 ft^2 23.2 m^2
Draft: 13" / .11 m bd up 4' / 1.22 m bd dn

Designed displacement: 1,875 lbs 850 kg.
Mast hgt above WL: 41' (est) m
Tankage:
Water: 20 gal. 76 liters
Fuel: 6 gal. 22 liters
Auxiliary power: one outboard

Type .... Full cabin
Foredeck .... Full foredeck
Cabins ...... three
Galley up/down down
Rudders ..... Daggerboard type
Rig .... Masthead sloop

Hirondelle 24

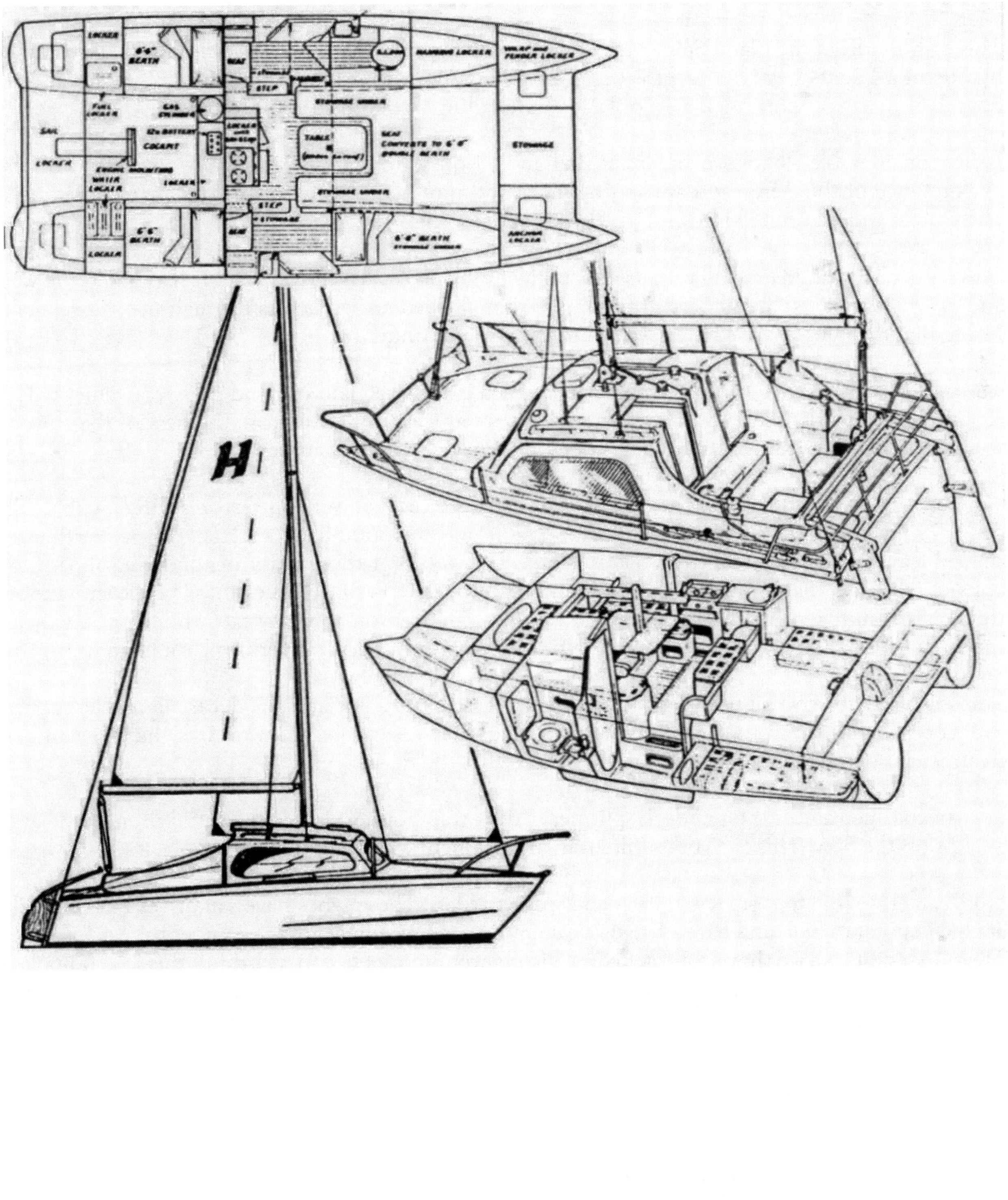

Island Spirit 37

Island Spirit, Class 4, Cruising/Racing

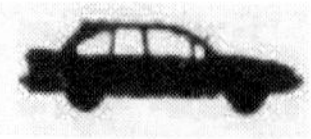

Designer: Phil Southwell

This early model of the Island Spirit was equipped with a Carbo spars Aero rig and was manufactured by Fortuna Yachts, SA. The yacht is designed and equipped for blue water. This is one handsome boat. It is nicely proportioned and looks like it is sailing fast even secured to the dock.

I did considerable sailing on this boat. It handled quite well and I was again, impressed by the performance of the Aero rig. I am getting more and more familiar with this rig having sailed on a Hirondelle and a Prout 45 also equipped with an Aero rig.

This particular boat, *Manx*, had won its class in the 1996 Capetown to Rio race. It handled quite well and I particularly liked the aft head arrangement that allowed engine access through the shower stall. I note that the later models have reverted to the engines under the bunks.

The boat has a U shaped galley-up arrangement which is very well laid out and quite convenient. The refrigerator is across the saloon from the galley, but at least it is away from the heat of the stove. I would also prefer to see a double sink. There is room for it in the cabinet.

The charter version features four staterooms while the owner's version features three, with a full owner's suite in the starboard hull. The saloon table and surround bench is L shaped rather than U shaped leaving room for real chairs. This is especially good for those planning on living aboard for considerable lengths of time. The overall ambiance of the saloon in excellent. The clever window treatment eliminates the need for the ubiquitous covers seen on so many cats while not destroying the spectacular panoramic view, one of the most intriguing aspects of a cruising catamaran.

I especially like the cockpit treatment and the access to the transoms and the decks via real level steps with good handholds. Getting around on the foredeck was reasonable and now with the new window treatment the upper deck is even more accessible as is the mast step.

For special mention is the foredeck split heavy duty trampoline with wide central promenade and anchor on the bow. This is a good utilitarian arrangement as it should be.

Underwater we find the usual mini keels and spade rudders. The engines have sail drives but the hulls are high buoyancy hulls and have a length/beam ratio above what most of the competitive yachts have. Their underwater shape shows a greater chine giving even more buoyancy. I think this is good policy knowing the stuff most owners drag aboard. This appears to be one yacht designed for that eventuality. Despite this lack of extremism in design, the vessel distinguished itself in the 1996 Capetown to Rio race.

Sailing summary

Island Spirit 37

1=Poor 2=Fair 3=Average 4=Good 5=Exceptional

	1	2	3	4	5
Motoring			3		
Docking			3		
Backing			3		
Sailing				4	
Weather helm/lee helm				4	
Windward ability				4	
Ease of tacking				4	
Visibility from the helm			3		
Tracking				4	
Seakindly/ride				4	
Underdeck Slamming			3		
Convenient Deck Layout				4	
Interior layout			3		
Adequate storage areas			3		

Vital Statistics

Island Splrt 37

LOA (length overall)	36' 1"	11.1 m
LWL (waterline length)	35' 9"	11.0 m
BMAX (maximum beam)	22' 1"	06.8 m
SA (sail area)	910 ft^2	84.2 m^2

Draft: 3' 9" / 1.11 m Keels
Designed displacement: 15,212lbs 6900kg.
Mast hgt above WL: 57' (est) 17.37 m
Tankage:
Water: 158 gal. 600 liters
Fuel: 79 gal. 300 liters
Auxiliary power: 18 hp. Twin diesel

Type .... Full Cabin
Foredeck .... Trampoline
Cabins ...... three/four
Galley up/down Up
Rudders ..... Balanced spade
Rig .... Fractional sloop

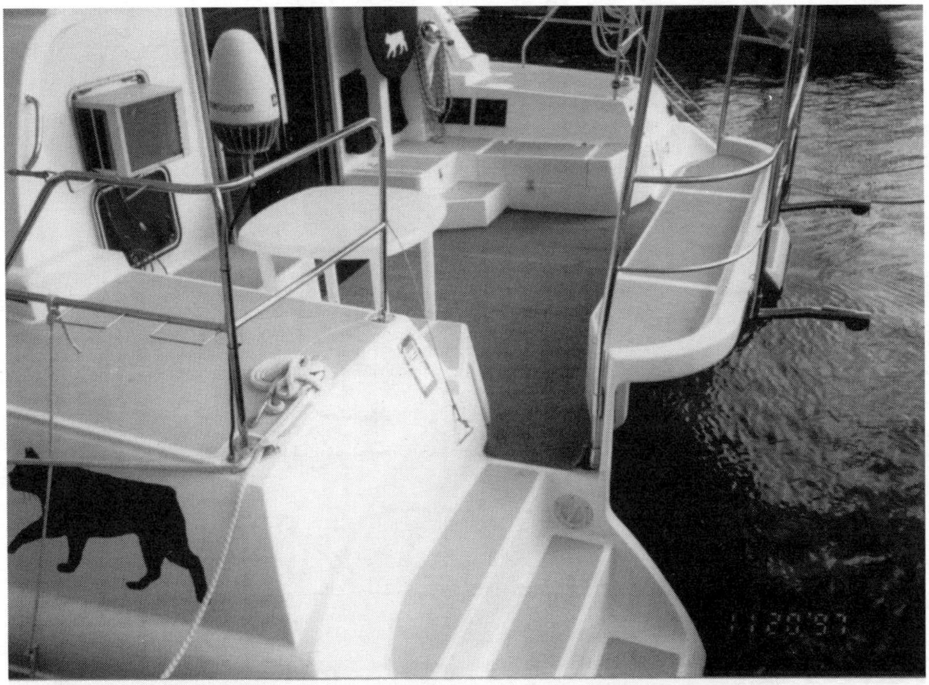

Note easy access to and from decks and from transoms.

Island Spirit 37

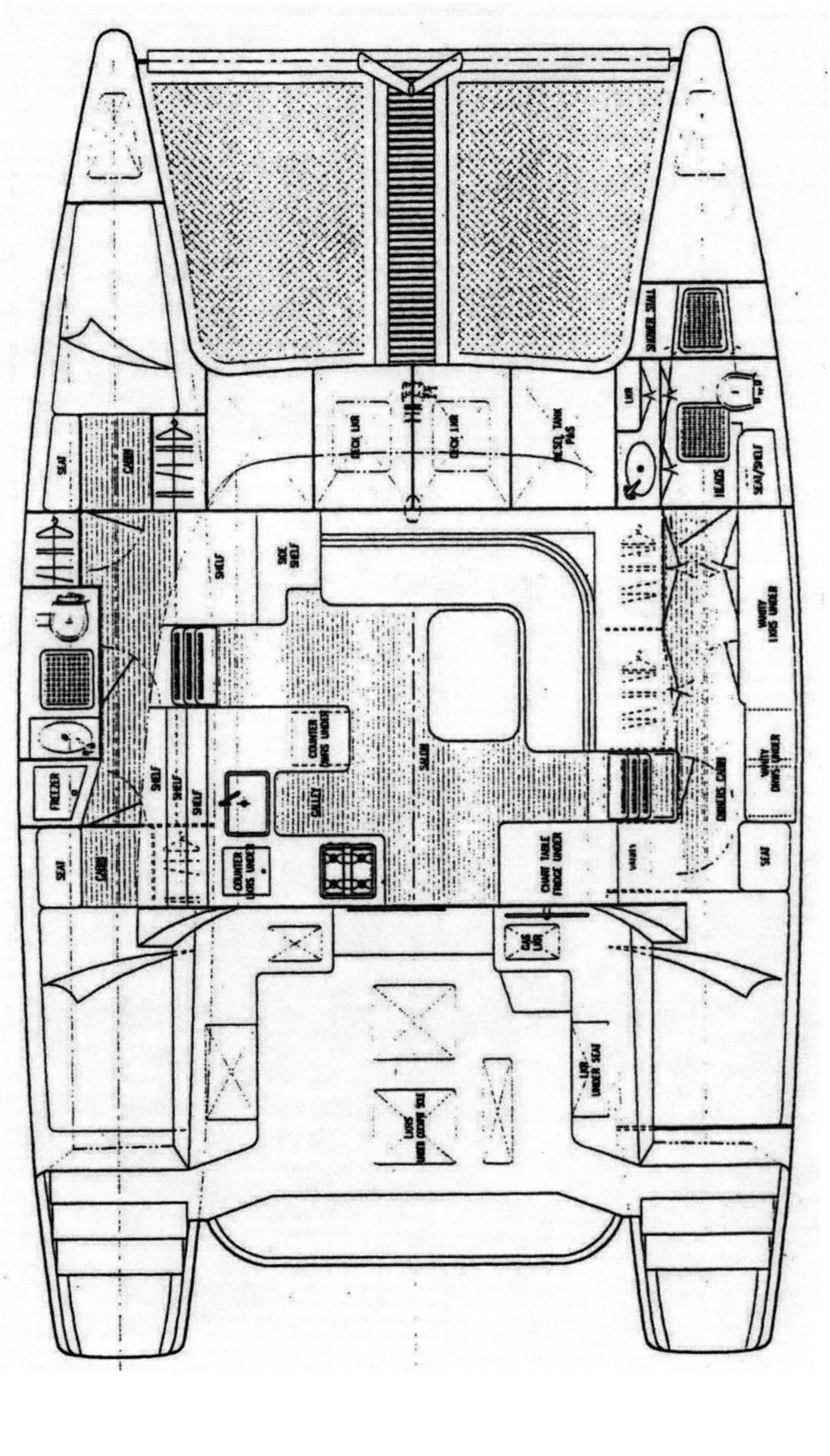

Lagoon 410 Class five, Cruising

Designer: Marc Van Peterghem, Vincent Lauriot-Prévost

The Strictly Sail Miami International Boat Show allowed me to inspect the new Lagoon 410 for the second time. The first time I saw it was at the Annapolis International Boat show at which the boat was so crowded with prospects I barely could get aboard. Miami offered more quality time.

The first view of the boat is startling. The rather vertical cabin window treatment is so unusual in a modern, French style catamaran it takes you off guard. After you gain composure, you realize that it is a bold step toward nautical sanity. Windows you can actually see out of and a configuration that gives much more headroom and open space in the main cabin plus a cabin roof you can walk upon and the final solution to those "greenhouse" windows. Rather than automotive or space ship windows, they are windows that look like they belong on a sea-going vessel. (I heard whispers that the windows are not strong enough because they are vertical. Let me assure you that anyone who says that hasn't a clue of what he is talking about.)

In fact, the entire layout, both inside and out is far more ship-like and a good step away from the overly rounded "Star Wars" approach to production cruising catamarans. After all, it is a boat, not an aircraft or an automobile. People operate a boat from the decks and the decks must be safe and convenient for such things as anchor handling, sail handling and docking. Cars and aircraft on the other hand are operated from the interior thus can afford the maximum in streamlining.

Once you get over the grandeur of the huge spacious luxurious saloon with real windows from which you can actually see the beautiful panorama available, you suddenly realize the diminutive little galley up arrangement simply is woefully inadequate for a cruiser. Born of charter boat heritage, where nobody really cooks, it pays only lip service to a proper galley. The sink has no counter space on the right side, there are virtually no storage lockers in the area. It does have two beautiful walk-past heads in the hulls and excellent privacy for the three cabins in the owner's version or four cabins in the charter version.

I feel the user friendly layouts both inside and out, especially the beautiful flat decks are a giant step in the right direction. The new design has considerably more buoyancy and carries it further forward than the old designs, a much-needed advancement for the modern reality of cruising.

So much for the good stuff. I found three problems, one minor, one serious and one very serious. The minor one being the puzzling anomaly of a teak toe rail! I suggest that a teak toe rail on this thoroughly modern vessel, without a stitch of archaic exterior wood trim is esthetically offensive, a maintenance headache and will eventually causes leaks into the foam core. In addition, there were bulls-eye leads for the roller furling screwed right into the deck! For shame! Why penetrate the deck and expose the laminate to leaks when for less money and less labor you can bolt on the lead blocks to the stanchions?

The serious problem is the lack of a rub-rail. Lack of a proper rub-rails are storm clouds on the horizon. I am baffled by those who wish to market to John Q. Public, but do not utilize research and feedback from those very same people. A proper rubrail is usually top of the list for most Mom and Pop cruisers. A rubrail is more important than a high top speed. One "salesman?" claims that the designer says that hitting a rubrail may cause delamination. If that be true, what kind of inferior layup do they have on that boat anyway? Does that mean hitting a rubrail could be more damaging than hitting a piling directly against the topsides? Or more damaging than scraping the topsides down the pilings as the wind blows you uncontrollably against the fuel dock? Think of the competitive disadvantage of a cruising boat offering with no rubrail when all the similar boats have them and the sales force is extolling their virtues?

The sales people who wish to market boats for private US use should understand the nature of our boating milieu which is quite different from the usual Med-Moor system used in the Mediterranean or the Caribbean.

Third, and truly serious, is the anchoring system.

For those of you who have not seen the boat, the anchor (single) tucks into an opening in the top of the underdeck in the upsweep just below where it meets the bridgedeck! That is correct, right where the waves smash with the greatest force when pushing the boat up to windward under power; meaning a constant salt-water bath for the windlass, the chain and the locker! There is no provision to get the anchor out to the bow in any usual way. The trampoline covers the anchor. The anchor is discharged from that hole-in-the-most-vulnerable-part-of-the-hull position and when an appropriate scope is laid out, a preset bridle is hooked on and enough additional scope let out to put the strain on the bridle. This means you must work through a small hatch at a level below your feet to raise or lower the anchor.

I can see this development clearly as the confluence of three factions: those who claim it moves the weight away from the bows and assures use of a bridle, the bean counters who claim it saves money and the marketing men who think they created something easy to use that they could sell. In reality, this triumvirate created a menace that will damage the boat, create a filthy dirty mess and prevent the proper use of multiple anchors when required. This is an anchoring system fraught with the possibility of tragedy, possibly a major tragedy.

First, how on earth do you clean this thing? Most people stand on the bow with a hose whilst they hoist the anchor. No chance here. The inbound chain is hidden under the tramp which means all the smelly sticky gooey mud goes right into the anchor locker which is right next to your bunks. Will the chain run out properly after this sticky mess dries? Marketing types must come to grips with the fact that much of anchoring takes place in the most evil smelling mud imaginable. The entire Chesapeake Bay, The Outer Banks of North Carolina, Coastal South Carolina and Georgia, most of Florida, the Gulf Coast states all have ubiquitous sticky mud bottoms. The Thames estuary and south of France have their polluted sticky mud anchorages also. From what I saw of the anchorage in the Tenerife, where every boat was covered with a thick oil slick, the rest of the places were nothing.

Is there a danger in sticking your hands down below decks to clip on a bridle and then releasing more chain? What happens if you accidentally drop the bridle? You are in the same compartment and in very close proximity to the windlass. You would not be the first person to be hurt with that combination. A dear friend of mine lost two of the fingers on her right hand in just such an incident.

Next, what happens in a windlass failure? A rather common occurrence. Sure, if your a charter customer you simply get on the VHF and scream. But if you are a private owner, then what? If you have an anchor properly deployed from the bow you have several emergency retrieval options not the least of which is using your halyard winches and a dock line to hoist the anchor or the time honored motoring up on the anchor and hauling it manually. Pretty tough to do that with this arrangement, but that is not the worst!

The worst is that quite often you will grind chain against your topsides in your anchoring and anchor retrieval. It is inevitable. Think over the process. You stop, lower your anchor and your boat begins to fall off, as always. When the anchor is deployed from the bow it is not much of a problem because even at an angle, the rode is away from the bow. With the anchor deployed from way back between the hulls, the rode is going to rub against the hulls and keels much of the time. How many of you have ever seen a boat that falls arrow-straight back when being anchored? Modern boats have an amazing tendency for anchor sailing based upon their automatic lee-helm quandary when sails are stowed. This virtually guarantees chafing the inside of the hulls.

In addition to the above, you have the problem of correct adjustment and retention of the bridle. If that is not correct, you will anchor-sail like mad because you are essentially anchored from near the middle of the boat. Trying to adjust scope means hauling up the anchor to a point high enough to release the bridle then letting out more scope to a place to re-hook the bridle, all the while anchor-sailing and scraping the devil out of the insides of the hulls. The thought of a bar-taught anchor chain leading up to the bridge deck between the hulls of an anchor-sailing catamaran is not a pleasant one.

In extreme conditions when you need multiple anchors, there is no safe or convenient way to get them to a central point or use the windlass with them. For those who ordinarily sail in places with strong currents where two anchors from the bow in a Bahamian moor are routine, there is no reasonable way to do it.

If you are a cruiser, you anchor most of the time. Most cruisers accept the fact that their anchor and anchoring system is the most important and most critical part of their boat. This cutesy, gimmicky system just doesn't cut it.

Lagoon 410

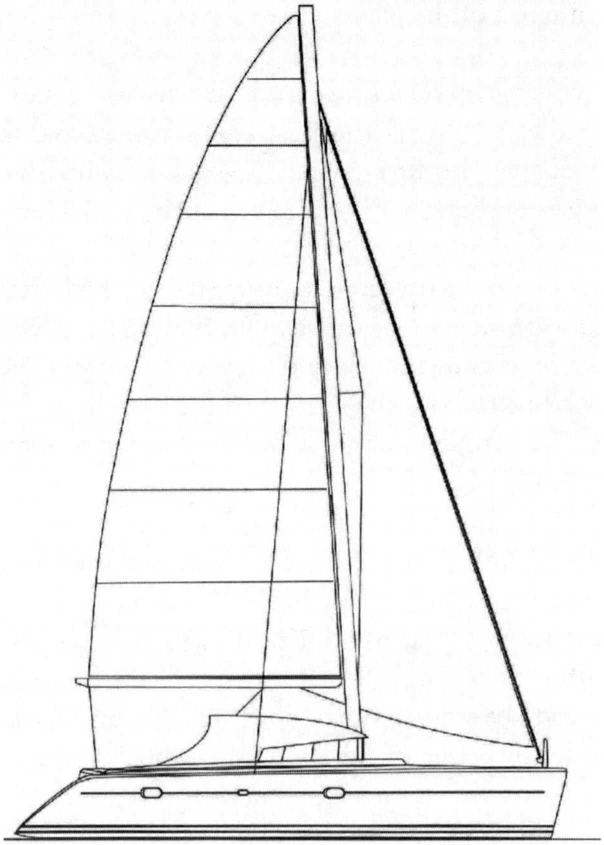

Vital Statistics

Lagoon 410

LOA (length overall)	40' 7"	12.37m
LWL (waterline length)	38' 3"	11.67m
BMAX (maximum beam)	23' 3"	7.09 m
SA (sail area)	958 ft^2	89 m^2
Draft: 3' 11" / 1.2 m	Keels	

Designed displacement: 29,300 lbs 9200kg.
Mast hgt above WL: 60' / 18.29 m
Tankage:
Water: 116 gal. 440 liters
Fuel: 52 gal. 200 liters
Auxiliary power: Twin 28 hp diesel sail drive

Type . . . . Full cabin
Foredeck . . . . Trampoline
Cabins . . . . . . Three
Galley up/down Down
Rudders . . . . . Balanced spade

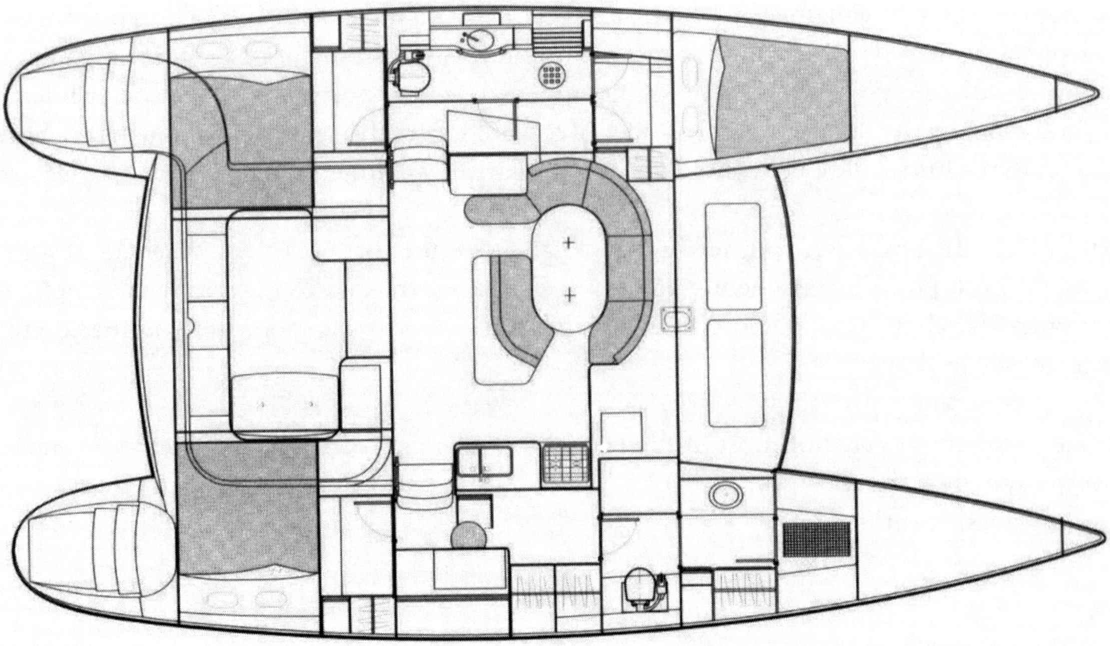

Lightwave 35, 38 Class 5, Cruising

Designer: Tony Grainger

Designed by Tony Grainger and manufactured by Overell Stanton Yachts, Australia, imported into the US by Southern Ocean Yachts in Miami, this catamaran is a newcomer to the American market but a well known vessel in Australia. It is designed and advertised as a blue water vessel.

It was one of the vessels that struck me as being better laid out and arranged than most of the competing swollen boats. By swollen I mean the tendency of designers and builders to force more accommodation into the vessel than it should realistically carry for its length and displacement.

This boat is a lot better than some of the other "wedding cake" style boats since it actually does have sufficient underdeck clearance and reasonable walk-able decks and a well thought out deck layout. What suffers is the helm, of which you must gopher-like, poke your head up above the top to see. Newer models have modified it somewhat featuring a double seat and a mini-Bimini for all around visibility and protection from the elements.

On the other hand, the interior is one of the better arranged boats when it comes to space utilization and ergonomics. The vessel I inspected had a galley-down arrangement with three double staterooms and a single head. Two staterooms and two heads are available as an option and I think that would be a much better arrangement in that size boat.

The saloon is enormous and there is room for some reclining chairs if desired. If you are really living aboard for any length of time, the ubiquitous central table with surrounding bench gets really old. You need a place to just flop and sprawl.

Engine room access on the port side is through the shower which is a really nice arrangement leaving a large equipment room with close to standing headroom. The perfect place to install watermakers or other equipment. The starboard engine is under the bunk in the rear cabin, except the two cabin model with two aft heads.

Other arrangements are possible, for those not chartering, it is most likely not a requirement to have three double cabins on a 35 foot boat Having mirror image head arrangements means that both engines with be available through doors in the shower compartment. This is of inestimable value. Easy servicing and separation of the engine and living compartments are an ideal arrangement. Accessing the engines from the interior rather than the transom steps also has a safety advantage.

One should not underestimate the need for sufficient comfortable space in an engine room. Aside from just being able to properly access all the machinery with room to swing wrenches, there should be room to store all the dirty tools and equipment required for maintenance. It is really nice and promotes proper maintenance when all your parts and supplies are neatly arranged right where you need them and you do not have to root through lockers all over the boat to find things.

Rub rails are an option at the moment, but according to the importer, most likely will be standard shortly. Rub-rails are critical safety equipment for short handed liveaboard couples.

One of the things that attracted me to this boat was the cockpit design and the access from the stern and from the sugar-scoop transoms. It is excellent flat walking access. There is no climbing across fortress like coamings and needing to step on seat cushions to access the vessel. Handling a dinghy in davits from the built in arch is made easier by the aft deck layout.

Vital Statistics

Lightwave 35

LOA (length overall) 34' 5" 7.46 m
LWL (waterline length) NA
BMAX (maximum beam) 21' 10" 06.9 m
SA (sail area) 700 ft^2 65 m^2
Draft: 3' 5" Keels
Designed displacement: 10,570lbs 4794kg.
Mast hgt above WL: 55' 16.8 m
Tankage:
Water: 88 gal. 333 liters
Fuel: 50 gal. 189 liters
Auxiliary power: Twin 18 hp diesel sail drives

Type .... Full cabin
Foredeck .... Trampoline
Cabins ...... three
Galley up/down down
Rudders ..... Balanced spade
Rig .... Fractional sloop

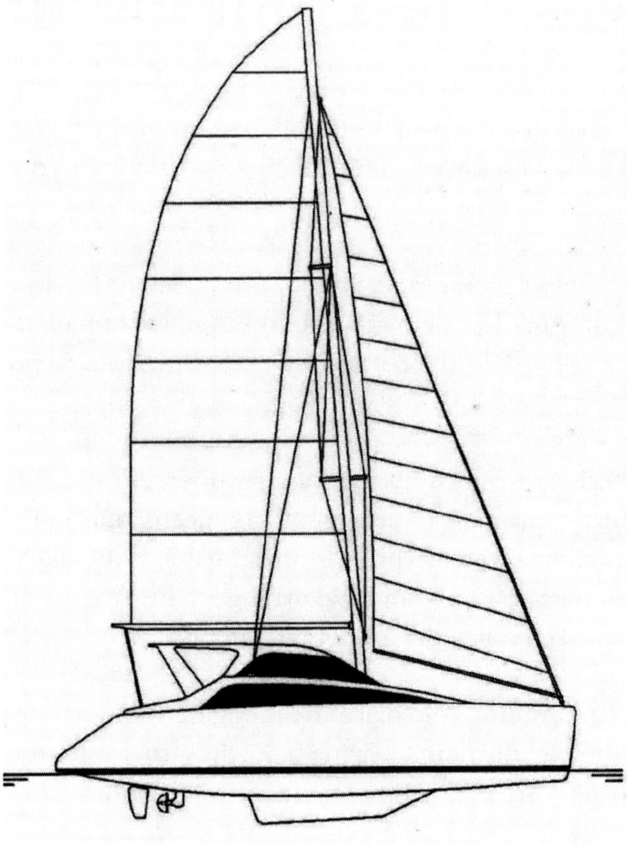

The aesthetics of the vessel are very pleasing. It handles it's multi-tiered look with pleasing proportions and does not have that boxy look of some catamarans in that size range.

Steering is Edson cable and the keels areintegrally molded integral with the hulls.
Rig is fractional sloop, with four shrouds, single headstay and double spreaders. Seems like overkill for that size vessel. One thing I really like is the way the mast base is accessible from the main deck.

Trampolines are nicely laid out and split by a promenade that also is a compression strut for the crossbeam.

The head compartment is palatial with a separate shower stall and the manual toilet is equipped with a legal holding tank.
(Lightwave 38 is identical except for engines and total displacement)

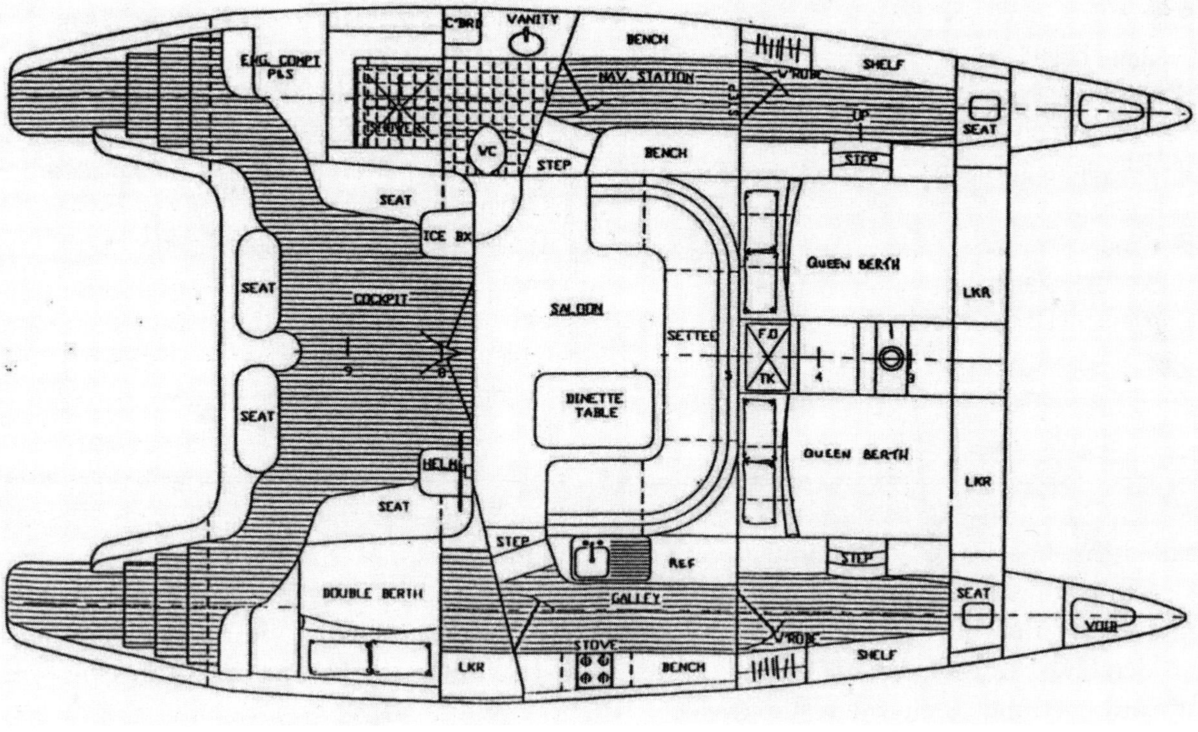

Maine Cat 30

Maine Cat 30, Class 3, Racing/cruisng

Designer: Dick Vermullen

Sailing summary

Maine Cat 30

1=Poor 2=Fair 3=Average 4=Good 5=Exceptional

	1	2	3	4	5
Motoring				4	
Docking				4	
Backing				4	
Sailing				4	
Weather helm/lee helm			3		
Windward ability			3		
Ease of tacking			3		
Visibility from the helm			3		
Tracking			3		
Seakindly/ride				4	
Underdeck Slamming				4	
Convenient Deck Layout			3		
Interior layout					5
Adequate storage areas				4	

Vital Statistics

Maine Cat 30

LOA (length overall) 30' 0" 9.14 m
LWL (waterline length) 29' 3" 8.91 m
BMAX (maximum beam) 18' 0" 5.48 m
SA (sail area) 500 ft^2 46.5 m^2
Draft: 2' / .60 m bd up 5' / 1.52 m bd dwn
Designed displacement: 6,000 lbs 2721 kg.
Mast hgt above WL: 46' 11" 14 m
Tankage:
Water: 54 gal. 204 liters
Fuel: 26 gal. 98 liters
Auxiliary power: Twin 9.9 outboard

Type .... Partial open bridgedeck
Foredeck .... Trampoline
Cabins ...... three
Galley up/down down
Rudders ..... Balanced spade
Rig .... Fractional sloop

I sailed this boat twice at Multihull Demo Days in Annapolis. I was impressed with its general seakindly nature and the layout of the controls. While we could not out-distance the super light speed demons, felt that the boat performed well and that in heavier air we would have closed the gap considerably. The boat did not pound, hobby-horse or display any bad habits. With twin screw outboards, getting back into our slip was a cinch.

There are some very interesting features that make this a comfortable boat. First, the accommodations are light and airy and there is no attempt to squeeze the proverbial five ponds into the four pound bag. This is especially true in the galley which has considerable counter space.

Overall, this is a nice boat that meets my criteria for a proper cruising boat.

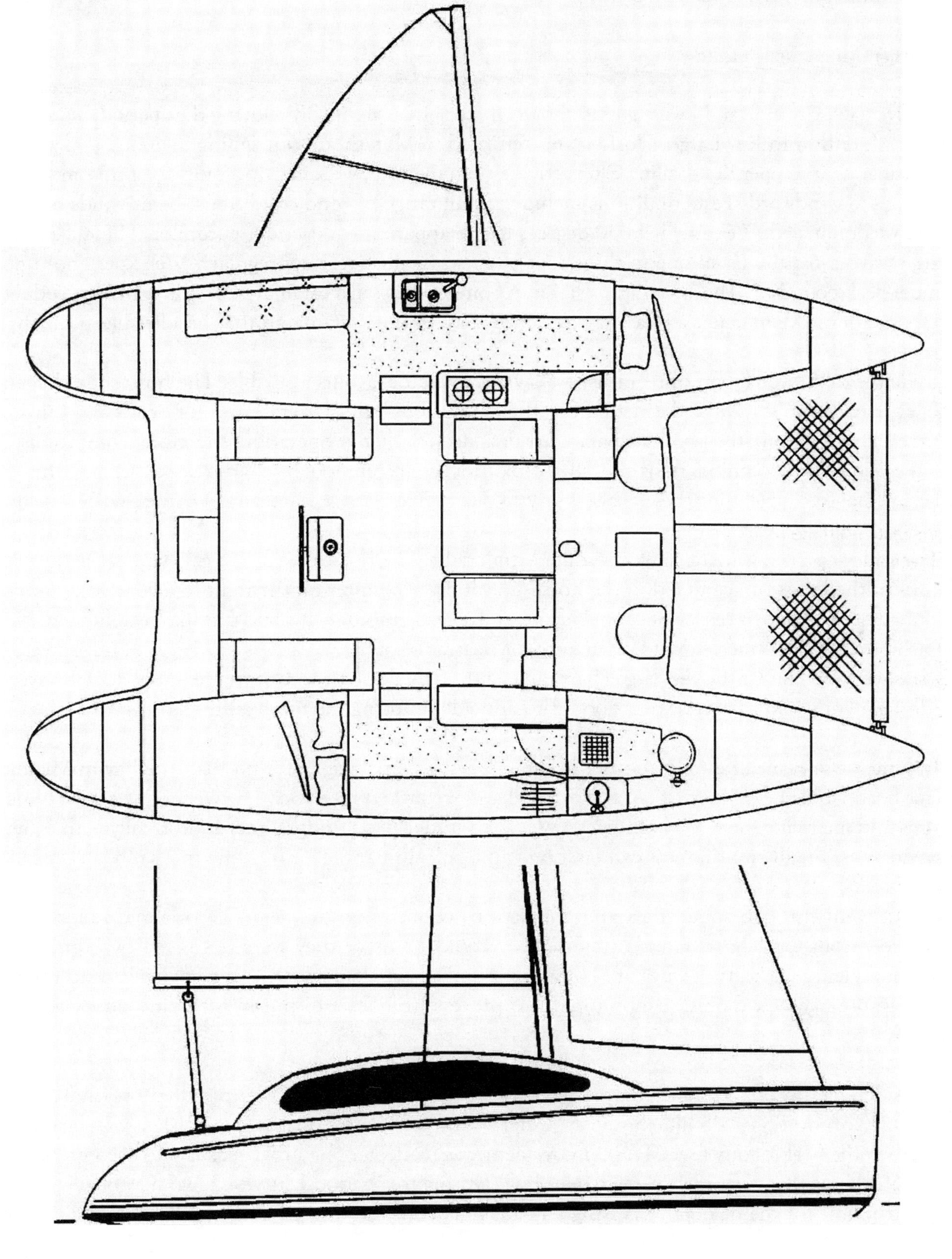

Manta 42

The Manta 38, 40, 42 Class 5, Cruising

Designer: Manta design team

For five days, Corinne and I were guests aboard the original Manta 38. During that period, we sailed from Marathon to Key Largo, Florida, and returned to Marathon. Our sailing included a race and participating in Upper Keys Sailing Club activities. In that trip, we sailed 150 miles. Of that mileage, half was to windward. Test conditions included a full range of wind conditions. Light winds on the nose predominated. We easily held 30 degrees to the apparent wind under autopilot. There were no large waves, just the usual one to three foot chop so familiar to Keys regulars. We experienced no under-deck pounding. This boat sails well. I don't often get to sail a catamaran with the instant rudder response of the Manta. It is a pleasant experience to sail a huge catamaran that handles like a dinghy.

Inside the cabin, you'll find one of the best owner/operator layouts available. The horseshoe shaped galley, navigation station and dinette fit in the main cabin. Hulls are reserved for heads and bunks. As you know, most of the large catamarans are designed for charter operation, therefore, compromises in layout and accommodations are all to luxuriously accommodate charter folks. Often, this is detrimental to people using these boats as private yachts. The Manta is definitely a fine owner's yacht.

The galley-up arrangement must get some extra credit. The horseshoe shape galley puts the double sinks to the left, stove centered in the horseshoe and the counter on the right in a good ergonomic configuration. There is sufficient counter space and storage cabinets It's laid out like real kitchen. No reaching across a burning stove to get utensils, or clearing a counter to access crockery. A huge freezer is stashed neatly under the very large chart table, although alternate locations are available. However, I do hear complaints from owners about the difficulty in using the top loading refrigerator.

In June, we day-sailed Kestrel, another Manta 38 (with a larger rig) around Biscayne Bay in Miami. This boat handled even better than the original and we tacked up tricky narrow channels with blind turns and menacing shoals on both sides. We sailed under fixed bridges, raced informally with other catamarans, finally quitting this exhilarating trip near midnight. This was one fine sailboat.

Manta went through some trials and tribulations concerning builders. The original molds were shipped from Canada to Florida. Before they could be used, they were destroyed by Hurricane Andrew. Henry Lucke, the owner of Manta hull number two, loaned the company his boat to use as a plug to make new molds. I monitored this process from start to finish, including surveying the molds.

Considerable negotiation and many trials and tribulations later, the group settled on Endeavour Yachts Corporation to build the boats. After a few completed boats, the decision was made to lengthen the boat to forty feet. This improvement was made, among other reasons, to accommodate the accessories that customers were demanding During this period I surveyed two new boats under construction for the owners. Surveying a boat as the construction surveyor entails many visits and inspections to the factory at unannounced times to inspect the materials and methods used during various construction phases. I also provided four purchasers full hull surveys of the older 38s as they came up on the used market. It must be noted that in all my personal experience with these boats, I have not found any structural failure related to design or construction.

From the 40 foot model the sterns were lengthened to create the Manta 42, ostensibly to improve performance and load carrying ability to support the ever growing demand of cruiser. Early in 2001, production was switched to Diversified Marine Group, Inc. in Palmetto, Florida.

Manta has some unusual and highly innovative options. First is the massive structure covering the cockpit. No question, it is the industry leader for rugged strength and fine materials. It not only incorporates a solid roof, but integrates the davits, radar arch, and mainsheet attachment point all in one fixture. However, it is not without its drawbacks. I feel it is too large, too heavy, and has too much windage for a bluewater vessel. For those actually considering the Southern Ocean and its routes, I would prefer a much more modest, lighter structure, and one that could be stripped of excessive windage in a survival storm.

The second innovation is the Berig Camberspar®. It is actually half a wishbone boom inside a sleeve in the sail. It maintains sail shape in the lightest zephyrs, eliminates the need for a whisker pole, eliminates the need for Genoa winches, and best of all, is self-tacking. In my opinion this is a big advancement in sail power. You get sophisticated shape and power while eliminating lots of work at tacking and the flogging headsails.

As with any innovation, there are drawbacks. The only valid drawback is that it is not as easy to stow the sail as is a roller furling sail. If you are primarily a dock sailor or weekender, roller-furling is handy. If you are primarily an offshore, heavy-duty sailor, or performance oriented, reefing a headsail by furling is usually a less than satisfactory situation, both for sail shape and sail longevity. Despite much hoopla by sail makers, rolling up a headsail as a method of reefing is not a real solution for offshore sailors.

The other drawback, not being able to roller reef, is not really a drawback. The Camberspar sail is able to reef but by dropping the Camberspar sail, you leave the headstay available to hank on a real storm jib if needed. As far as needing to go forward is concerned, most of those queries come from monohull sailors (when the risks of going forward in nasty conditions presents a valid question).

No question about it, going forward on a skinny, slippery foredeck, usually with poor footing because of bow clutter and heeled over at 30 degrees or so with green water sweeping across the bow is to be avoided at all costs, thus the perennial question and the roller furling solution. Once you have sailed a well found cruising catamaran, those reservations melt away with the awesome actual experience of

Sailing summary

Manta 38 (40, not specifically listed)

1=Poor 2=Fair 3=Average 4=Good 5=Exceptional

	1	2	3	4	5
Motoring			3		
Docking			3		
Backing			3		
Sailing				4	
Weather helm/lee helm				4	
Windward ability				4	
Ease of tacking				4	
Visibility from the helm				4	
Tracking				4	
Seakindly/ride				4	
Underdeck Slamming				4	
Convenient Deck Layout			3		
Interior layout			3		
Adequate storage areas			3		

Vital Statistics

Manta 38 (add 24" to LOA for 40)

LOA (length overall)	38' 0"	11.58 m
LWL (waterline length)	37' 0"	11.28 m
BMAX (maximum beam)	20' 9"	6.32 m
SA (sail area)	967 ft^2	294 m^2
Draft: 3' 0"		Keels
Designed displacement:	7,493 lbs	3399 kg.
Mast hgt above WL:	64' (est)	19.5 m

Tankage:
- Water: 120 gal. 454 liters
- Fuel: 40 gal. 151 liters

Auxiliary power: Twin 18 hp diesel saildrives

Type .... Full cabin
Foredeck .... Hermaphrodite, Trampoline
Cabins ...... Three
Galley up/down Up
Rudders ..... Balanced spade

safely, and I stress safely, being able to go forward and change a headsail in virtually any conditions, without heeling, without the bow plunging through green water. Most of the foredeck danger and trauma is eliminated on catamarans such as the Manta.

On a level playing field, with lots of room on both sides, with a sturdy center walkway, with a gentle sea-kindly motion, hanking on a headsail is a piece of cake.

Manta meets one of my most stringent evaluating criteria for a cruising boat: all four corners of the boat must be visible from the helm. Maneuvering a 21-foot x 42-foot boat requires good visibility. With the twin diesels so far apart, maneuvering options are available to you that are not even available to twin-screw power boats.

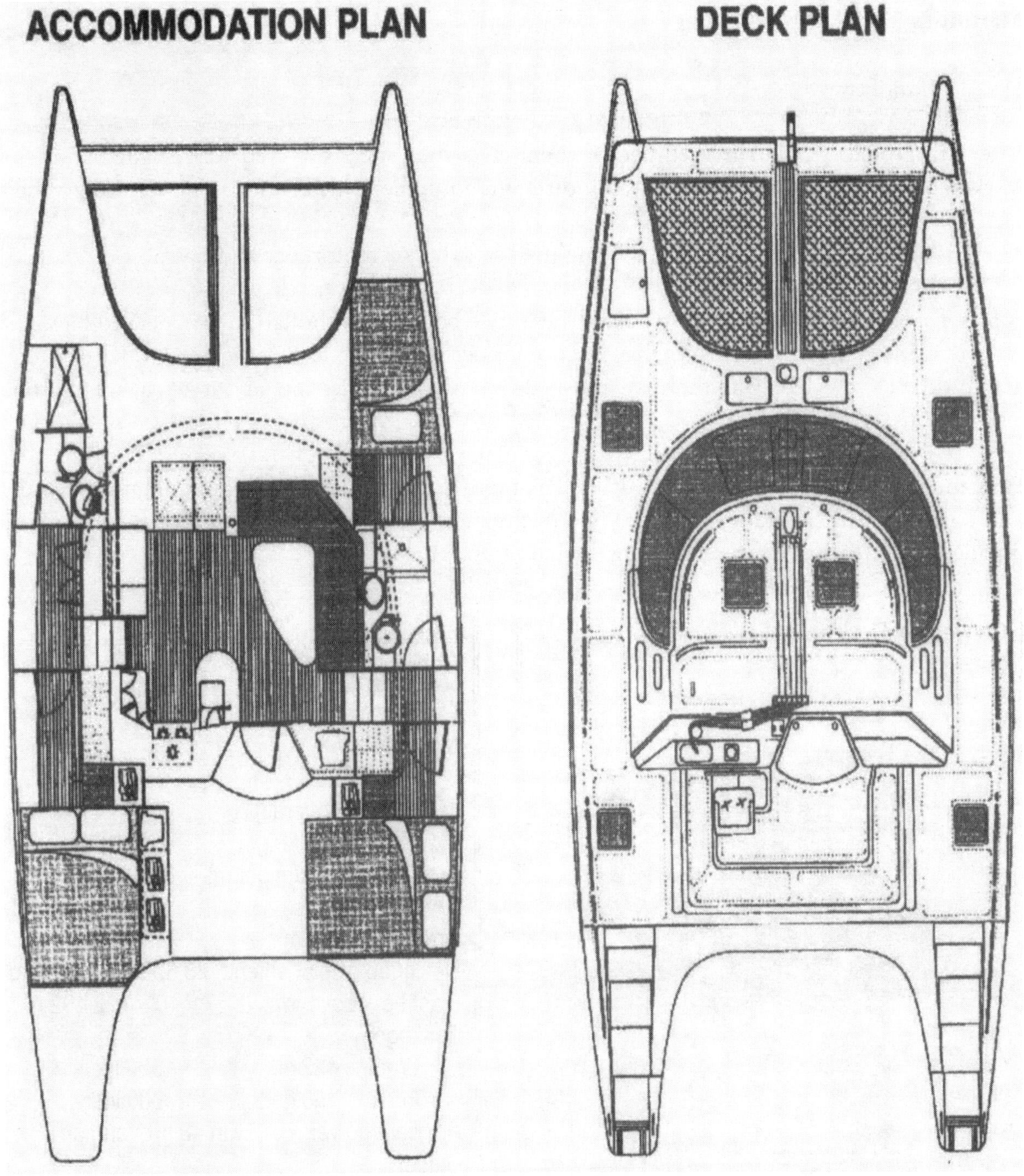

The above diagrams are from the latest Manta 42 brochure. Since there are only minor differences in the vessel which essentially depict the continuing evolution of the yacht I felt in the spirt of this book I simply could not include all the diagrams of past vessels. They are available from Manta or from your surveyor of choice.

Marquises 53

Marquises 53 Class 5, Cruising

Designer: Jean Berret

Interesting boat. I have surveyed two of them. The first one I surveyed was plagued with sloppy workmanship, poor quality control and indifferent equipment installation. The second was better.

There is one in the Caribbean that is being used strictly as a motorboat and has no rig.

A reasonable sailor but as with so many of these style boats, even though they appear to have all the accouterments of high performance they just do not perform up to expectations. Good, but not great as would be expected. But then we must always ask the question about what are reasonable expectations?

Later models have adapted to other Foutaine Pajot roof line which features a built-in gutter/hand grip and shade over the windows. Perhaps one of the most sensible innovations in cruising boats in a decade. The later models were also lengthened to 56 feet.

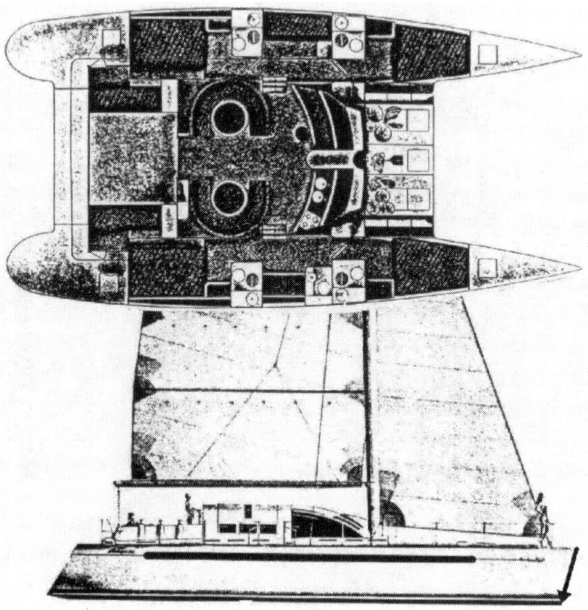

Vital Statistics

Marquises 53

LOA (length overall) 53' 0" 16.3 m
LWL (waterline length) 52' 15.8 m
BMAX (maximum beam) 27' 0" 8.15 m
SA (sail area) 1722 ft^2 160 m^2
Draft: 4' 6" / 1.4 m Keels
Designed displacement: 28,652 lbs 12997 kg.
Mast hgt above WL: 71' 10" 21.92 m
Tankage:
Water: 422 gal. 1600 liters
Fuel: 158 gal. 600 liters
Auxiliary power: Twin 48 hp Yanmar Diesel

Type .... Full cabin bridgedeck
Foredeck .... Trampoline
Cabins ...... Six
Galley up/down Up
Rudders ..... Balanced spade

Moorings 3800 Class 5, Cruising

Designer: Alexander Simonis

The Moorings 3800 is a refreshing move towards sanity in the cruising catamaran field. Having been a voice in the wilderness for years, I finally found a company that actually listened to their customers and designed according to customer suggestions.

Its things like flat, well protected decks and walkways with toe rails and handholds, easy access to the cockpit without the necessity for clambering over wall like bulkheads, like convenient low stress dinghy access that can be operated by less than expert cruisers. Like a clean uncluttered cockpit.

The Moorings 3800 incorporates almost all of the features I have promulgated as necessary for a proper cruising boat for many years. There are only two deficiencies, in my opinion, they are lack of a rub rail and a difficult to service and operate anchoring facility.

The salesman suggested, "if you want a rub rail, we will install one, no problem." he went on to say that the boats used in the Caribbean and South Pacific were always anchored out or on moorings so it was not important." I countered by saying, "of the used boats being sold from the charter fleets, the ones without rub rails usually have topsides scarred far worse than those with rub rails." As far as the neatly tucked into the foredeck anchor is concerned, as long as you stay in the protected anchorages of the tropical trade wind charter routes and within VHF radio contact of home base, you can get away with that system. Once you leave on your own, you better have not only a greater knowledge of anchoring, but proper equipment that can be deployed, rigged and serviced from an accessible location.

The moorings also offers an interesting charter/ownership plan. It will not make anyone any money but it will defray some of the ownership costs. The best part about the plan is that it allows people who might otherwise be overwhelmed by the logistics of cruising boat ownership a real opportunity to sail some of the world's prime waters with minimal ownership concerns and in their own yacht.

The vessel itself is a study in how to design a boat that is aesthetically pleasing yet meets the essential requirements for cruising sailboat safety and convenience. It is built with easy access to flat decks with lots of handholds and well developed life lines and toe rails. On deck, walkways are secure and there are plenty of places to enjoy sunbathing. The helm is both convenient and protected. Really innovative features include slated sunshades over the forward windows which are actually steps leading up to the top deck. The top deck itself is flat enough to be a secure platform to handle the mainsail. An arch arrangement that holds the traveler and the mainsheet is yet another innovation. Seems a large percentage of charter people complained about those items taking up space and creating a hazard and nuisance in the cockpit. The arch top arrangement eliminates those complaints quite nicely.

Moorings 3800

Below decks, it is a three double cabin arrangement, ergonomic "U" shaped galley up, with a matching "U"shaped saloon. One nice feature is that the starboard hull can be an owners suite while the port hull has the head between the two cabins. This is a spacious boat that has not been ruined by squeezing every possible inch for bunk space Visibility from the interior is very good.

Underwater we find the ubiquitous but quite satisfactory for a cruising boat, mini-keels. Rudders are skeg mounted, the strongest, most secure way to protect rudders except having them attached to the end of a full keel. Props are mounted through integral molded struts. Hull/beam ratio is modest, giving lots of weight carrying ability and only losing a slight bit of mythical top speed performance.

Sail rig is also modest. Taking a hint from those who deal with extreme sail plans and get the feedback from people who have difficulty raising or lowering an extreme roached mainsail, the sail plan is powerful but conservative.

I spent some time aboard the vessel at Boat shows and at Demo Days, hosted by Performance Cruising, Annapolis. However, I have not spent sufficient time under sail to accurately comment on the vessels sailing characteristics thus I will defer comment to a later time.

Photo: 1 Moorings 3800 shows the deck accessibility from the stern of the vessel. It also highlights the clever arch arrangement that contains both the traveler and the mainsheet. Well supported lifeline corner posts and stanchions are noteworthy as is the dinghy lifting device.

Photo: 2, Moorings 3800 forward view of flat foredeck and top deck, window treatment steps, aesthetic appeal and opening ports for ventilating forward cabins, wide side decks, sturdy pulpits large mooring cleats on forward cross beam.

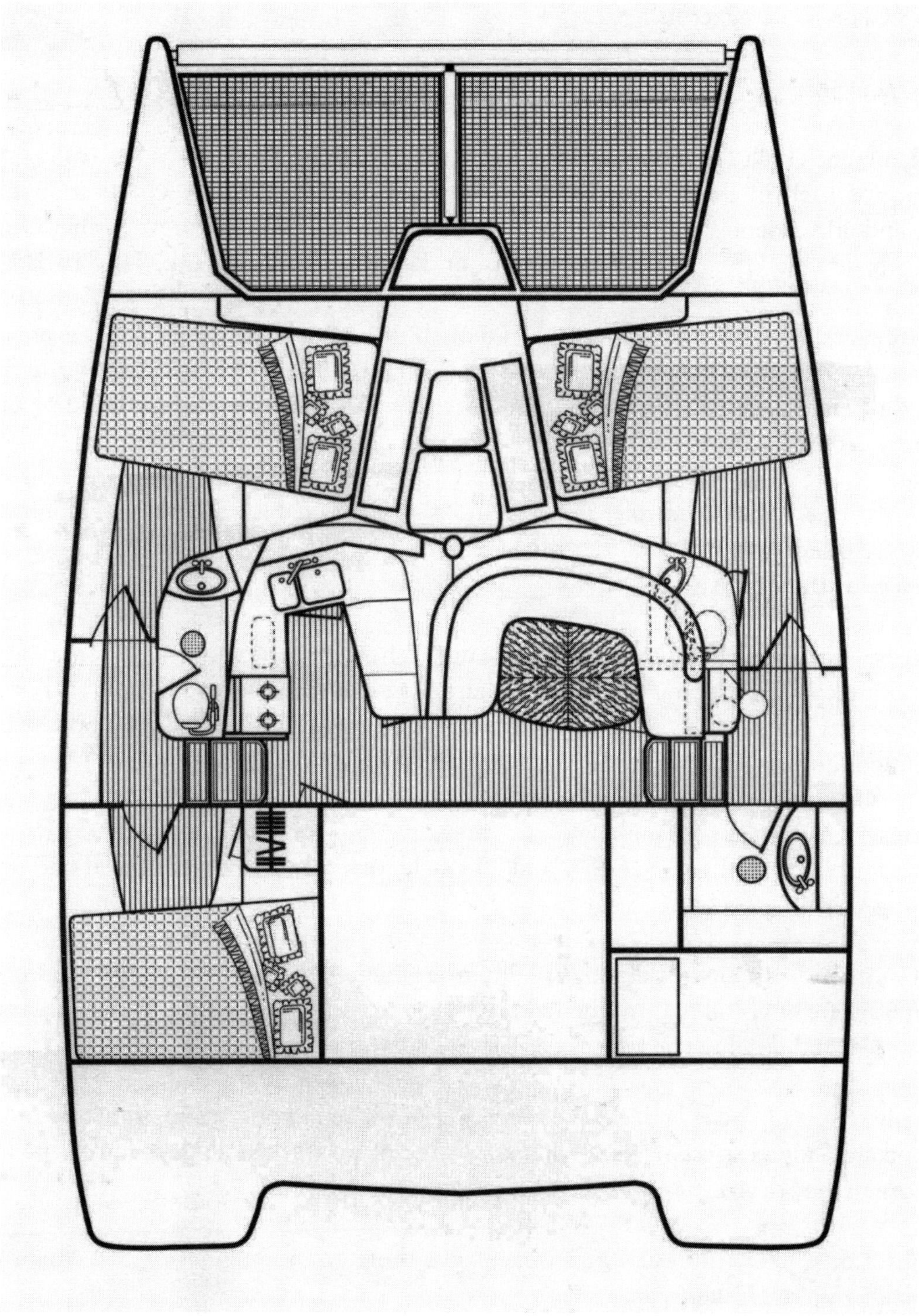

Accommodation plan shows unusual but effective location of forward bunks. The twin horseshoe arrangement works well. The extra wide beam allows that extravagance but may be detrimental to handling, sailing and underdeck slamming.

Outremer 43

Outremer 43 Class 3, racing, cruising

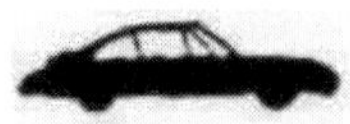

Designer: Gerard Danson

Outremer 43, a French built boat with just a few units trickling into the United States really has made quite a name for itself in the league of high performance cruising boats. Unquestionably both performance and price oriented they show that it can be done. The manufacturer, Atelier Outremer, has just recently set up a US dealer, Gregor Tarjan on Long Island, New York, 516-818-3113.

The classification and design intent for this vessel an offshore racing boat and this boat is an example of one that has the majority of its compromises for performance. It has very fine hulls without the fat sterns needed to carry the weight of engines and other heavy accessories. It is designed to win races with a reasonable degree of comfort.

To this end, the one I sailed had tiller steering and a single Yamaha 9.9 four stroke outboard for auxiliary power. According to the manufacturer, several options for power are available including the usual twin diesel with sail drives.

Outremer is one fine looking boat, having a hollow curved sheer line with a flush deck and a smallish cabin house. It presents the appearance of traveling fast even at rest. It has two double and one single cabin/stateroom, a typical French central table with a bench surround in the saloon and a rear facing galley up configuration.

Sleeping cabins are an interesting combination of three different types. One double atwhartships, one double for and aft in the hulls and on fore and aft single forward. The vessel I sailed had one head tactfully arranged in an aft cabin.

Visibility from interior is quite good and visibility from the helm is reasonable. Wheels wood most likely increase the arc of visibility from the helm. As with any tiller steered boat, there are trade offs for simplicity, feel and weight. The boat is daggerboard equipped.

Decks are easily traversed and side decks are very adequate. Toe rails and rub rails are accomplished by a perforated aluminum extrusion as seen on many other boats at the hull/deck joint. Life lines are standard as are dinghy davits.

Deck and interior storage is especially adequate since there are no engines or other high volume equipment taking up the locker space.

Anchoring facilities are adequate for the vessel and the fractional sloop rig is similar to most modern catamarans.

Sailing summary

Outremer 43

1=Poor 2=Fair 3=Average 4=Good 5=Exceptional

	1	2	3	4	5
Motoring		2*			
Docking		2*			
Backing			3*		
Sailing				4**	
Weather helm/lee helm				4**	
Windward ability				4	
Ease of tacking				4	
Visibility from the helm				4	
Tracking				4	
Seakindly/ride				4	
Underdeck Slamming				4	
Convenient Deck Layout				4	
Interior layout			3		
Adequate storage areas			3		

Vital Statistics

Outremer 43

LOA (length overall)	43' 0"	13. 8 m
LWL (waterline length)	42' 0"	12.8 m
BMAX (maximum beam)	23' 0"	07.0 m
SA (sail area)	932 ft^2	86.6 m^2

Draft: 17" / .234 bd up 6' 7" / 2.00 m bd dwn
Designed displacement: 10,000 lbs 4500 kg.
Mast hgt above WL: 49' 3" 15.02 m
Tankage:
Water: 53 gal. 200 liters
Fuel: 26 gal. 100 liters
Auxiliary power: one 9.9 outboard

Type .... Cabin with partial bridgedeck
Foredeck .... Trampoline
Cabins ...... Three
Galley up/down Up
Rudders ..... Balanced spade
Rig .... Fractional sloop

Common problems and price history are not available at this writing except to say that due to the nature of construction and fitting out, advertised prices are considerably lower than similar sized boats.

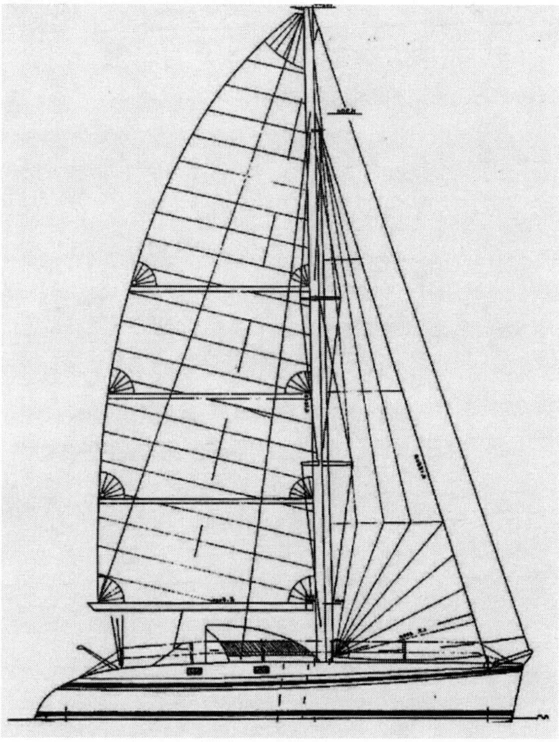

I have only inspected two of them thus far thus only giving me a brief introduction, to these interesting vessels.

*This will change with twin diesels. The vessels I have sailed were outboard powered.

**Insufficient sailing time for a full evaluation

Outremer 43

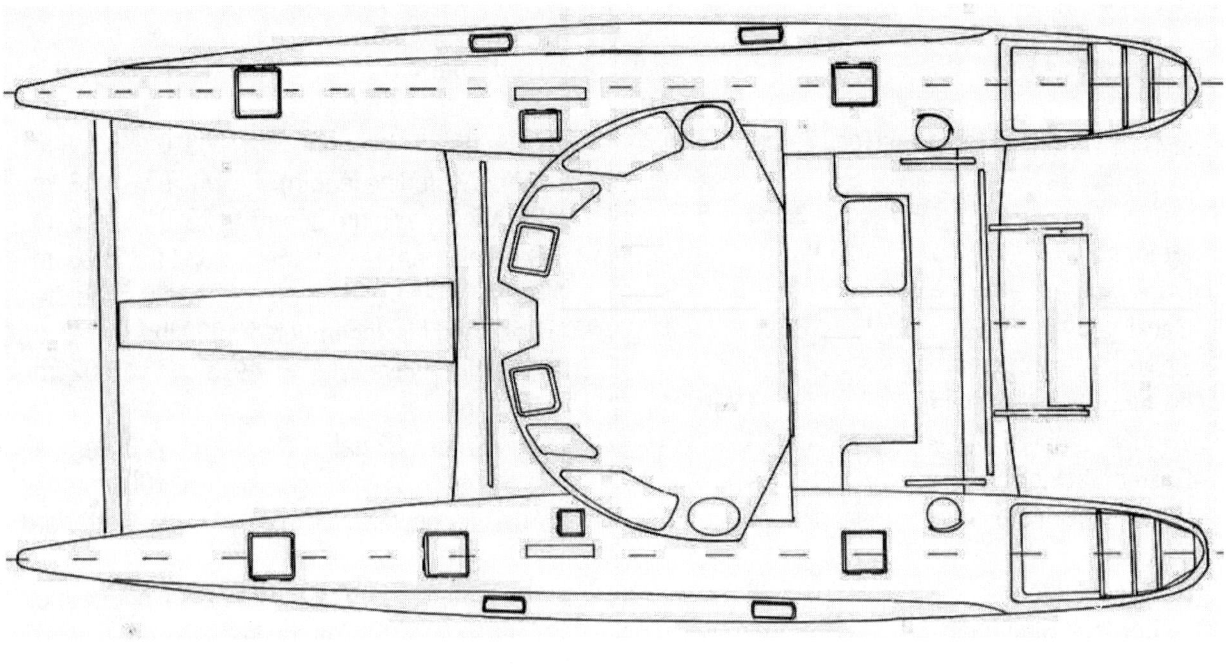

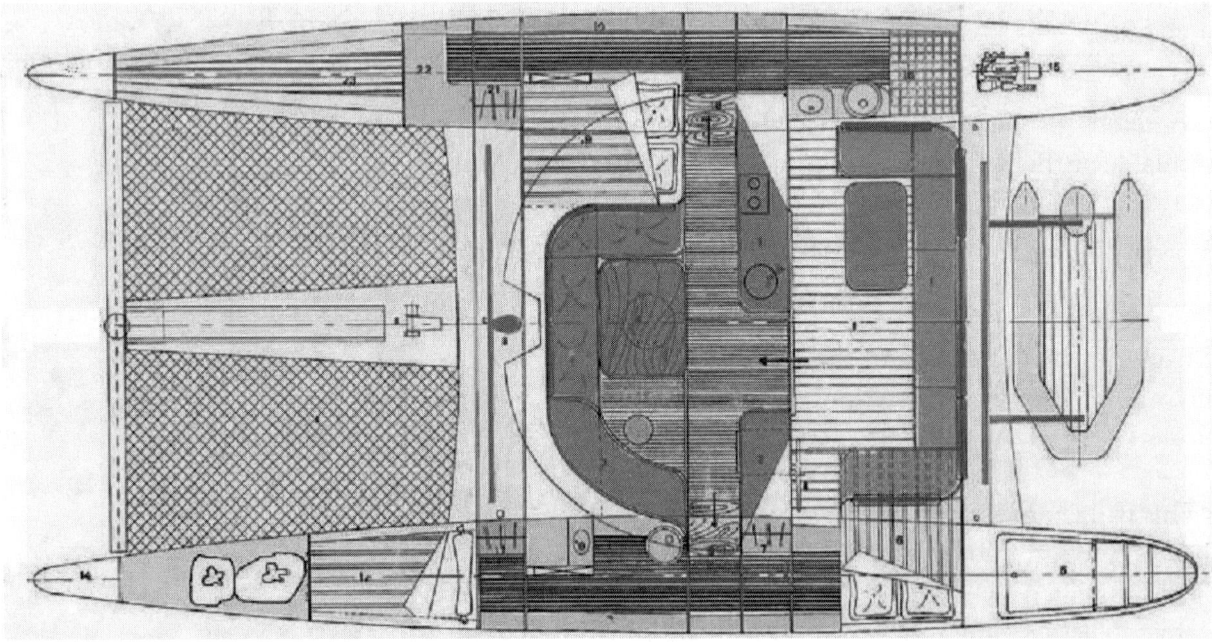

PDQ 32, Class five, Cruising

Designer: Alan Slater

Sailing summary

PDQ 32

1=Poor 2=Fair 3=Average 4=Good 5=Exceptional

	1	2	3	4	5
Motoring				4	
Docking				4	
Backing				4	
Sailing			3		
Weather helm/lee helm				4	
Windward ability			3		
Ease of tacking			3		
Visibility from the helm			3		
Tracking					
Seakindly/ride					5
Underdeck Slamming					5
Convenient Deck Layout				4	
Interior layout					5
Adequate storage areas				4	

Vital Statistics

PDQ 32

LOA (length overall)	31' 7"	9.60 m
LWL (waterline length)	31'	9.40 m
BMAX (maximum beam)	16'	4.90 m
SA (sail area)	443 ft^2	41 m^2

Draft: 3' 2" Keels
Designed displacement: 7,200 lbs 3273kg.
Mast hgt above WL: 45' / 13.7 m
Tankage:
Water: 47 gal. 177 liters
Fuel: 27 gal. 102 liters
Auxiliary power: Twin 9.9 outboards in wells

Type . . . . Full cabin
Foredeck . . . . Trampoline
Cabins . . . . . . Two
Galley up/down Down
Rudders . . . . . Balanced spade

PDQ 32

This is one slick boat. Aside from sailing nicely, it has an exceptional internal layout. It has a center cockpit arrangement without many of the deficiencies often associated with center cockpit layouts. The cockpit is spacious, protected by a permanent hard top and provides good visibility. Decks have reasonable access, the forward bulkhead having vertical windows allows both an interesting forward seat and good headroom in the cabin. This arrangement also eliminates one catamaran bugaboo, the greenhouse effect from heavily slanted windows. There are two good sized sail lockers forward with anchor lockers in both forepeakes.. There is a crossbeam with gull-striker, trampoline and a three wire, masthead rig allowing a full batten main with large roach.

In the cockpit are the twin 9.9 four-stroke outboards. They raise and lower from cleverly designed compartments under the seats. This is a convenient arrangement which keeps the engines outside any accommodation area and makes utilizing gasoline as safe as in your car. The batteries fit in that same compartment making the switches easy to access and the wiring short and to the point.

The interior layout is most ingenious. Starting at the stern, there are large lazarette lockers and the steering system is accessible by removing the lazarette floorboards. Between the two cabins, forming a walkway into the cockpit, is a central locker containing the fuel tank, bilge pump and pump switch-over valve. The two mirror image aft staterooms have ventilation from three sides and a deck hatch. Each stateroom has a queen-sized bed its own hanging locker.

The saloon has a squared horseshoe shaped bench seat with central center table wich can be converted to a bunk if required. It comfortably seats six people on the three sides facing the bench and could seat two more on the forth side using folding chairs.

With the galley on port side and the head on starboard, there is never a user conflict. The head is large, no problem using all the facilities properly and the boat has a built-in holding tank outside the accommodation area. The galley is a real delight. It is hard to conceptualize that this galley with about 12 feet of counter space, an eye-level front opening fridge and lots of locker space is on a thirty-two foot boat. Definitely can fit more than one person in this galley.

Under the boat you find very little in the way of impediments to water flow, just some minor splashing from the motor mounts. This boat has the highest bridgedeck clearance of all the boats in its size range. It has rather buoyant hulls and there is little or no slamming. The boat has fixed low-aspect ratio keels and spade rudders.

As expected, motoring and maneuvering is a delight. Twin screws do it every time. With the engines just about on the center of rotation, we experienced no cavitation or ventilation, normal problems with outboards mounted far aft or directly on the transom. The boat sailed up to expectations, about what is normal for a thirty-two foot cruising catamaran. Do not expect double digit speeds in most conditions, it simply is too short and too heavy to do it. You simply cannot compare cruising boats to racing boats for performance. Having said that, I can further state that higher the wind speeds the greater your advantage over other types of cruising boats with similar waterline lengths. Double digit speeds are routine in high wind conditions.

My only criticism is the same as most of the present crop of catamarans: the spade rudders. The influence of the racing crowd and the bean-counters has spade rudders as a universal choice. However, in the real world of cruising, there are myriad obstacles to overcome. Lobster pots, discarded fishing lines, plastic bags and other assorted debris that inextricably finds its way into your rudders. Snagging junk in a spade rudder means a trip overboard. Running aground most often results in a bent rudder shaft or damage. A skeg rudder neatly solves both problems.

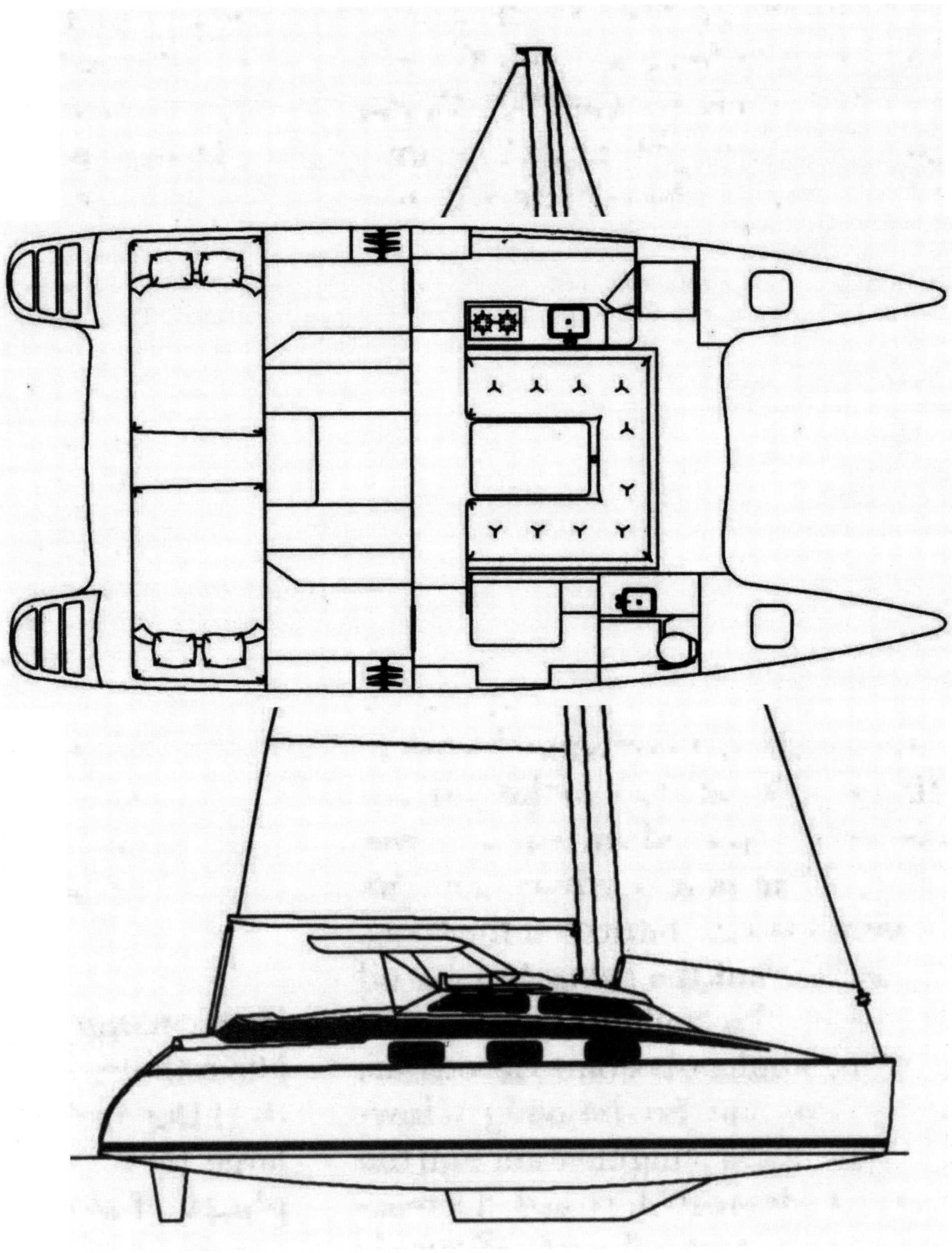

PDQ 36

PDQ 36, Class 4, cruising/racing

Designer: Alan Slater

Classification and/or design intent
Limited ocean and offshore use. PDQ stands for: pretty darn quick!

Aesthetics of the vessel are good. A sloped foredeck design, center divided trampoline, moderately low profile cabin top. Reversed transoms with steps. Three cabin layout, two equal cabins with double beds on bridgedeck forward of saloon. These cabin have exceptional headroom height because of the sloped deck. The cabin in the port stern is available with a wide single or bunk beds. The galley down is on the port side and navigation area on starboard. Typical U-shaped saloon seating with central table. Galley down on port side and is very large and spacious with an exceptional amount of clear counter space. The refrigerator is at eye level.

The vessel has an exceptional "bathroom" with a stall shower and smoked acrylic shower door. The facilities equal those found ashore except for the marine toilet. Holding tanks and "Y" valves are standard equipment.

Decks/toe rails/rub rails/life lines/ The boat has a combination rub rail/toe rail of perforated aluminum, a handy place to secure child nets, fenders, snatch blocks and other gear. There are anchor lockers on each hull foredeck and there is sufficient room to mount a windlass if required. There are bow rollers on the bow of each hull appropriate for anchor storage. Cleats are correctly sized.

The rig is a masthead sloop with single upper shrouds, twin backstays and double lowers. The headstay is attached to a bridle. This arrangements negates the need for an extensive trussed forward crossbeam. There are no spreaders.

Annapolis boat show time is a wonderful time for boat testing. Thus it was with PDQ. I hitched a ride on several demo sails. I skippered occasionally. For those unfamiliar with Annapolis, it is hectic, crowded and dock space is at a premium.

Getting into our allocated slip meant traversing a long channel that was only inches wider than the boat. Spinning on the twin 9.9 Yamaha outboard engines and backing to the dock impressed our now attentive audience. Applause came from onlookers at the restaurant looming above us. The patrons considered our performance live entertainment!

After Annapolis, three PDQ sailed to Florida. Intermittently, we sailed side by side. Through the Florida winter and the Miami Boat Show, we had many opportunities to cruise together and more opportunity to sail offshore in southern waters.

The results of all this observation and sailing PDQ show that PDQ has above average windward ability for a cruising catamaran without center or dagger boards. I could find little to criticize about tacking ability. The Bimini top and the sloping foredeck hamper sail handling a small amount.

The sloping foredeck concept is not one that I particularly favor. The reasons being that it makes it difficult and in some instances like the Banshee or the Wildcat, actually dangerous to walk upon. It limits the use of the deck as a sundeck but mostly it eliminates one of the primary catamaran attractions, the seated eye level forward looking windows. This sloped deck feature is moderated somewhat in the PDQ 36 by having reasonable side decks and a very workable foredeck and trampoline area.

With wind speeds in the 5 to 10 knot range, under sail, the PDQ 36 would achieve close to wind speed anytime the wind was forward of the beam. Tacking was routine, we never lost any. Jibing was routine catamaran jibing, little if any, special attention and no special equipment needed. There was only occasional mild bridgedeck slapping, despite occasional high waves, or power boat wakes. Power boat wakes are really worse than natural waves. Equipment such as tracks and winches were well laid out and easy to operate. All equipment met my test for a class 4, cruising racing, catamaran.

First introduced in the late 1980s and undergoing gradual metamorphoses, I have had many opportunities to survey PDQ 36 in various states of repair and disrepair and to observe cruisers using the boats in their natural element. From my observations this is a remarkably trouble free boat. Later models offer additional accessories such as the LRC model (long range cruiser) that has twin diesels. There is a new cockpit arrangement that has a continuous bench seat across the stern. There is also a hardtop for the cockpit option.

The outboard engine arrangement in the cockpit under the bench seats is a practical and convenient way to mount the engines. The history of using these engines has been mostly satisfactory. The advantages are that the engines and fuel are all outside where there is neither excess danger nor odor as opposed to diesel engines located under bunks that have the affinity for being two more family members whom you must hear, smell and deal with their latent heat. In addition, you must get under the bunks for routine inspection and maintenance and drag containers of hot diesel oil through the entire boat at service time. The danger part comes from having through-hull fittings of intake, exhaust and shaft log. Actually, carrying a spare 9.9 engine is less expensive, has less weight, is less intrusive and exchanging one is a repair anyone can make any time anywhere.

Additionally, the diesels use valuable under-bunk space normally reserved for dry goods such as blankets, linens, clothing, pillows and other light bulky items. Losing that space, there is really nothing to replace it. I also found that access to the engines severely restricted. I could not even check the oil on the starboard engine because it was in such an inaccessible location.

All things considered, the PDQ 36 is a soundly constructed, reasonably laid out moderately priced choice in a cruising catamaran.

Sailing summary

PDQ 36

1=Poor 2=Fair 3=Average 4=Good 5=Exceptional

	1	2	3	4	5
Motoring				4	
Docking				4	
Backing				4	
Sailing				4	
Weather helm/lee helm			3		
Windward ability				4	
Ease of tacking			3		
Visibility from the helm			3		
Tracking				4	
Seakindly/ride				4	
Underdeck Slamming			3		
Convenient Deck Layout			3		
Interior layout			3		
Adequate storage areas		2*			

Vital Statistics

PDQ 36

LOA (length overall) 36' 5" 11.09 m
LWL (waterline length) 34' 4" 10.45 m
BMAX (maximum beam) 18' 3" 5.56 m
SA (sail area) 490 ft^2 45.52 m^2
Draft: 2' 10" Keels
Designed displacement: 8,000 lbs 3628 kg.
Mast hgt above WL: 46' 9" 14.25 m
Tankage:
Water: 50 gal. 189.25 liters
Fuel: 36 gal. 136.26 liters
Auxiliary power: Twin 9.9 outboard standard
Twin diesel in LRC model optional
Type .... Full cabin
Foredeck .... Trampoline
Cabins ...... Three
Galley up/down Down
Rudders ..... Skeg mounted

* I surveyed two boats that had so much equipment installed that it used every single storage facility in the boat. There was not a single locker left in which one could store a case of beverage.

This boat has many compromises for high performance. Properly loaded, it does indeed provide that performance. However, it simply does not have the carrying capacity to load it up with every possible shore-side convenience and accessory. In the LRC model, the weight of the engines in the aft pulls the stern down. The addition of a large RIB inflatable with a big engine pulls the stern down far enough to immerse the backstay chainplates. This is not a good situation.

The PDQ 36 is a wonderful vessel. However, people considering the purchase of this, or any other similar catamaran, must realize its limitations.

PDQ 36

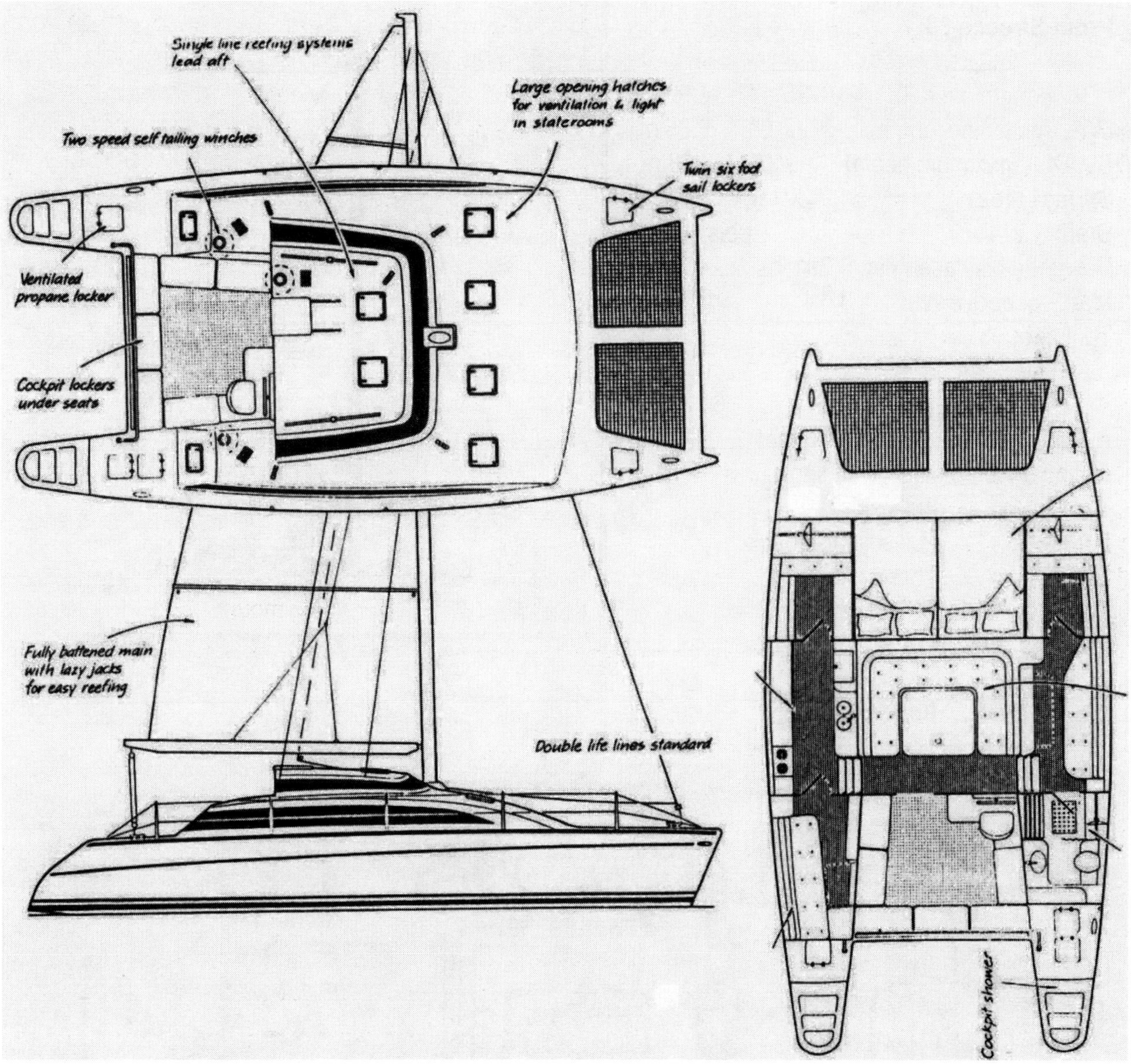
Single line reefing systems lead aft
Large opening hatches for ventilation & light in staterooms
Two speed self tailing winches
Twin six foot sail lockers
Ventilated propane locker
Cockpit lockers under seats
Fully battened main with lazy jacks for easy reefing
Double life lines standard
Cockpit shower

Prout Sirocco 26

Prout Sirocco 26 Class 5, Cruising

Vital Statistics

Prout Sirocco 26

LOA (length overall)	25' 11"	7.99 m
LWL (waterline length)	22' 6"	6.85 m
BMAX (maximum beam)	12' 9"	3.88 m
SA (sail area)	301 ft^2	28 m^2
Draft: 2' 4"		Keels
Designed displacement:	5,291 lbs	2400 kg.
Mast hgt above WL:	37' 4"	11.35 m
Tankage:		
Water:	44 gal.	200 liters
Fuel:	22 gal.	60 liters

Auxiliary power: one outboard
Optional 7.5 hp diesel with sonic drive
Type . . . . Full cabin
Foredeck . . . . Full solid foredeck
Cabins . . . . . . Three
Galley up/down Down
Rudders . . . . . Transom hung, skeg protected
Rig . . . . Fractional sloop

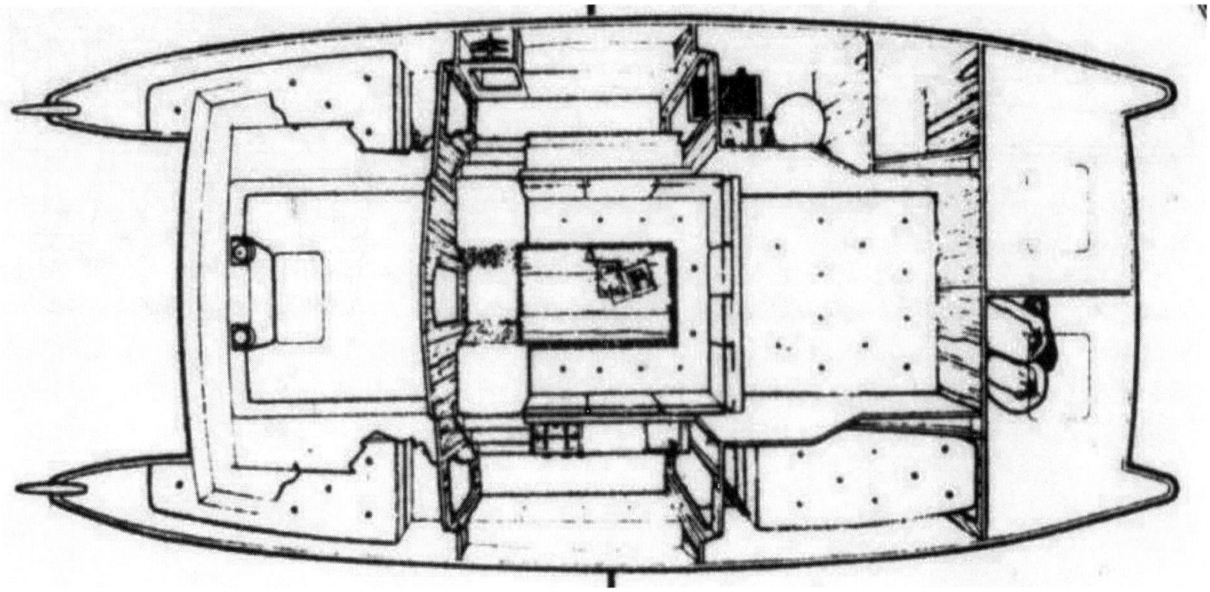

Prout Quest 31 Class 5, Cruising 

My first experience with one was when Corinne and I delivered it from Symons place on Long Island, N.Y. to the Annapolis, Maryland, boat show. Bill had specifically told us that there was a twenty-five gallon fuel tank installed. I topped the tank off from Jerry-cans as part of our make ready, and off we went. Twenty-five gallons being sufficient to get us to Annapolis. Somewhere near the middle of Delaware Bay, we ran out of fuel. The tank was not twenty-five gallons as stated, but merely fifteen gallons. Corinne and I then sailed that boat up the Cohansey River, a winding, narrow, swift, muddy waterway to the dock at a little fish camp for fuel.

My next experience with one was in the 1978 YACHTING ONE-OF-KIND Regatta at Annapolis, where, at the request of Bill Symons, I skippered one in the regatta for a new owner. Like so many British boats, the rigs are ultra conservative. Therefore, do not expect sparkling performance in light wind regions. Properly loaded, this boat will perform quite well when there is sufficient wind.

Since those days, I have surveyed three others. Not overloaded, the boat sails quite well. It tacks nicely, as do most of the Prout boats. In my opinion, it, like most boats in this size range, it is on the cusp of being too small for serious long term liveaboard cruising. I must say, however, I met several people, living aboard them and doing it. They are happy campers.

Most of these boats came equipped with diesel auxiliary and Sonic leg. Occasionally, you can find one with an outboard. The outboard powered boats generally float considerably higher due to the difference in weight.

Prout Quest 31

Sailing summary

Prout Quest 31

1=Poor 2=Fair 3=Average 4=Good 5=Exceptional

	1	2	3	4	5
Motoring			3		
Docking			3		
Backing			3		
Sailing			3		
Weather helm/lee helm				4	
Windward ability		2			
Ease of tacking			3		
Visibility from the helm				4	
Tracking				4	
Seakindly/ride				4	
Underdeck Slamming			3*		
Convenient Deck Layout				4	
Interior layout			3		
Adequate storage areas			3		

* depends on load in forward lockers

Vital Statistics

Prout Quest 31

LOA (length overall)	31' 10"	9.45 m
LWL (waterline length)	28' 6"	8.69 m
BMAX (maximum beam)	14' 3"	4.34 m
SA (sail area)	443 ft^2	41 m^2
Draft: 2' 6"		Keels
Designed displacement:	7,000 lbs	3175 kg.
Mast hgt above WL:	35'	10.7 m
Tankage:		
Water:	gal.	liters
Fuel:	15 gal.	57 liters

Auxiliary power: 12 hp diesel with sonic leg
Type .... Full cabin
Foredeck .... Full decked
Cabins ...... Three
Galley up/down Down
Rudders ..... Skeg mounted, transom hung
Rig .... Mast-aft sloop*

*The mast aft rig has both advantages and disadvantages. The advantages lie in the fact that all the sail handling lines are in the cockpit and the tabernacle can allow the mast to tip forward, using the main boom as the leverage pole. For those who believe that the head sails drive the boat and that roller furling sails are a convenience, this rig is the ultimate.

Disadvantage lies in the fact that the geometry of rigging is reversed, thus leaving it almost impossible to get a tight headstay and seriously over stressing the backstays when trying. Be careful when tuning this rig. I have observed serious damage to boats from over tightening stays and shrouds.

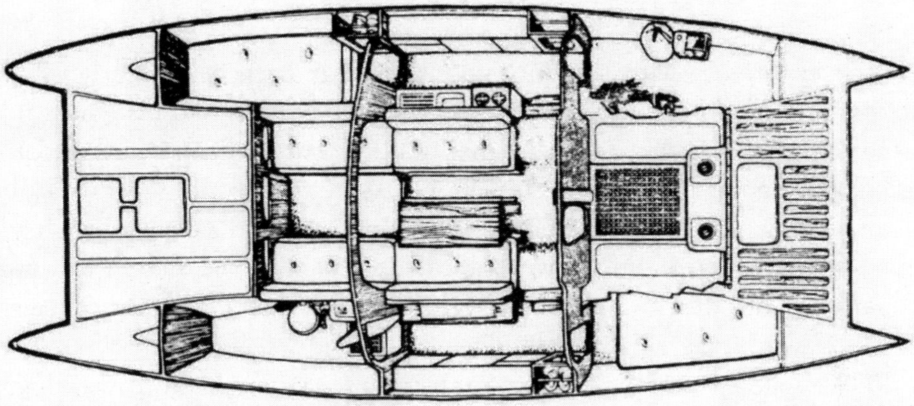

Quest 31

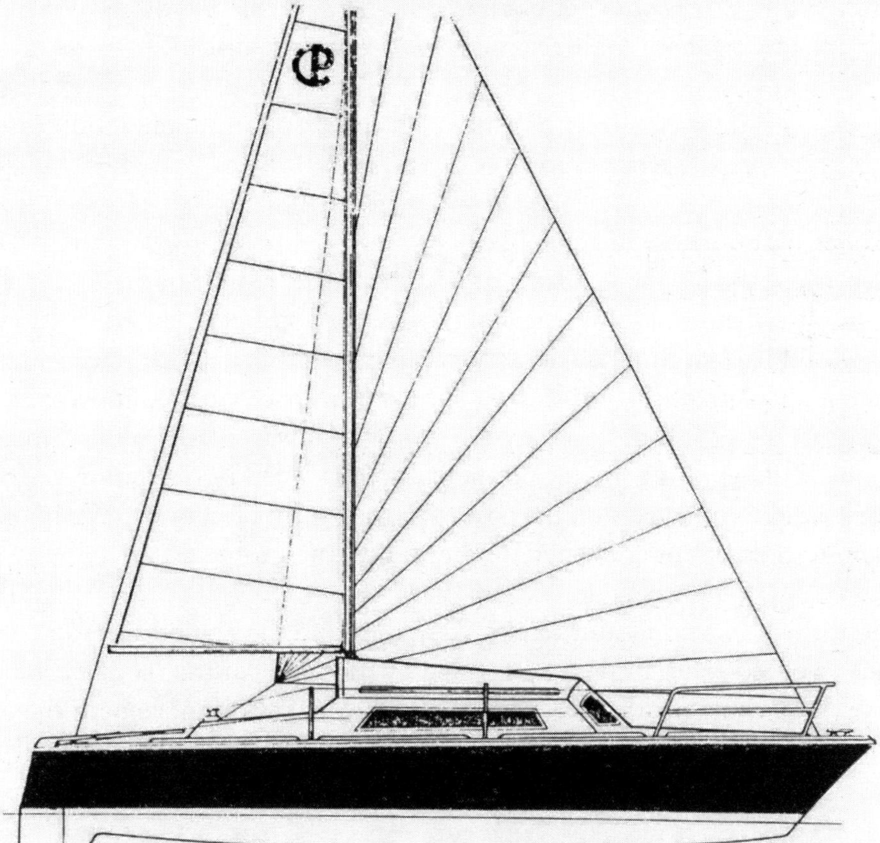

Prout 34, 35 Class 5, Cruising

My first experience with the Snowgoose was way back in the early 70's when I delivered one to Montauk, Long Island, New York for a race. Bill Symons had his hands full with other boats, and I was in charge of that one. Following that first exposure, I delivered three others on the East Coast of the U.S., one going through the Okeechobee Waterway to the West Coast of Florida and then as far North as Panama City. Make no mistake, this is a good sailing boat. It tacks well, handles nicely, responds well to the helm. However, don't expect this boat to be a windward wonder. Between the mast-aft rig, the distribution of buoyancy and the low aspect ratio keels, windward performance suffers. Nevertheless, this a sea kindly vessel. There are myriad sprinkled around the harbors of the world. We sailed part of the Bahama winter sailing season in company with Conrad and Josie on *Pussy Cat*, a Prout, Snowgoose 34. Conrad did a remarkable job of making that boat sail. (That boat was kidnaped by vigilante soldiers on the West coast of Africa.)

I have also surveyed a total of thirty eight others, done some modification on one and had a very embarrassing experience on one.

We were just offshore of Cape May, New Jersey, having lowered the sails, preparing to come in the inlet, when the engine suddenly quit with a screeching shuddering bang. It sounded really serious. We hoisted the jib and sailed right to the dock at South Jersey marina. I looked it over as best I could, but could find nothing obviously wrong. Since it made such a ferocious noise, we assumed it was something internal and very serious. The yard mechanic looked it over, he could find nothing either, except that it would not even try to turn over, it was frozen solid! . Since it was a brand new Volvo engine, Bill Symons got a replacement engine loaded in his station wagon and drove to Cape May, While I did some surgery on the boat to prepare for replacement. We hooked the engine to the main boom to hoist it out and when we did, found the problem.

The forward motor mount had broken, allowing the engine to drop just enough so that the front pulley wedged solid in the "vee" formed by the shape of the nacelle. The engine fit in the nacelle so closely, you could not discern by eye the slight downward movement the engine had taken when the mount broke, but there was sufficient friction to stop the engine dead. Talk about embarrassing . . .

Prout vessels have two items of difference from most other vessels that need to be considered: The mast aft rig, and the double ended hull shape.

The mast aft rig has both advantages and disadvantages. The advantages lie in the fact that all the sail handling lines are in the cockpit and the tabernacle can allow the mast to tip forward, using the main boom as the leverage pole. For those who believe that the head sails drive the boat and that roller furling sails are a convenience, this rig is the ultimate.

Disadvantage lies in the fact that the geometry of rigging is reversed, thus making it almost impossible to get a tight headstay and seriously over stressing the backstays when trying.

Double ended theory as applied to these catamarans, places the center of buoyancy close to the center of gravity, which is close to the center of the boat. Therefore, when waves come up from astern, they pass a significant distance under the boat, before they begin to lift, thus lifting from the center of the boat and eliminating the potential for pitchpole. Boats with broad transoms, have the center of buoyancy further aft, thus they appear to want to drive their bows under when large waves come up astern because the buoyancy is so far aft it picks the stern up first. On the double ended Prout boats, when waves come up from astern the boat either lifts level or tips backwards slightly, depending upon loading.

The double ended concept is a workable concept. The disadvantage is that it is hardly feasible to mount anything heavy, such as a dinghy in davits, on the transom. In addition, for the devotees of the "weight is great" anchoring method, having a slew of heavy style anchors with miles of chain in the forward anchor lockers will dramatically change the handling of the boat as well as the underdeck pounding.

Prout 34, 35

Prout 34, 35

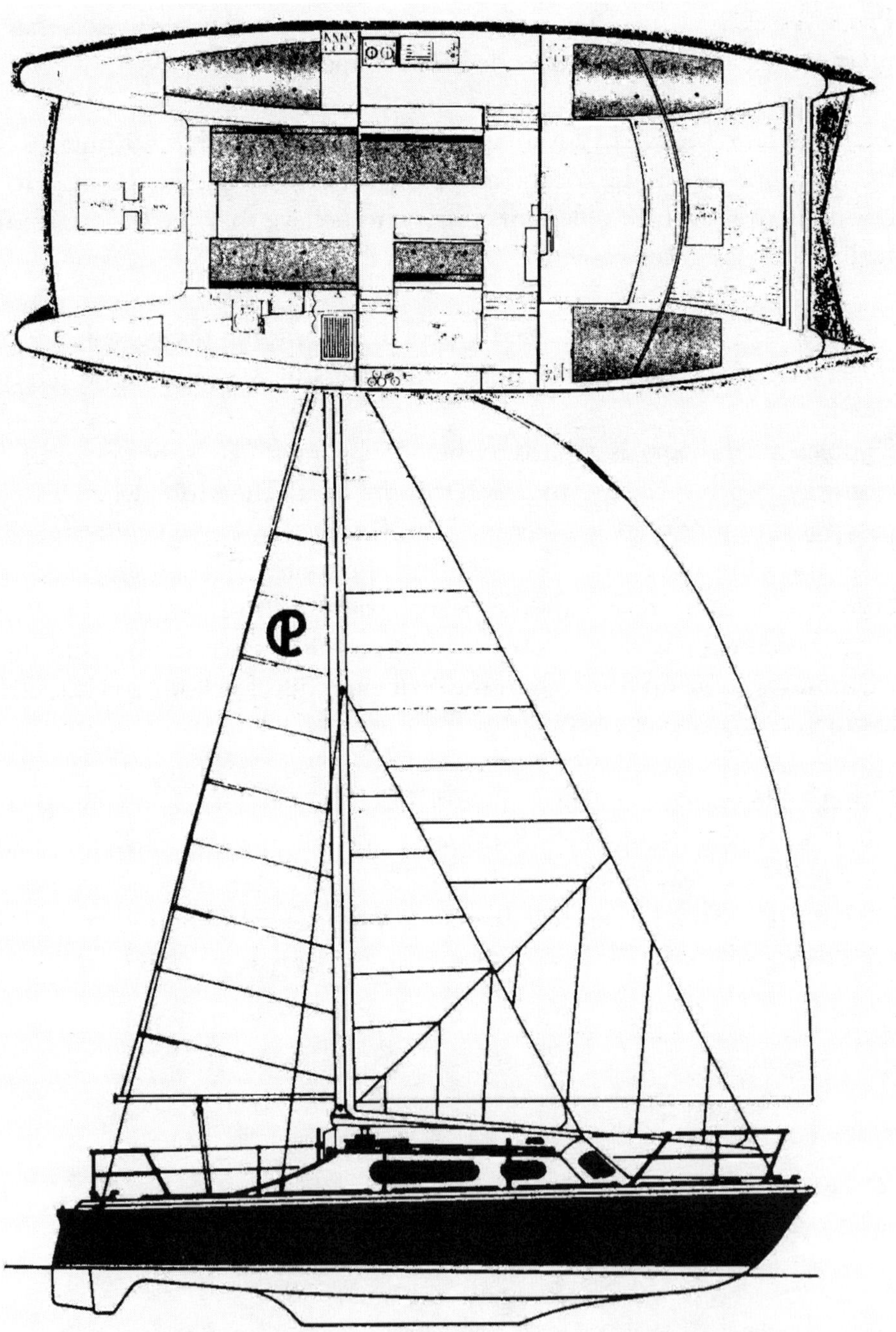

Prout 34, 35

Sailing summary

Prout , 34, 35

1=Poor 2=Fair 3=Average 4=Good 5=Exceptional

	1	**2**	**3**	**4**	**5**
Motoring			3		
Docking			3		
Backing			3		
Sailing			3		
Weather helm/lee helm			3		
Windward ability		2			
Ease of tacking				4	
Visibility from the helm				4	
Tracking			3		
Seakindly/ride				4	
Underdeck Slamming			3		
Convenient Deck Layout				4	
Interior layout			3		
Adequate storage areas			3		

Vital Statistics

Prout, 34, 35

LOA (length overall)	34' 3"	10.4 m
LWL (waterline length)	30' 6"	9.30 m
BMAX (maximum beam)	15' 3"	4.6 m
SA (sail area)	500 ft^2	46.4 m^2
Draft: 2' 9"	Keels	
Designed displacement:	8,000 lbs	3630 kg.
Mast hgt above WL:	45' (est)	13.7 m
Tankage:		
Water:	110 gal.	416 liters
Fuel:	36 gal.	136 liters
Auxiliary power:	Diesel with Sonic drive leg	

Type . . . . Full cabin
Foredeck . . . . Full solid foredeck
Cabins . . . . . . Three
Galley up/down Down
Rudders . . . . . Skeg mounted transom hung

Prout Snowgoose 37 (Class 5, cruising)

This vessel is in reality an updated version of the well proven Snowgoose 34 and 35. I have surveyed twenty six and sailed on thirteen others. The basic difference between the two boats is that the 37 is enough larger to really use the inner forestay for a sail and it is so rigged. The standard Snowgoose is a good sailing, seakindly vessel. Some later versions, like the Snow Goose Elite, that has expanded the interior to try to compete with some of the roommarans I have certain reservations about.

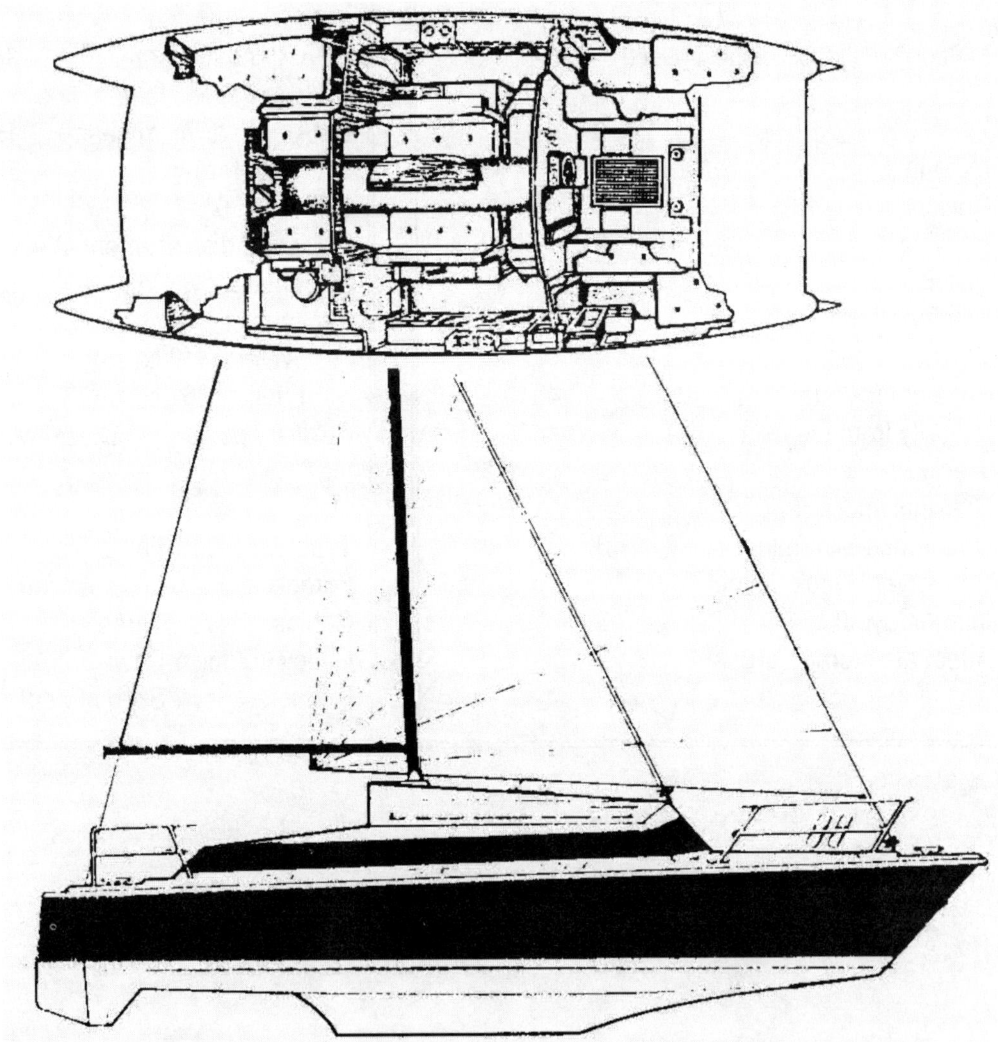

Prout Snowgoose 37

Sailing summary

Snowgoose 37

1=Poor 2=Fair 3=Average 4=Good 5=Exceptional

	1	2	3	4	5
Motoring			3		
Docking				4	
Backing				4	
Sailing				4	
Weather helm/lee helm				4	
Windward ability			3		
Ease of tacking			3		
Visibility from the helm				4	
Tracking				4	
Seakindly/ride			3		
Underdeck Slamming				4	
Convenient Deck Layout			3		
Interior layout			3		
Adequate storage areas			3		

Vital Statistics

Snowgoose 37

LOA (length overall)	36' 8"	11.17m
LWL (waterline length)	33' 0"	10.06m
BMAX (maximum beam)	15' 3"	4.65 m
SA (sail area)	570 ft^2	53 m^2

Draft: 2' 8" / .85 m Keels
Designed displacement: 9,600 lbs 4355 kg.
Mast hgt above WL: 52' (est) 16 m
Tankage:
Water: 70 gal. 265 liters
Fuel: 226 gal. 83.27 liters
Auxiliary power: Diesel with Sonic drive leg

Type .... Full cabin
Foredeck .... Solid full foredeck
Cabins ...... Three
Galley up/down Down
Rudders ..... Skeg hung transom

Prout Snowgoose 37

I have always unhesitatingly suggested the Snowgoose 37 as an ideal blue water cruiser. It is big enough and powerful enough to do the job, yet not so big that it cannot be easily handled by a cruising couple nor so big it cannot fit into marina slips or travel lifts.

Prout 50

Prout Quasar 50 Class 4, Cruising/racing

Designer: Prout design team

A Rare Privilege.

April 15th 1999 when so many were struggling with their tax returns, I had the privilege of taking what I consider a well designed catamaran on sea trials. The Prout Quasar 50.

It was a perfect day for sea trials. The wind was howling a steady 20 to 25 with occasional gusts to 27 and 30. There were white caps in the bay south of Peanut Island which extended right up to our haul-out slip at Cracker Boy yard, West Palm Beach, Florida. The Palm Beach inlet was filled with breakers and it was interesting to watch the boats go up to the end of the breakwaters and then turn around. None of the private boats ventured out, only the pilot boat went out, but he had too!

There was plenty of equivocation between the owner the broker and the buyers as to the prudence of sailing out into that maelstrom. The decision made, since the purpose for sea trials is to check for any weak spots, bad habits, performance and a host of other details, we plunged forward through the breakers.

This boat has a roller furling mainsail hauled out by an electric anchor windlass mounted in the cockpit, a really sensible arrangement as it is more powerful, more versatile and less expensive than an regular installed major chandlery company electric winches. Both head-sails are roller furling. We had no problem furling or unfurling sails in that wind although we never did get to unfurl the Genoa.

At this point, I must explain that in cruising catamarans I have always been a critic of excessive beam, extremely fine bows and big fat transoms that put the center of buoyancy aft. My sailing, surveying and delivery experience leads me to believe that those are the characteristics that lead to the baffling mediocre performance seen in so many modern cruising catamarans. I also feel most manufacturers squeeze in too much accommodation for the size of the boat, which is what leads to the overloading syndrome. This boat, in my opinion, is properly sized for the amount of accommodation.

Prout and others have steadfastly adhered to the double ended philosophy with the center of buoyancy close to the center of the boat. This allows waves either from forward or astern to lift the boat in a more level attitude rather than pitching the boat forward.

It works! The Quasar 50 is considered a "narrow beam" boat by modern pundits who tout their own wide beam wares. The boat is 47' LWL x 23'6" BMAX making it exactly the classic length/beam ratio measurement of 2:1 used since Polynesian times and used by some of the most successful boats ever built. (I must remind my readers that the reason for going to the extra wide beam in cruising catamarans was to create considerably more volume and space in the same length boat, not for safety or performance.)

Bullying our way through the breakers, the first thing you notice is the absence of any slamming or pounding. It seems almost incredible, but we never slammed the bridgedeck even once during the entire trial period. This is a very rare and unusual occurrence. Most boats, both monohull and multihull will pound somewhat, some worse than others. Planing powerboats will kick your kidneys out. Not only did we not pound, but we also remained at a more level attitude than most boats. There was no pitching. We did take some spray as far back as the cockpit, but never any green water. I stood on the foredeck, balancing myself by holding the headstay, patiently watching the action of the boat.

Of course, part of this performance is due to the fact that the boat is 50 feet long. Length helps any boat, even poorly designed ones.

With full main and full staysail we sailed at ten knots with no effort. As far as proper trim or trying to maximize speed or anything else we might normally do, forget it. With a crew of complete greenhorns to the boat and a basic survival instinct we simply did what we could out there in the wind and waves. Since my job was to search for structural imperfections, I spent most of my time crawling around the bilges and normally inaccessible areas looking for defects not normally seen except under load. (A good way to get really seasick.)

As with most cruising catamarans, the feather light helm on all points of sail, great tracking, faithful responsiveness and great seakindliness came through in spades.This boat just has a little more of each.

Back inside the bay, dealing with just wind, I could determine that the very long straight keels and the oversize skegs on the rudders were responsible for the sluggish tacking. The cutter rig also exacerbates this problem, a problem often solved on competing catamarans by the spade rudder, mainsail driven, modern rigs which tack almost like monohulls. On the other hand, long shallow keel, skeg mounted rudder features give a true cruising boat safety and sailing features worth every bit of the tacking disadvantage.

Overall I surveyed four of these vessels, one had a unique tuna tower on the aft section. I sailed many miles aboard these boats and I believe been aboard almost all that are in the United States.

Sailing summary

Prout 50

1=Poor 2=Fair 3=Average 4=Good 5=Exceptional

	1	2	3	4	5
Motoring					5
Docking			3		
Backing				4	
Sailing				4	
Weather helm/lee helm				4	
Windward ability				4	
Ease of tacking			3		
Visibility from the helm			3		
Tracking				4	
Seakindly/ride					5
Underdeck Slamming					5
Convenient Deck Layout			3		
Interior layout			3*		
Adequate storage areas				4	

* *various layouts exist*

Vital Statistics

Prout 50

LOA (length overall) 49' 0" 14.95m
LWL (waterline length) 45' 0" 13.73m
BMAX (maximum beam) 23' 6" 7.16 m
SA (sail area) 1178ft^2 110m^2
Draft: 3' 4" 1.02 m Keels
Designed displacement: 20,000lbs 9090kg.
Mast hgt above WL: 63' (est) 19m
Tankage:
Water: 100 gal. 454 liters
Fuel: 200 gal. 910 liters
Auxiliary power: Twin 27 hp diesel

Type .... Full cabin
Foredeck .... Trampoline
Cabins ...... Three to five
Galley up/down Down
Rudders ..... Skeg mounted
Rig .... Masthead cutter

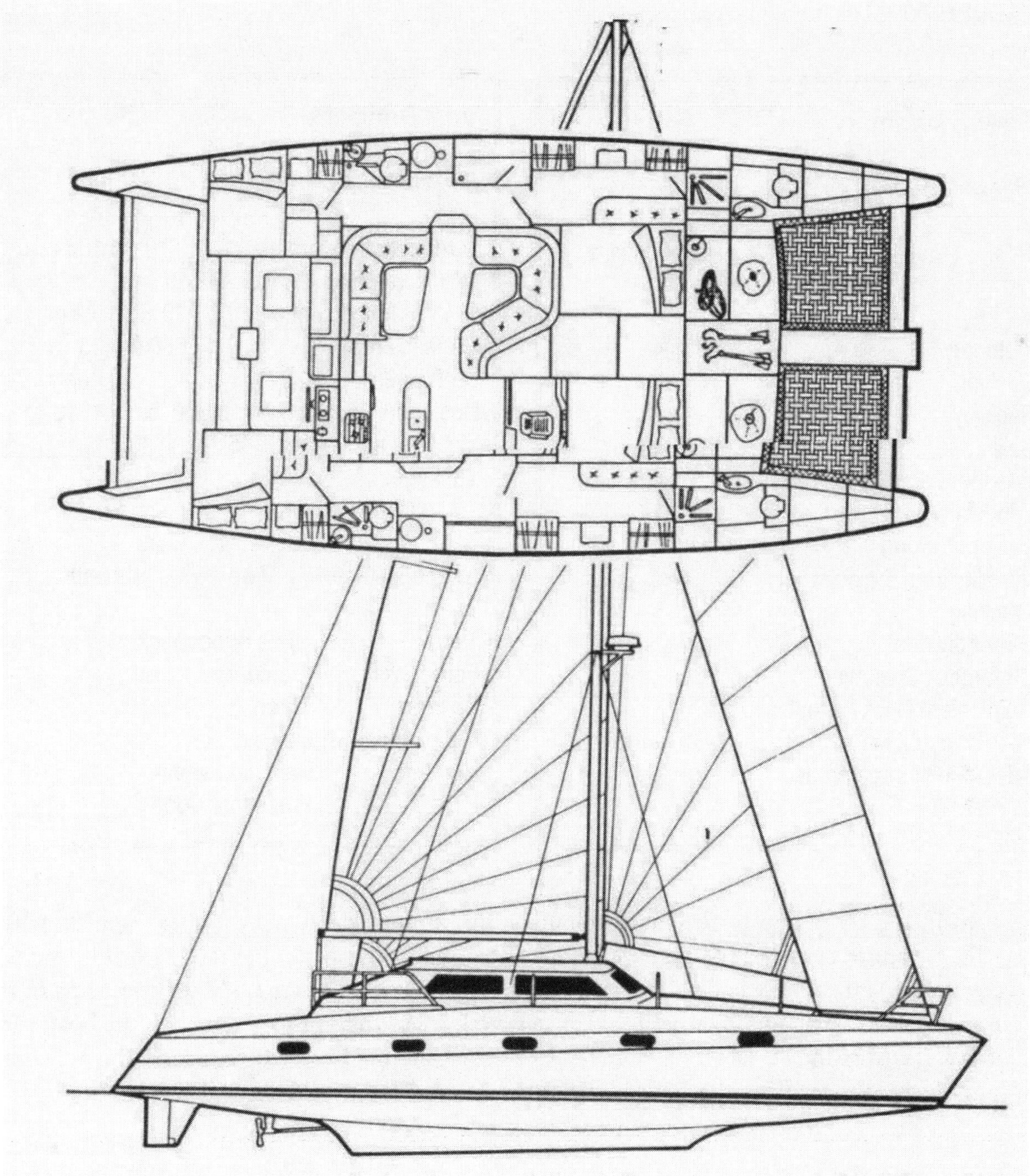

Seawind 1000 Class 3, racing/cruising

Sailing summary

Seawind 1000

1=Poor 2=Fair 3=Average 4=Good 5=Exceptional

	1	**2**	**3**	**4**	**5**
Motoring				4	
Docking			3		
Backing				4	
Sailing				4	
Weather helm/lee helm				4	
Windward ability				4	
Ease of tacking				4	
Visibility from the helm					5
Tracking				4	
Seakindly/ride			3		
Underdeck Slamming			3		
Convenient Deck Layout			3		
Interior layout			3		
Adequate storage areas			3		

Vital Statistics

Seawind 1000

LOA (length overall) 33'0" 10 m
LWL (waterline length) ' ft m
BMAX (maximum beam) 19' 5" 5.9 m
SA (sail area) 680 ft^2 63 m^2
Draft: 2' 11" .90 m Keels
Designed displacement: 8,800 lbs 4000 kg.
Mast hgt above WL: 45' (est) 13.7 m
Tankage:
Water: 120 gal. 450 liters
Fuel: 32 gal. 120 liters
Auxiliary power: Two 9.9 outboard

Type .... Open bridgedeck
Foredeck .... Trampoline
Cabins ...... Three
Galley up/down Down
Rudders ..... Balanced spade
Rig .... Fractional sloop

The Seawind 1000 is one of the more popular imports into the United States from its native Australia. there are ten in the Miami area alone. They have an exceptional race record in the US. However, many of them have had a three foot extension added to the sterns. As with so many of the newer production catamarans, they need that extra waterline length to properly balance the boat, carry the loads and change the LB ratio to a more favorable one. Another boat that reinforces the point that you cannot stray too far from the classic LB ratio of 2:1 without paying a penalty.

Seawind 1000

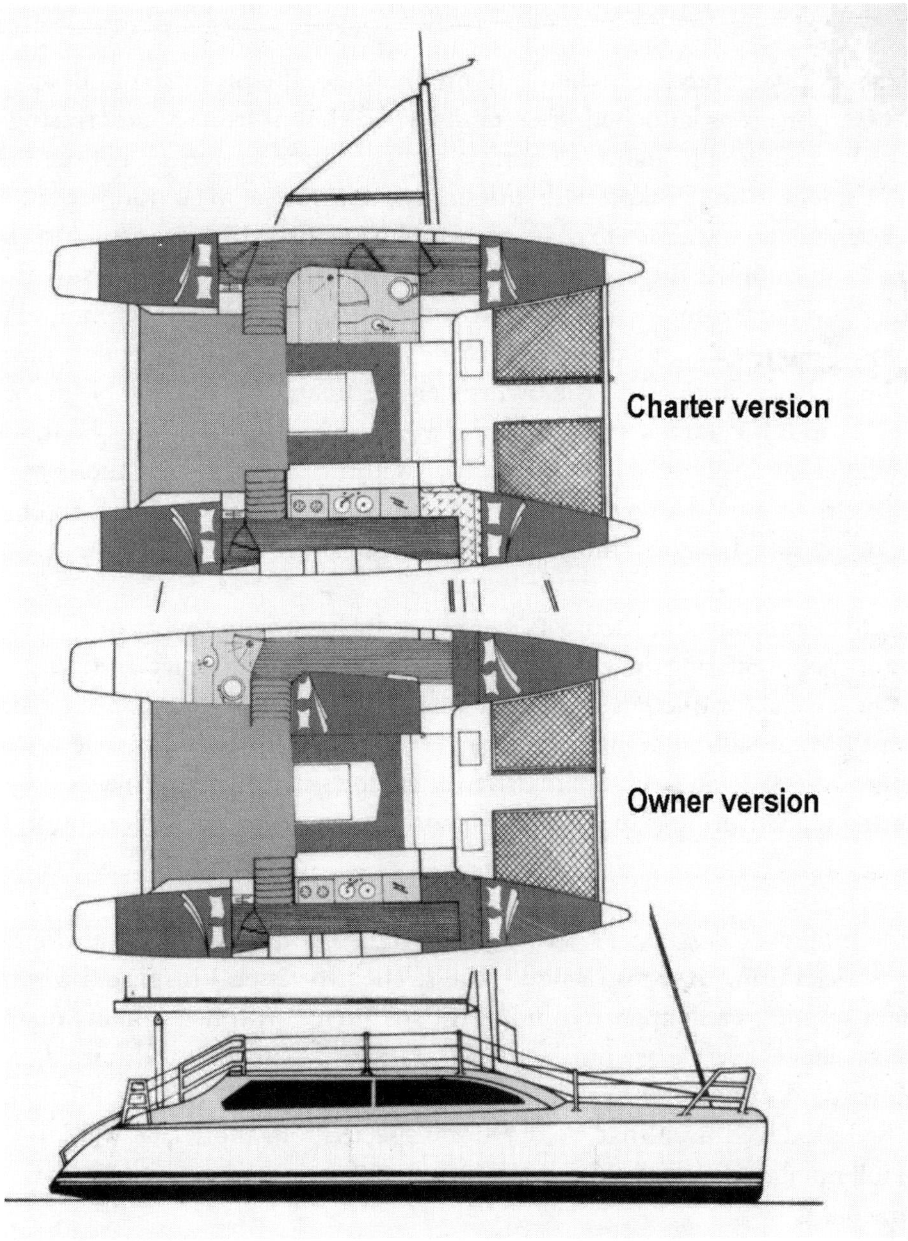

Victory 35

Victory 35, Class 5, Cruising

Designer: Bill Hepburn

I first saw a Victory catamaran back when hull number one, the prototype built at Bennet Brothers yard in North Carolina, was launched. It certainly was different from other cruising catamaran offerings. It had exceptionally high freeboard, a thick, rounded foredeck between the hulls, and a different arrangement in the cockpit with a double seat at the helm station. Inside, it had standing headroom throughout, an owners suite complete with luxurious bathroom, (too lush to call it a head) a queen size bed and complete dressing room. That was on one side. The other side (hull) was equally luxurious with a queen size bunk in the aft cabin, a double in the forward cabin, and a wash basin in between cabins. I inquired about the design while speaking with the designer, naval architect Bill Hepburn of Pine Island, Florida and the owner Len Feldman of Brooklyn, New York. The answers were refreshingly candid. Feldman said: "I want a women to step aboard this boat, look around and say, 'Yes, I could live on this boat." Hepburn added, "Without compromising the boat's ability to sail." Did they achieve their goal? The owner's of hull number one, *Primo* formerly in charter by Passage Charters in Cortez, Florida, report satisfaction with both usage and maintenance.

From hull number two onward, the boats were built by Endeavor Yachts in Clearwater. As a sales incentive, Victory began offering three days of personalized sailing instruction (like giving away a free dingy). I gave the lessons for one purchaser where we spent most of the time practicing docking and a small amount of time sailing. One of the things I learned from that episode is that the standard centrally mounted diesel engine with Sonic drive leg can be skillfully used to access tight corners, but if a lot of marina maneuvering is in your future and you are not a skillful operator, I would suggest twin diesels.

I followed the progress of two later Victory cats as the surveyor on the job at the Endeavour yard, making visits to each during the construction process. The two vessels I inspected were built according to the designer's scantlings and appeared properly constructed with high-quality materials. Both had what I would consider above average finish. In 2000, Victory switched producers. Diversified Marine Group in Palmetto, Florida currently (2002) produces the Victory. I was the surveyor on the job for hull number two produced from that factory and inspected four more that were in various stages of production. Hull number two, *Blo Bashi*, did a single handed shakedown cruise from Grand Bahama Island to Curacoa, WI.

Victory 35

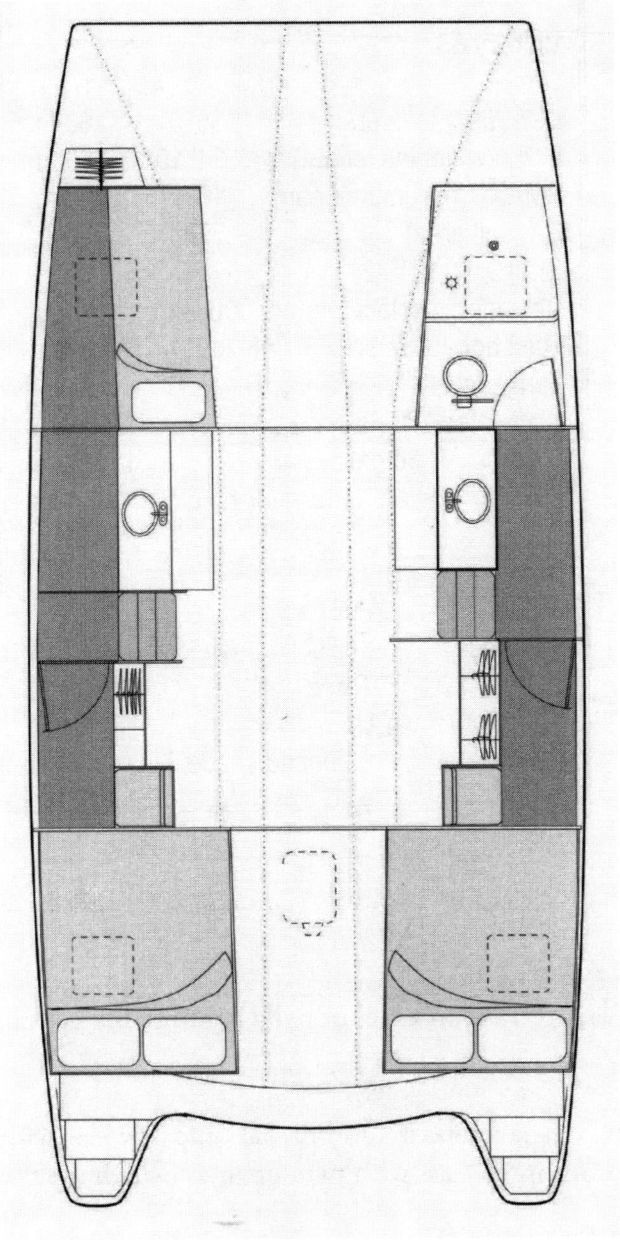

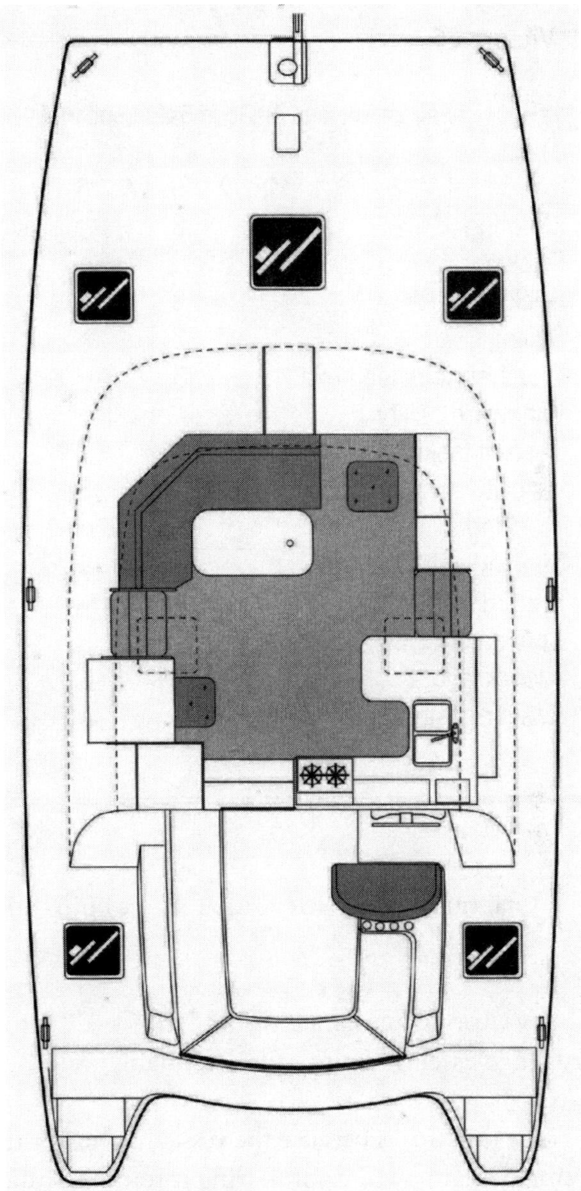

Victory 35

Sailing summary

Victory 35

1=Poor 2=Fair 3=Average 4=Good 5=Exceptional

	1	**2**	**3**	**4**	**5**
Motoring			3		
Docking		2*			
Backing			3		
Sailing			3		
Weather helm/lee helm			3		
Windward ability		2			
Ease of tacking			3		
Visibility from the helm					5
Tracking				4	
Seakindly/ride			3		
Underdeck Slamming			3		
Convenient Deck Layout					5
Interior layout					5
Adequate storage areas					5

Vital Statistics

Victory 35

LOA (length overall)	35' 0"	10.68 m
LWL (waterline length)	34' 10"	10.63 m
BMAX (maximum beam)	16' 0"	4.88 m
SA (sail area)	570 ft^2	70.3 m^2
Draft: 2' 10"		Keels
Designed displacement:	9,000 lbs	4082 kg.
Mast hgt above WL:	48' (est)	14.64 m
Tankage:		
Water:	80 gal.	303 liters
Fuel:	36 gal.	136 liters
Auxiliary power:	one 27 hp diesel w/Sonic leg	

Type .... Full cabin
Foredeck .... Full solid bridgedeck
Cabins ...... Three
Galley up/down Up
Rudders ..... Balanced spade

The total number of Victory catamarans built and sailing as of this writing, (January, 2002) is: thirty-three (hull #133)

The Victory Catamaran certainly fits well into Class 5. It is an exceptionally comfortable coastal cruising vessel that also sails reasonably well.

* To be fair, this is because the vessel has higher than average freeboard and has a Sonic drive leg with a single engine while comparing it to other catamarans in its class with twin engines (which are the best possible arrangement for docking.)

Wildcat, Class 3, racing/cruising

Designer: Jeff Schionning

At the Miami Boat Show I spent considerable time looking at a new offering built in South Africa, the *Wildcat*. It certainly is an attractive looking boat with many pleasing aesthetic features. Personally, I could see it with a smaller cabin, stripped out interior and outboard power as a reasonable class 3 racer/cruiser or sport boat. Being offered as a solid class four-cruiser /racer or class five cruiser, I must take issue with many features I consider unsafe or at the least, inconvenient.

Perhaps I have not stated this truism sufficiently often. The truism is that boats, unlike automobiles or airplanes, are operated from outside the vessel. Automobiles and airplanes benefit from streamlining mostly because they move at high speeds. Sport boats and racers acquire minimal benefits from streamlining because their design intent is to forego any compromise that might interfere with maximizing potential top speed. Cruising boats are not like that. There is precious little to gain from streamlining at cruising sailboat speeds but lots of safety and utility to lose.

Any cruising vessel must be so configured that you can easily and safely reach any part of the topsides or deck area of the vessel. This is necessary for the obvious routine reasons of docking and anchoring. This is especially poignant for cruising couples who seldom have the advantage of the local football team to manhandle the boat around docks or to rush forward on a slippery rounded deck with no handholds to toss or receive a dock line.

Decks

First, the decks are extremely rounded, giving no flat area anywhere upon which a person may sunbath on a towel. Perfect for the racer to reduce windage and cut down on weight, deadly for the less agile trying to dock the boat. Continuing around the deck, one notes that the stern deck quarters are so severely sloped both downward and outward as to not only be potentially dangerous but making it difficult to get on and off the boat safely. Even in the benign conditions of the boat show it was necessary to put mats down to keep people from slipping as the clambered up from the dock across the sloping decks and into the cockpit. I mused over what it would have been like if we had a typical rainy day at the show. That area also loses any usefulness as deck area. I always ask potential purchasers of cruising boats to visualize the routine of handing up groceries from the dinghy, or even a low dock for that matter, during a rain.

There is yet another aspect to this severe slope and that is the impossibility of putting anything on deck, even temporarily. One of my clients, buying a Crowther designed Offshore 44 with very much the same characteristics finally sold it complaining bitterly that his children's toys kept sliding off the boat and getting lost. I suggest that anyone handing up mask and snorkel, flippers or your day's catch had better not lay it on that deck for even an instant.

Going further around the deck, we note there is no toe rail. On many boats, the toe rail is molded into the deck, on some it is an extrusion and on others it is a decorative wooden rail, but it is there. A toe rail is an important safety and convenience feature that should be on all cruising boats, with rare exceptions.

Arriving at the foredeck you note that it is so steep you hesitate to try and walk down, even in good weather on a new boat. What will it be like when the weather is bad, the boat moving and the non-skid a little worn? I predict that any owners will install walking lines down the center of the deck as seen in some flush deck catamarans that have considerably less slope. This is a vital navigation area to handle sails, anchors, docklines etc. Its going to be tougher than it should be for a cruising couple to launch and retrieve a spinnaker with that deck configuration.

Cockpit

Finally back in the relative security of the cockpit, you find one of the strangest anomalies yet on a cruising boat, that is the cockpit sole is not level but slopes forward! This forward sloping area will undoubtedly cause confusion and accelerated falling accidents. The anomaly continues inside the cabin where the cabin sole slopes aft. If it is difficult to negotiate at show dockside, how is it going to be in the open ocean? Sloping interior decks are not good. It is especially confusing in areas normally level, like the floor of your home. Imagine living in a house in which the floors sloped?

Helm

The helm does have good visibility, but at what price? The helmsman's seat is perched up on top of a pole almost completely out of the cockpit. The reason it is so high is because of oversize main bulkhead, often called the catamaran curse. It got that name from the fact that in many catamarans, no one but the helmsman can see forward. This massive wall feeling is exacerbated by the forward sloping cockpit sole. From appearances it seems that this is one of the very rare catamarans where the helmsman could actually be hit by the boom! That also brings up the extreme difficulty of installing a Bimini top, an obvious necessity in tropical climes.

Summary

This may be a great boat to tear across oceans with because it has all the "speedophile" influences. However, the second truism is, to quote Eric Hiscock, "even the most intrepid cruiser spends ninety percent of his time at anchor." Therefore, I suggest those looking for true cruising boats weigh carefully the features described in this article. Please do not think of this critique as overly harsh. For a class 2 or perhaps class 3, those very features of slopes and round areas would receive lavish praise for both their weight reduction and wind drag reduction. Extreme streamlining in that case would be beneficial.

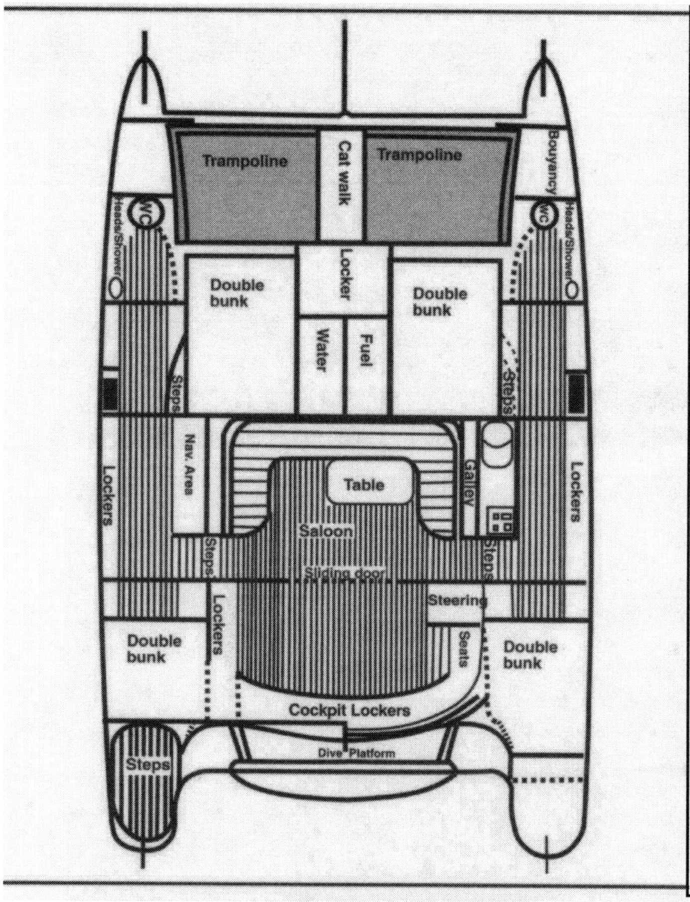

Vital Statistics

Wildcat 35

LOA (length overall)	34' 5"	10.50 m
LWL (waterline length)	32' 10"	10.00 m
BMAX (maximum beam)	26' 7"	8.07 m
SA (sail area)	807 ft^2	75 m^2
Draft: 2' 7"		Keels
Designed displacement:	9,972 lbs	450 kg.
Mast hgt above WL:	55' 9"	17 m
Tankage:		
Water:	106 gal.	400 liters
Fuel:	53 gal.	200 liters
Auxiliary power:	Two 8hp OB, option 2 diesels	

Type .... Full cabin
Foredeck .... Trampoline
Cabins ...... Three
Galley up/down Down
Rudders ..... Balanced spade

Photo: Wildcat foredeck
Note: This extremely steep foredeck will need a hand railing down the center to be usable.

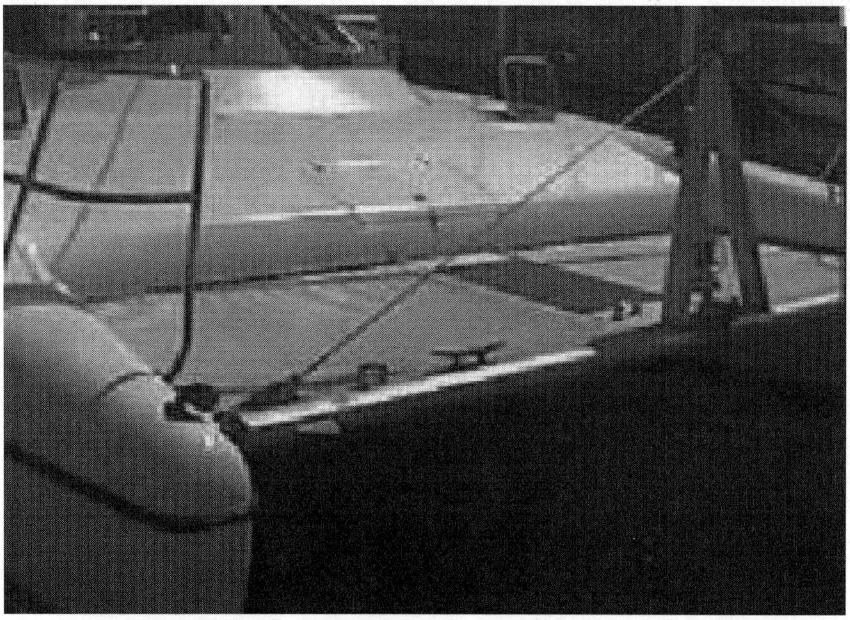

Photo: Wildcat starboard stern
Note: The extreme slopes to be negotiated in order to enter the cockpit
Note the extreme position of the helm.

Book 3, Part 2, Catamarans not in production

In this section, I will critique 34 catamarans of different manufacture with which I have significant experience. As in part 1, "Catamarans in Production," there will be a certain amount of crossover because the company may be producing boats, but not that particular model. Each section will be listed alphabetically with a cross reference by length.

This is a tough call because many companies have temporarily discontinued certain models in favor of newer models but still have the molds for the earlier models. Often, they will mold a vessel on order, thus making categorization problematic. Prout is one such company.

It is not feasible to list every catamaran I have surveyed thus I have culled through my list and critiqued those most likely to be of interest to those looking for new or previously owned boats. Specifically missing from the listings are the following vessels:

Admiral 38, 49
Banshee 37
Catana 472
Comanche 32
Corsair 3600
Cracksman 28
Flicka 36
Flying Feather 44
G Force 36
Karmin 35
Kelsall 40
Logical 44
Mayotte 500
Ocenanic 30
Outremer 40
Polycon 40
Prout Manta 38
Prout 38, 39
Sea Cat 35
Sea Wind 24
Sea Wind 40
Simpson 35
Shuttlecat 35

Catamarans not in Production

Americat 30 . Class 5, Cruising

1. No longer in production.
2.Long history. Started as Intercat, morphed through several different builders, eventually became Endeavourcat 30.
3. Number of units actually built and sailing, unknown

Americat 30

.... *Designer: Robb Ladd*

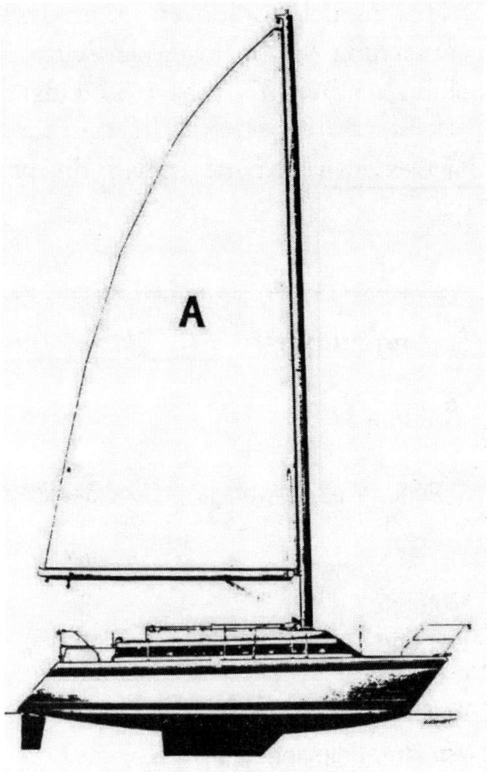

Vital Statistics

Americat 30

LOA (length overall)	30' 0"	9.14 m
LWL (waterline length)	29' 6"	8.99 m
BMAX (maximum beam)	14' 4"	4.37 m
SA (sail area)	430 ft^2	39.9 m^2
Draft: 2' 0"		Keels
Designed displacement:	6300 lbs	2857kg.
Mast hgt above WL:	44' (est)	13.4 m
Tankage:		
Water:	60 gal.	227 liters
Fuel:	50 gal.	189 liters
Auxiliary power:	one	outboard

Type .... Full cabin
Foredeck ..... Solid foredeck
Cabins ...... Two
Galley up/down Down
Rudders ..... Balanced spade
Rig .... Cat

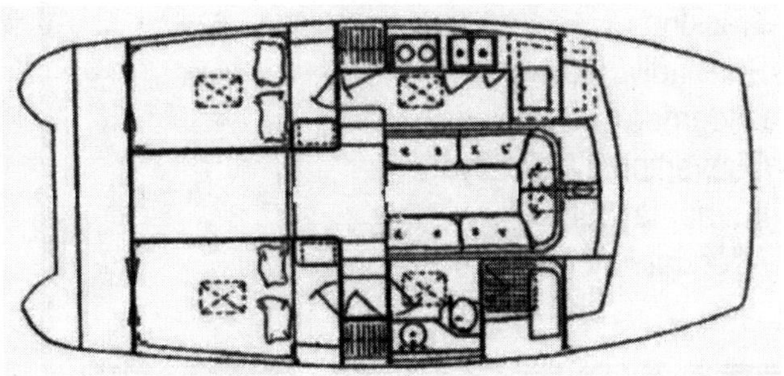

Antigua 37 . Class 5, Cruising

Designer: Michel Joubert/Bernard Nivelt

1. This model no longer in production but the company is doing very well. Very popular manufacturer with many models, circa 2002 building mostly larger boats. Other discontinued models include Maldives 32, Tabago 35, Fidgi 35, Louisiane 37, Cassamance 44.

I have extensive experience with this particular vessel and the entire line of Fountaine Pajot.

Sailing summary

Antigua 37

1=Poor 2=Fair 3=Average 4=Good 5=Exceptional

	1	**2**	**3**	**4**	**5**
Motoring			3		
Docking			3		
Backing			3		
Sailing			3		
Weather helm/lee helm			3		
Windward ability			3		
Ease of tacking				4	
Visibility from the helm				4	
Tracking				4	
Seakindly/ride				4	
Underdeck Slamming				4	
Convenient Deck Layout			3		
Interior layout			3		
Adequate storage areas			3		

Vital Statistics

Antigua 37

LOA (length overall) 37' 0" 11.3 m
LWL (waterline length) 36' 0" 10.9 m
BMAX (maximum beam) 19' 6" 06.0 m
SA (sail area) 738 ft^2 68.6 m^2
Draft: 2' 0" Keels
Designed displacement: 9,918 lbs 4498kg.
Mast hgt above WL: 53' (est) 16.15 m
Tankage:
Water: 62 gal. 280 liters
Fuel: 33 gal. 150 liters
Auxiliary power: Twin 18 hp diesels

Type .... Full cabin
Foredeck ..... Trampoline
Cabins ...... Three
Galley up/down Up
Rudders ..... Balanced spade
Rig .... Fractional sloop

Antigua 37

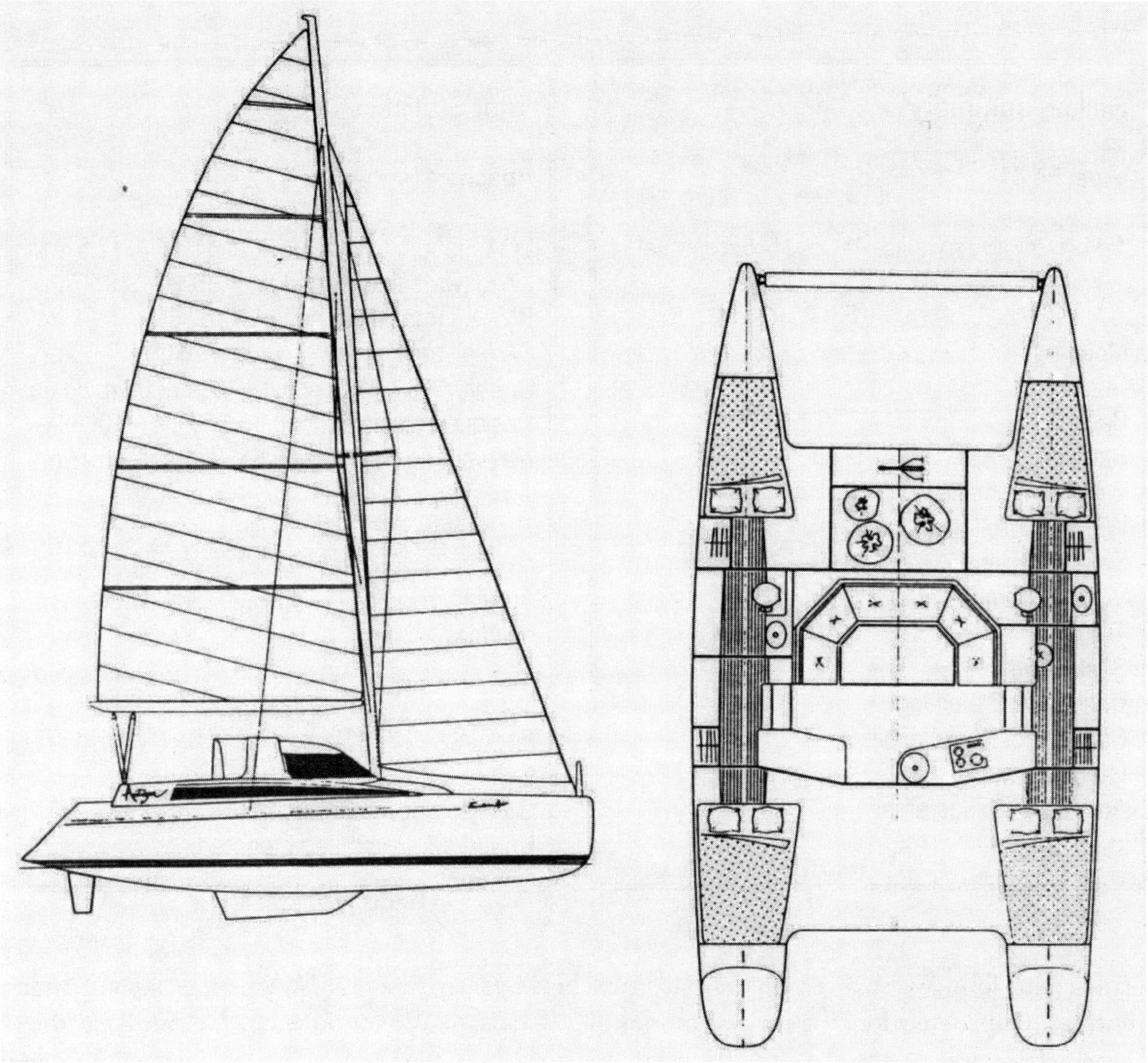

Apache 41 Class 3, Racing/Cruising

Designer: Rod Macalpine-Downey

Sailing summary

Apache 41

1=Poor 2=Fair 3=Average 4=Good 5=Exceptional

	1	**2**	**3**	**4**	**5**
Motoring			3		
Docking			3		
Backing				4	
Sailing				4	
Weather helm/lee helm				4	
Windward ability				4	
Ease of tacking			3		
Visibility from the helm			3		
Tracking			3		
Seakindly/ride				4	
Underdeck Slamming				4	
Convenient Deck Layout		2			
Interior layout		2			
Adequate storage areas			3		

Vital Statistics

Apache 41

LOA (length overall)	41' 0"	12.5 m
LWL (waterline length)	36' 0"	10.97 m
BMAX (maximum beam)	19' 6"	5.94 m
SA (sail area)	710 ft^2	66 m^2
Draft: 3' 0"/.91 m bd dn	6' 9"/2.08 m bd up	
Designed displacement:	13,350 lbs	6,068 kg.
Mast hgt above WL:	55' (est)	16.7 m

Tankage:
Water: gal. liters
Fuel: 6 gal. 22 liters
Auxiliary power: diesel or outboard options

Type .... Full cabin
Foredeck ..... Trampoline
Cabins ...... Three
Galley up/down Down
Rudders ..... Skeg mounted

I sailed with John Walsh a few times on his Apache 41, *Chessypeake Cat*. We used to go out "mono hunting," deliberately looking for hot monohulls to outsail. We found them, we outsailed them. *ChessyPeakeCat* was equipped with outboard motors in wells in the sterns. That arrangement was quite unsatisfactory and later model sported a variety of different drive types including one that had regular inboard/outboard units stuck through the transoms.

Years later, I spent much time with Dave and Loretta Rosenberg, aboard *Boomerang,* an Apache 45. Corinne and I spent most of a Winter season sailing with them, through the Intracoastal, the Keys and the Bahamas. I got to know that vessel very well. Dave had purchased the boat in Holland and sailed it transatlantic to Florida.

Later, when *Boomerang* changed hands and sailed back to Holland, I did the survey for the new owner. Lo and behold, after several years of cruising the Mediterranean it wound up back in Florida again and again I surveyed it for the new owner. Last I saw of it was in 1999 heading up to Chesapeake Bay. Being able to compare the effects of aging on the same vessel first hand is a good learning experience.

I have spent time around two other Apache 41's, but not to the extent of *Boomerang*. To the best of my knowledge, Boomerang now lives in the Chesapeake with its diesel removed and twin Honda outboards installed.

Note: The Apache was originally advertised as a 40. It was quickly changed to 41 to conform with American marketing practice. It was later extended to a 45 by the addition of four feet to the transom and a rear crossbar and trampoline.

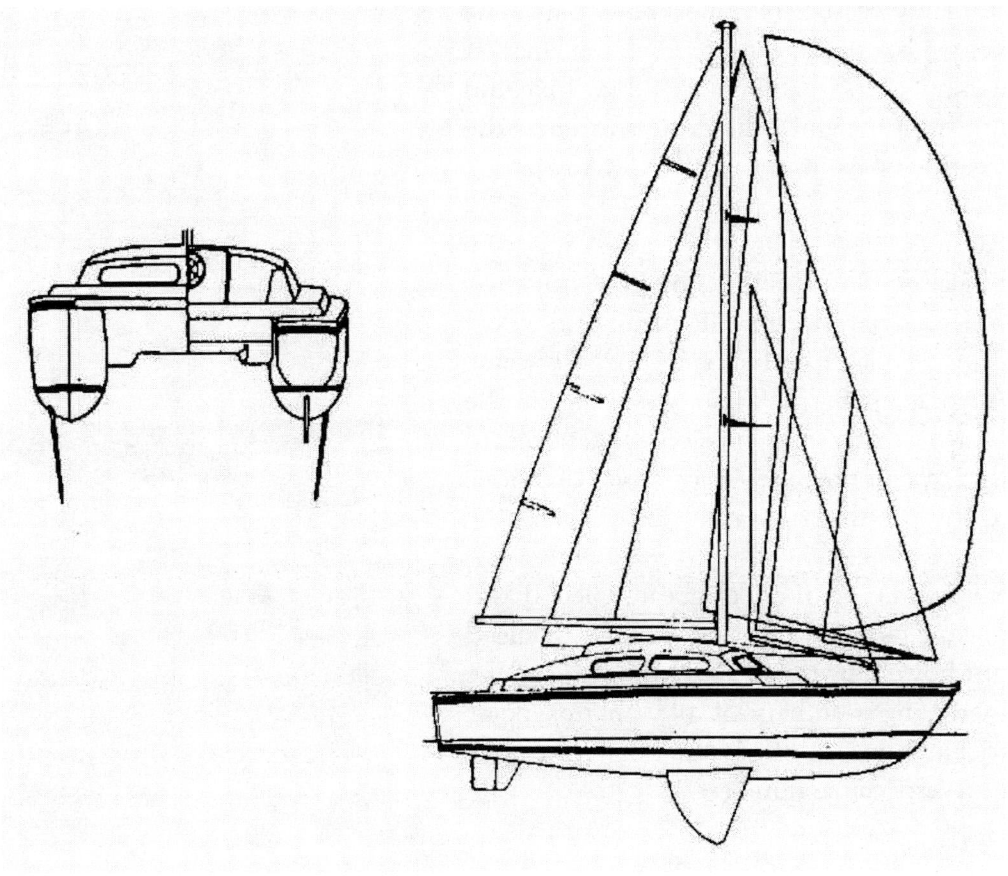

Beneteau

Bleu II (Beneteau 35) Class 4, cruising/racing

Designer: Philippe Briand

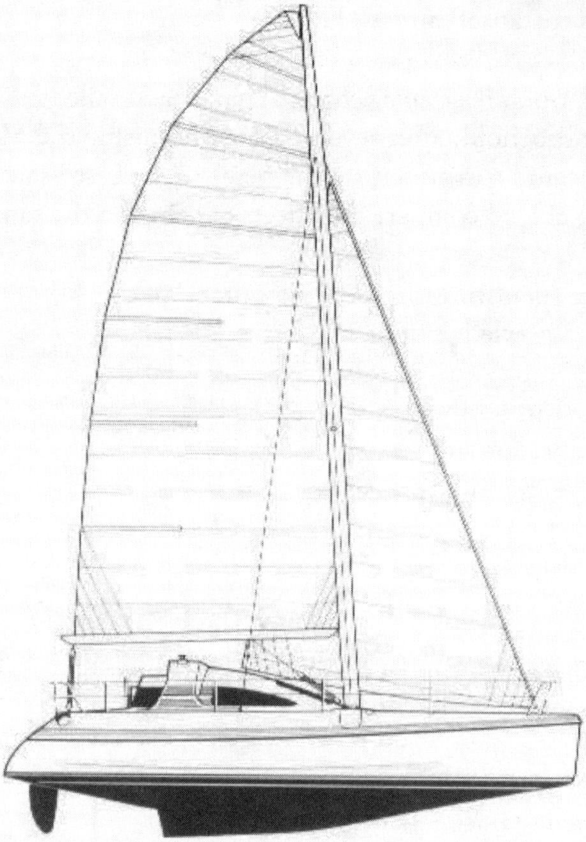

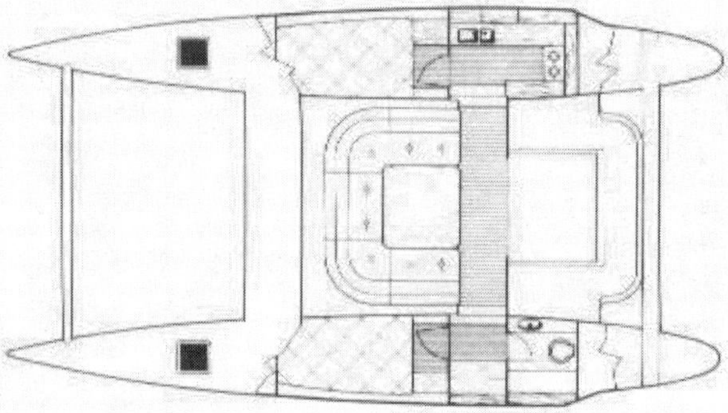

When I first saw this boat in Annapolis about 1986, I was ecstatic. Finally, one of the major manufacturers got into the act! It is gorgeous! Looks like it flew right in out of Star Wars. It had everything, looks, equipment, layout, twin diesels, twin wheels, etc. Then something happened? Marketing suddenly ceased. No one knows why. The boat just disappeared from the market without even an attempt to sell it. We could not even convince the dealers to enter the existing boats in our local CCMA regattas.

Rumors flew that the cause was the recalcitrant local importer of Beneteau boats who was fiercely anti-catamaran and a very powerful player in the local Annapolis scene. At that time, Beneteau was new in the United States and possibly susceptible to that type of pressure. I have never been able to confirm that rumor.

Meanwhile, in the Caribbean, Bleu II was making its mark. Beneteau has not divulged just how many they built but judging by the increasing numbers of old charter boats trickling up into the United States, it must be a respectable number.

In later years, I did manage to get to survey a couple of them as they were imported into the US as used charter boats. Unfortunately, those particular boats suffered very badly from poor maintenance thus giving a distorted view of the boats.

Bleau II is an interesting boat. It has mirror image hulls with little 8 hp Yanmar diesels and the vessel is demountable. It has a fine ergonomic layout for what it is. The twin helm on this boat make good sense since they are in the cockpit and behind the main bulkhead.

Sailing summary

Beneteau 35 Bleu II

1=Poor 2=Fair 3=Average 4=Good 5=Exceptional

	1	**2**	**3**	**4**	**5**
Motoring			3		
Docking			3		
Backing			3		
Sailing				4	
Weather helm/lee helm				4	
Windward ability			3		
Ease of tacking			3		
Visibility from the helm				4	
Tracking				4	
Seakindly/ride				4	
Underdeck Slamming			3		
Convenient Deck Layout			3		
Interior layout			3		
Adequate storage areas		2			

Vital Statistics

Beneteau Bleu II

LOA (length overall)	35' 2"	10.40 m
LWL (waterline length)	30' 0"	08.90 m
BMAX (maximum beam)	20' 0"	05.95 m
SA (sail area)	701 ft^2	65.2 m^2
Draft: 3' 0"	.90 m	Keels
Designed displacement:	7,500 lbs	3402 kg.
Mast hgt above WL:	50' (est)	15.24 m

Tankage:
Water: gal. liters
Fuel: gal. liters
Auxiliary power: Twin 8 hp diesel

Type .... Full cabin
Foredeck ..... Trampoline
Cabins ...... Two
Galley up/down Down
Rudders ..... Balanced spade
Rig .... Fractional sloop

Catalac 8m, 9m, 10m

Designer: J. Winterbotham & T.M. Lack

Catalac – the "Energizer Bunny" of catamarans

Catalac cruising catamarans were in production from the late 1960s until the mid 1980s. More than 500 vessels were built and almost all are still sailing. In fact, there has been a flourishing charter fleet of Catalac catamarans in the Mediterranean since the 1970s. They keep going, and going, and going.

Catalac comes in four sizes: 8 meter (27'), 9 meter (30'), 10 meter (34'), and 12 meter (41'). The most common is the second oldest, the Catalac 27; the fewest units built were the latest model, the 10 meter cat. Catalac ceased production after business owner Tom Lack was seriously injured in an automobile accident.

An interesting insight into the life and times of a sailboat business rests with this vessel. As good as this vessel is, sales in the U.S. were exceptionally difficult. The importer/dealer back in the 1970s had his problems. On behalf of Tom Lack, I spent a considerable time at the Miami Boat Show (which in those days was held in Coconut Grove) observing, investigating, and interviewing. What I found was intriguing.

"I would not buy that boat no matter what the price," exclaimed one show-goer. "It's going to be difficult enough to explain buying a catamaran, let alone a 'Catalac.' How am I ever going to...and why should I have to...explain to my fellow club members that 'Catalac' is not made by General Motors or that 'Catalac' is not a play on words by some sleazy operator," he added emphatically.

What I had discovered – the name confusion – I believe not only prevented Lack from being the premier cruising sailboat seller in the U.S., but also hurt the acceptance of cruising catamarans in general. It was more than a decade later that Eric Martel and his enthusiasm for the French catamarans created the impetus for the sales growth in today's multihull market.

I believe that the 27 and 30 are among the best cruising catamarans ever produced. Not only are they strong, sturdy, and sea- kindly, they have superb ergonomic layouts.

My opinion is based on sailing a 27-footer 1,000 miles to windward from Florida to St. Thomas, USVI. It was the boat I single-handed the boat from Annapolis to Florida, then through the Bahamas. I also sailed and surveyed more than a dozen others.

Besides the exceptional layout, what I like about the boat is its construction and clever use of space. I have not found a 27 or 30 with any hull failure due to building or scantling difficulties. However, that does not mean things are perfect. The older models had transom hung kick-up rudders. The rudder blades were made out of flat plate metal. Since they had no chord, they had no lift and, at speed, they ventilated badly.

The rudder heads and tiller bar were made from mild steel. These rusted badly in a few years and created an aesthetic disaster aside from the obvious safety problem.

The chintzy locker hinges were made out of some kind of aluminum alloy. If they didn't break, the hinge pins would migrate out, leaving you with two pieces and a pin if you were lucky enough to retrieve the parts before they found their way overboard.

Thinking of buying a Catalac and zipping along at double-digit speed? Forget it! These boats are definitely not speedsters and windward ability is poor. On the other hand, they are so stable and safe that years ago Lack authorized a £10,000 reward to anybody that could document a Catalac with one hull out of the water. To date, no one has claimed that reward.

The boat will carry heavy loads far better than many others due to a wider hull-beam ratio, deep rocker design extreme deadrise hulls with hard chines, and big, powerboat- like transoms. They also handle extremely well. Unlike others of that generation, they tack securely without backwinding the jib and handle smartly around docks, easily turning in their own length – especially those with the outboard connected to the steering system.

Another of the features I really like are the windows. They are not only the ones that give the least problems from leaking, but they also have the best ventilation arrangements. The windows tilt back to allow ventilation in the rain or can be completely removed for maximum airflow.

The essential differences between the 27 and the 30 are the interior and deck layout – the hulls are the same. One compelling advantage of the 30 is the removable bulkhead between the forward cabin bunks which creates an enormous cabin with a king size bed and individual dressing areas and hanging lockers. The aft head on the 30 is also more convenient then the forward head on the 27.

These boats have been built with several engine combinations. Originally, they were powered with Viere two-cycle engines that ran in either direction. For reverse, you stopped the engine started it again in the other direction. They were the essence of simplicity with the fewest number of parts and lightest weight of any engine. However, the American market would not accept them and most were replaced by twin eight horsepower Yanmar diesels. Outboards were an option on all 8m and 9m models. Most 8m Catalacs are outboard powered and most 9m are twin diesel.

Sailing summary

Catalac 8m (27)

1=Poor 2=Fair 3=Average 4=Good 5=Exceptional

	1	2	3	4	5
Motoring				4	
Docking				4	
Backing				4	
Sailing			3		
Weather helm/lee helm			3		
Windward ability		2			
Ease of tacking				4	
Visibility from the helm				4	
Tracking				4	
Seakindly/ride				4	
Underdeck Slamming				4	
Convenient Deck Layout					5
Interior layout			3		
Adequate storage areas				4	

Vital Statistics

Catalac 8m (27)

LOA (length overall)	27' 0"	8.23 m
LWL (waterline length)	25' 6"	7.77 m
BMAX (maximum beam)	13' 8"	4.22 m
SA (sail area)	275 ft^2	25.5 m^2
Draft: 2' 0" / 0.67 m		Keels
Designed displacement:	6,283 lbs	2850kg.
Mast hgt above WL:	39' (est)	11.89 m

Tankage:
- Water: 70 gal. — 265 liters
- Fuel: 20 gal. — 76 liters

Auxiliary power: Outboard or twin diesel

Type . . . . Full cabin
Foredeck . . . . . Solid full foredeck
Cabins . . . . . . Two
Galley up/down Down
Rudders . . . . . Transom hung, kick-up
Rig . . . . Masthead sloop

Catalac 9m (30)

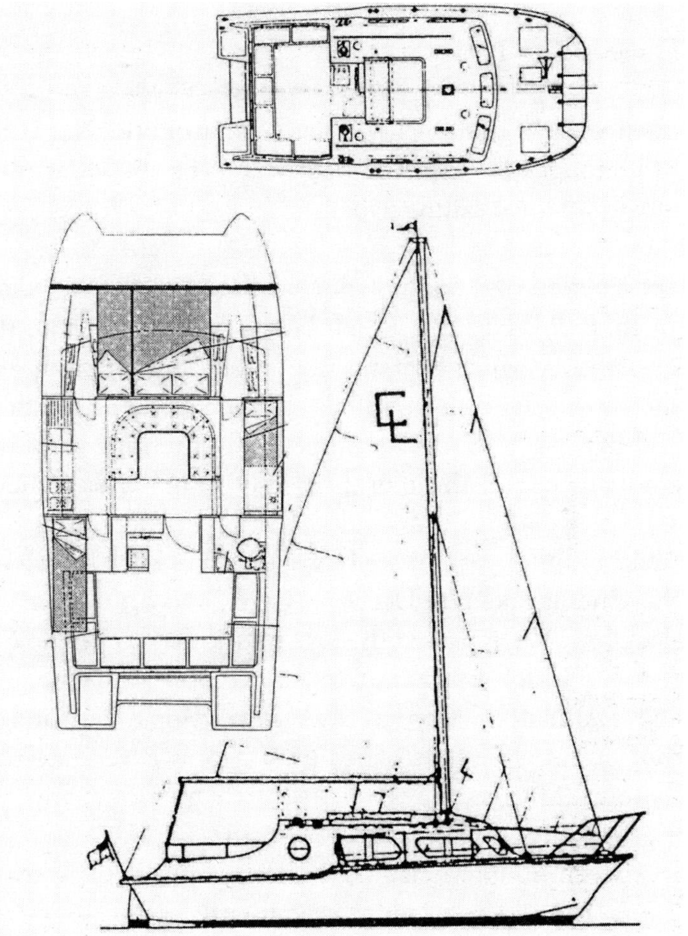

I surveyed over a dozen of these boats, some several times at change of ownership. I am impressed by the ongoing quality and endurance of these vessels. One particular boat has crossed the Atlantic three times. I have also sailed extensively on them having delivered two and performed sea trials on several others.

The interior layout in my opinion, is the best of the catamarans in that size range. Most owners remove the panel which separates the forward into two cabins, thus creating a king-sized bunk with his and hers sides. The created stateroom in addition to a huge bed has separate lockers for each side, good forward visibility and good ventilation.

The galley-down arrangement has lots of counter space. Many of the boats I surveyed used the aft cabin as a pantry or storeroom.

The head being in the starboard aft quarter makes it very convenient for all users without the cross traffic so prevalent on small boats. Most have twin diesels but some are outboard powered.

This is a vessel that really can carry a load but do not expect it to sail fast or go to windward well.

This is the original Catalac. The 8m has the same hull, just a different deck molding and interior arrangement. Being lighter, the 8m is also a little faster. However, the 9m has a far better layout. Because of the similarity between the vessels and their age and availability at this writing, I am not going to detail this vessel completely.

Catalac 41 (12m)

Catalac 12m (41) Class 5, Cruising

Designer: Tom Lack

I delivered a Catalac 41, *Turning Point*, from the North Shore of Long Island, to Annapolis, Maryland in October of 1987. The boat was to be shown at the Annapolis Boat Show. As expected, it was mostly a motor sailing trip.

For a period of time through New York City, we sailed alongside the beautiful fifty-five foot gold plate ketch: *Pacific High.* As we left New York Harbor, into the prevailing southwesterlies and the swell, it was obvious which boat was not only faster, but more seakindly. *Pacific High* would lunge forward off the waves, pounding and taking green water across the decks. Turning Point, just loped along nice and dry, no underdeck slamming, no lurching or rolling, just a steady upright forward motion. We gradually pulled ahead of them and in three hours they were hull down over the horizon.

At 12 minutes past midnight we left Cape May, New Jersey for the trip up the Delaware Bay, through the Chesapeake and Delaware Canal and on to Annapolis, Maryland, for the sailboat show. It was slack low. We followed the tide up the Bay to The C & D Canal, which connects the Delaware Bay to the Chesapeake Bay and continued down the Chesapeake to my home on the Magothy River, just above Annapolis, Maryland. We were trying to outrun hurricane Gloria, which was coming at us from the South. As we approached the Magothy River entrance, we could see the storm coming up the Bay. By four PM that day, we were securely hurricane anchored off my dock on the river.

In sixteen hours we had covered one hundred and forty odd miles, meaning that we averaged 8.75 knots. I would consider that eminently satisfactory considering the boat and the conditions. Tom Linton, champion Shark sailor and my crew for the trip, had mixed emotions. There was little "feel" of sailing, but for those who truly relish comfort and convenience, this is the boat.

After the hurricane, I used Turning Point's monster size center mounted three speed winch to get all my anchors back up out of the mud, but that is another story.

The twin screw configuration and the good visibility from the cockpit combine to make this a really handy boat around docks. With no effort at all you can put this boat anyplace you like.

I have had some input into the Catalac 41, *Allison II*, which was developed for my good friend and client, Dr. Robert Holst. That Catalac 41 was to be laid out and equipped in such a fashion that all systems were to be manual, repairable anywhere and that the engine drive would be completely removable and serviceable without haulout. His intention was to make cruising in remote regions without any yacht facilities a way of life and his vessel had jolly well better be prepared.

The Catalac 12m has an inside steering station.

A sistership, *Heart Light*, Originally owned by Hal McGinnes, survived the infamous Queen's Birthday typhoon. (A story in itself)

Sailing summary

Catalac 12m (41)

1=Poor 2=Fair 3=Average 4=Good 5=Exceptional

	1	2	3	4	5
Motoring				4	
Docking					5
Backing			3		
Sailing			3		
Weather helm/lee helm			3		
Windward ability		2			
Ease of tacking			3		
Visibility from the helm			3		
Tracking			3		
Seakindly/ride				4	
Underdeck Slamming			3		
Convenient Deck Layout			3		
Interior layout			3		
Adequate storage areas					5

Vital Statistics

Catalac 12m (41)

LOA (length overall)	40' 10"	12.45 m
LWL (waterline length)	36' 10"	10.97 m
BMAX (maximum beam)	17' 6"	5.26 m
SA (sail area)	685 ft^2	61 m^2
Draft: 3' 3" / .99m		Keels
Designed displacement:	18,500lbs	8392kg.
Mast hgt above WL:	46' 2" (est)	14 m
Tankage:		
Water:	120 gal.	542 liters
Fuel:	96 gal.	363 liters
Auxiliary power:	Twin 30 hp diesels	

Type .... Full cabin
Foredeck ..... Full solid bridgedeck
Cabins ...... Five
Galley up/down Down
Rudders ..... Skeg mounted
Rig .... Masthead sloop

Catalac 41 (12m)

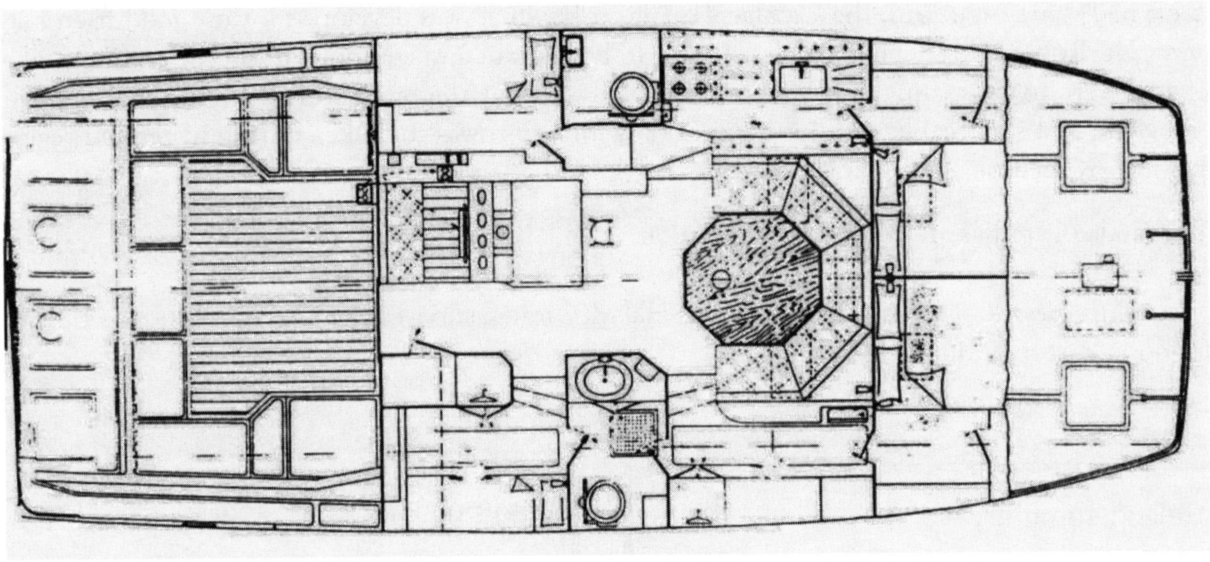

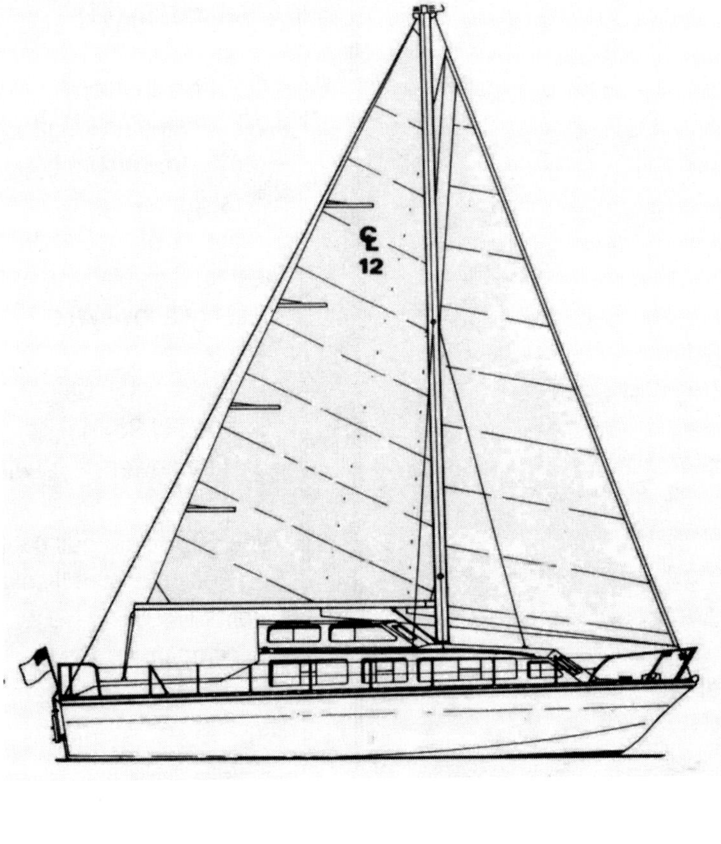

Catfisher 28 Class 5, Cruising

Designer: Terry Compton

This is a most interesting vessel. For a considerable period it was the most luxurious of all cruising catamarans. Despite its diminutive length, it has the mass, the volume, the deck area and the feeling of a much larger boat. Under power, with its twin screws, it is the equal or better of any powerboat at maneuvering and docking. Under sail, it's a little different story. (Some had other power options)

On one delivery, I managed to get a ketch rigged 28 footer up to seven and a half knots on the speedometer. (No GPS was available at that time) That is faster than any comparable monohull with a 24-foot waterline, especially the other Fisher yachts in that size range. It is probably 20 to 30 percent faster than the speed achievable from the comparable monohull Fisher, under the same conditions. These conditions were, broad reaching in 20 knot winds from the Northwest in modest waves. I was sailing along the coast of New Jersey offshore. I have since spoken to other Catfisher owners who claim satisfaction with the performance under sail. I certainly was not satisfied.

Later model changes included additional length to 32 feet, mostly to stop the excessive hobby-horsing of the 28 footers. The short waterline, high center of gravity and canoe stern all contributed to the hobby-horsing, but changing the stern shape and the waterline length seemed to do the trick.

The standard boat came with a double main sheet arrangement, which is simple, expedient, and has a minimum of both gear and holes drilled through decks and bulkheads. This arrangement removes the need for vangs, preventers and travelers. However, it is a miserable arrangement to try to tack! I have never heard anyone complain. For that matter, I have heard no comment at all. I suspect that says something about how people sail the boat or how much the boat is sailed. Actually, Catana uses that arrangement on some models.

Based on the present, circa 2002 cruising boat demand and cruising environment, this is one vessel I would really like to see back in production.

Catfisher 28

Sailing summary

Catfisher 28

1=Poor 2=Fair 3=Average 4=Good 5=Exceptional

	1	2	3	4	5
Motoring				4	
Docking				4	
Backing				4	
Sailing		2			
Weather helm/lee helm			3		
Windward ability		2			
Ease of tacking		2			
Visibility from the helm			3		
Tracking			3		
Seakindly/ride			3		
Underdeck Slamming				4	
Convenient Deck Layout					5
Interior layout					5
Adequate storage areas				4	

Vital Statistics

Catfisher 28

LOA (length overall)	28' 0"	8.53 m
LWL (waterline length)	25' ft	7.65 m
BMAX (maximum beam)	13' 1"	04.2 m
SA (sail area)	460 ft^2	47.3 m^2
Draft: 3' 1"		Keels
Designed displacement:	8,000 lbs	3629 kg.
Mast hgt above WL:	44' (est)	13.4 m
Tankage:		
Water:	84 gal.	318 liters
Fuel:	48 gal.	174 liters

Auxiliary power: Twin hydraulic drive
Type . . . . Full cabin
Foredeck . . . . . Full solid foredeck
Cabins . . . . . . Two
Galley up/down Down
Rudders . . . . . Skeg mounted
Rig . . . . Masthead sloop/Ketch option

1. The company is no longer in production, however, the company had an excellent reputation for fine craftsmanship during its production years.

2. There is an inside steering station cleverly incorporated in this vessel. It is galley-down with two large double cabins and one single.

3. There were alternate power and rig choices. The several that I delivered were ketch rigged. More than likely, the sloop rig will sail better.

Catfisher 28

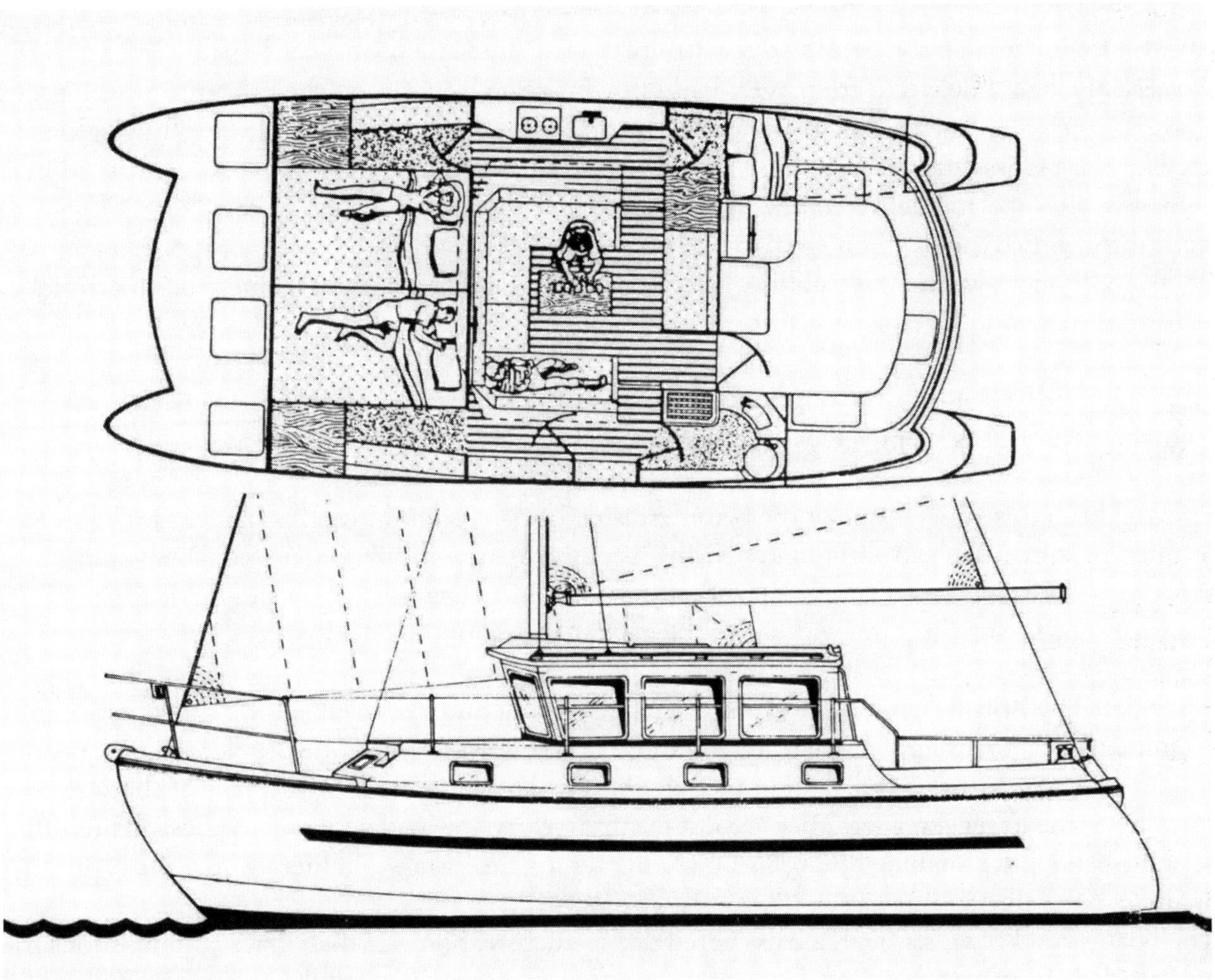

Cherokee 35

Cherokee 35 Class 5, Cruising

Designer: Rod Mcalpine-Downey

It is not too often a person gets the opportunity to view the same boats over a span of many years. *Panache,* the boat I delivered from West Palm Beach to St. Croix sits on the broker's dock, for sale. *Pywacket,* the Cherokee I sailed in companion with, for most of two cruising seasons throughout the U.S. East Coast and the Bahamas, I surveyed for the third time just two weeks ago, sailed in company with one and delivered one from West Palm Beach to St. Croix, USVI. To recap, we went through two gales and one hurricane on the way down. When the winds dropped below 30 knots, we thought we were in paradise. Needless to say, we did a lot of motoring, but we arrived alive with tremendous respect for that vessel.

Now, 12 years later, I have the opportunity to survey the same vessel after its many trials and tribulations as well as keeping tabs on several other Cherokee including *Susan M. Jackson.*

Frankly, I feel this vessel is one of the better cruising boats ever built. It has sufficient hull volume to carry an enormous load without destroying its performance. It sails better than you would expect a boat of this size to sail. The fact that it is only thirty five feet long belies the true size of the vessel. To be sure, it is not without problems, but it is a lot of boat also.

It is palatial by British standards, with most models having four real double cabins, two heads and an immense saloon/galley/nav station area and a big cockpit. There is one problem that owners should be aware of that is the nature of the double spreader rig. This rig has a pair of check stays running to the lower spreader. They are not meant to carry any load while sailing, just to limit the forward bend and or dampen pumping. Over tightening these stays will literally tear the boat apart.

The boat is solid fiberglass with a balsa cored deck structure. Some of these have gotten serious rot and needed extensive rebuilding. Get a good surveyor to go over it carefully. Not all the bulkheads are fully tabbed into the structure which is a weakness that was corrected in later models.

The cockpit, while roomy is also very deep and in a three stair step arrangement which is not particularly convenient and a requires considerable climbing around.

Many of these boats also had twin hydraulic drives powered by a single engine in the cockpit. At the historic period these were built it was a clever way to save weight and expense. However. The test of time has not been kind to the system and one by one they are being replaced with either outboards or twin sail drives. The engine being in the cockpit gave superb access for servicing. Other items were cleverly located as well. The fuel tank is in a compartment just forward of the mast and in front of the saloon curved seat bulkhead.

As with most of the boats from this era, it is built like a tank and is immensely strong. For those of you looking for a sturdy cruising boat in that size range take a look and see if it fits your needs. Some of the things to look out for are corrosion in the forward crossbeam and leakage in the hydraulic systems.

There are a couple of problems, endemic to all the Sailcraft boats. First, are the windows. They are chronic leakers and if you leave them open two things happen: a) you crack your shins on them and, b) you will tear them off with the jib sheets. Second, the deck and cockpit layout is not particularly handy, nor are the sheeting arrangements. Because of the stair-step arrangement of the deck, clambering around at sea is difficult as there are no handholds or balancing points. However, many of these boats have been retrofit with fixed acrylic ports and have had centerboards removed and replaced with fin keels.

The early model Cherokee had a single, pivoting centerboard, hoisted from a shin-smashing winch in the cockpit. This arrangement proved a chronic headache and was replaced by low aspect ratio keels on most of the boats and all the later models. It has skeg hung transom mounted rudders of incredible strength. The skegs go all the way through the hull to the top deck for support. A very impressive strong way to handle the skegs.

Cherokee 35

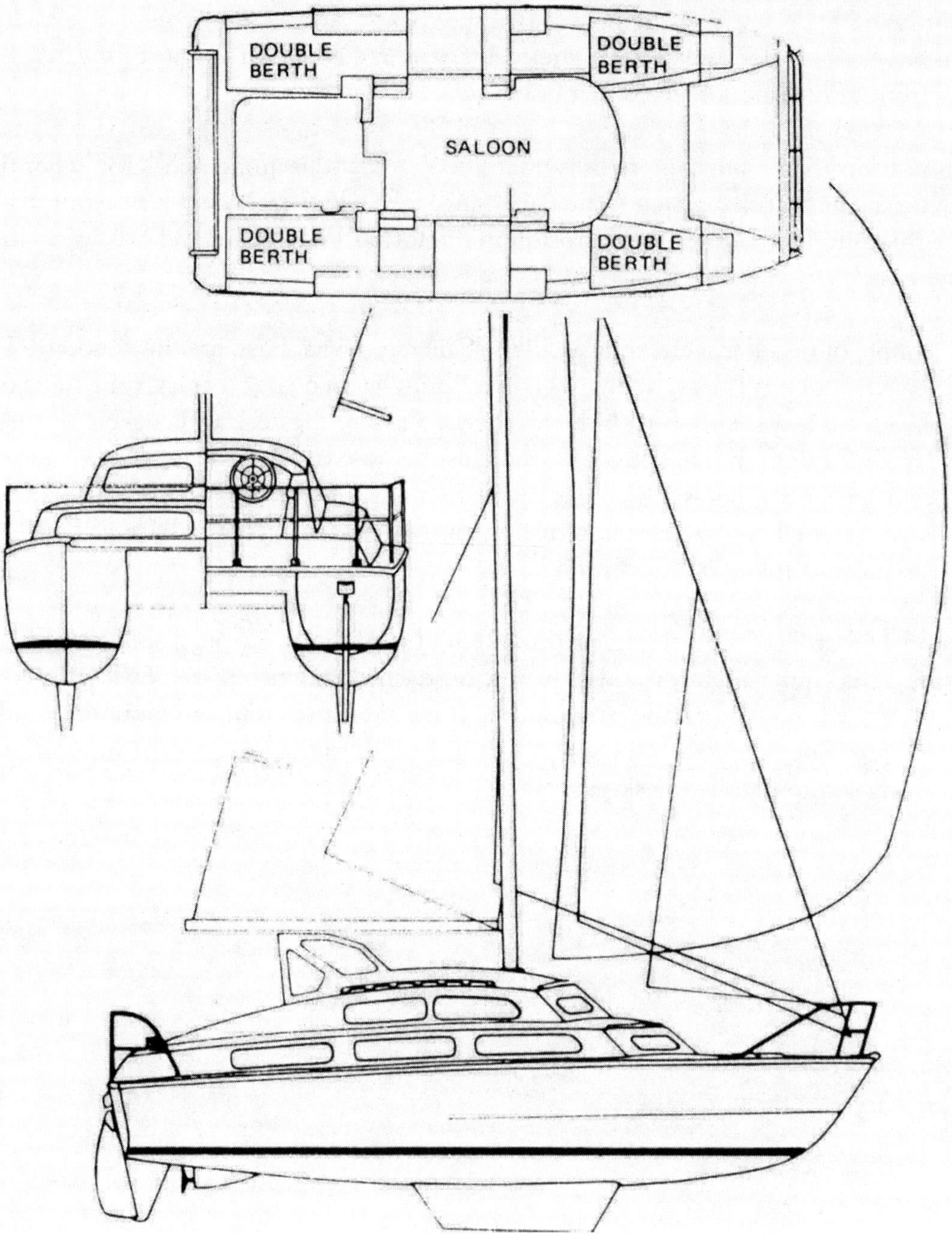
DOUBLE
BERTH
DOUBLE
BERTH
SALOON
DOUBLE
BERTH
DOUBLE
BERTH

Sailing summary

Cherokee 35

1=Poor 2=Fair 3=Average 4=Good 5=Exceptional

	1	**2**	**3**	**4**	**5**
Motoring				4	
Docking					5
Backing				4	
Sailing			3		
Weather helm/lee helm			3		
Windward ability			3		
Ease of tacking		2			
Visibility from the helm		2			
Tracking				4	
Seakindly/ride				4	
Underdeck Slamming				4	
Convenient Deck Layout		2			
Interior layout				4	
Adequate storage areas			3		

Vital Statistics

Cherokee 35

LOA (length overall)	35' 0"	10.67 m
LWL (waterline length)	31' 0"	09.45 m
BMAX (maximum beam)	16' 6"	05.3 m
SA (sail area)	615 ft^2	57.13 m^2
Draft: 3' 6" / 1.07 m		Keels
Designed displacement:	11,122 lbs	5,045 kg.
Mast hgt above WL:	51' (est)	15.5 m
Tankage:		
Water: 120 gal.	454 liters	
Fuel: 90 gal.	340 liters	

Auxiliary power: Std. Thornycraft or Perkins w/twin hydraulic, drive various options

Type .... Full cabin
Foredeck ..... Trampoline
Cabins ...... Four
Galley up/down Up
Rudders ..... Skeg hung

A January Bash To Windward,

Saga of the Cherokee 35, *Panache.*

New Year's Day, 1987 dawned sans hangovers. Fred Feldman, my right-hand man; Pat Field of the 30-foot Catalac, *Camelot;* and fellow Upper Keys Sailing Club member and *Southerly* owner, Lee Mielke, were all assembled at the boat we were to deliver to St. Thomas, U.S. Virgin Islands: a Cherokee 35 catamaran named *Panache.*

Corinne and Miriam (Pat's wife), were busy provisioning the boat. Corinne, my wife of 32 years and heroine of countless sailing sagas, had politely-but-firmly told me to "go fly a kite," when I had tried to get her to go on this trip with me. Seems the female mind has a better memory for hardship and travail than my feeble brain. She remembered quite vividly the difficulty of pounding 1000 miles to windward in winds that usually averaged 30 knots. The wives left us on Friday, we left early Saturday. You must realize I am not at all superstitious. That old story about not leaving on a Friday means nothing to me. But, just in case, we left on Saturday.

The forecast touted "Northeast winds 5-10 knots." Usually, you don't cross the Gulf Stream in any kind of norther, even a light one. Since we had a schedule to keep and wanted to take advantage of the north-westerlies that were forecast for the following days, we left West Palm Beach headed for Northwest Providence Channel, the passage through the Bahama Islands.

Surprisingly, our Gulf Stream crossing was uneventful. There was only one small patch of large waves, somewhere around mid-stream. The usual horror of crossing in a norther was not to be found. By 8 a.m. the next morning we were abeam of the Berry Islands and impatiently awaiting our north-westerlies to drive us as far east as possible before starting the inevitable slog to windward when we hit the trade winds.

Generally, winds had been light and the favorable (for us, that is) forecast kept us motoring along, using up fuel we had not planned on using. I'd intended replacing the fuel we were using somewhere in the eastern Bahamas in order to have enough to insure getting to the Virgin Islands without a problem. San Salvador, being farthest east, was our primary destination for this purpose.

Suddenly the wind increased and backed to the south. I shrugged. "Certainly unusual to have southerlies this time of year," I speculated. "Probably won't last long," I continued. Then things got worse. We started to see 40 and 50 knot gusts on our wind speed indicator. I grumbled about the deteriorating weather and changed course for Nassau, the only possible shelter we could attain at night, under storm conditions. With the wind coming from the south, we had some protection from the chain of islands and reefs that run from Nassau towards the northeast, ending at the tip of Eleuthera. Conditions got even worse, the wind backed further until it was southwest, removing any lee and any chance of fetching Nassau. Gusts started to go over 60.

As we changed course to the east, a particularly vicious gust caught the edge of the roller-furling headsail and ripped it. With each succeeding gust it ripped more. There was not even the slightest chance of getting the sail off the headstay, since it was a luff-groove type and would have meant unfurling the sail completely in order to lower it. With mountainous waves, pitch black darkness, and winds now gusting to 70 knots, there was not a chance.

Now we really had problems. There was no way of running before the storm because we had Great Abaco Island and its assorted reefs and shoals directly to leeward. The only possibility was to continue on to the east and try to get into the lee of Eleuthera. The storm simply did not want to abate. Gusts of 70 knots were interspersed with sustained winds in the sixties. We inched forward under bare poles and engine, about two or three knots and 60 or 70 degrees off the wind and waves. Some-time just before dawn, an enormous wave, at least twice the size of the others and sporting a double-crested breaking head, smashed into us from directly abeam. Part of it broke into the cockpit and ricocheted around the coamings until it found its way into the open door and soaked the carpet in the entranceway. About that time, I began to recap my various deep sea storm experience and reaffirm my concepts of the seakindliness of cruising catamarans. There is no question in my mind that had I been on any of the monohulls I have sailed on thus far, that wave would have totally overwhelmed the boat putting it completely underwater. If hatches had been open it would have sunk like a stone and if crew were not fastened to the boat with life harnesses, they would have been swept off like so many bowling pins. There is also no question in my mind that had I been on any of the trimarans I have sailed, the boat would have capsized. Yet here we were, except for the minor annoyance of a wet carpet and a couple of loosened secondary bulkheads, we were practically comfortable!

Peak winds dropped to the sixties as we rounded the northeast cape of Eleuthera. We did attain sufficient lee to feel more moderate wave action. Hopefully, we could fetch Little San Salvador off Cat Island before night fall in order to anchor and remove the remnants of the roller-furling jib that was now streaming out behind in long chevrons and shaking the headstay the way a cat shakes a mouse.

Suddenly, there was a sharp cracking sound, like the snap of a 22 rifle. I rushed up on deck and found a small screw had hit the windshield. Puzzled, at first, about where it had come from, I suddenly realized that it was from the sections of the roller-furling headstay which was now well into the process of self destruction. Then it began raining screws, most of which we miraculously caught either on the foredeck or on the dinghy that was turned upside-down and lashed to the trampoline.

On the horizon was the tip of Cat Island. ETA around the point and into a real lee was 14:00. It was dead downwind, the wind was still in the fifties and it had been blowing for 36 hours and by then had veered to the northwest. I was very apprehensive at the possibility of having huge breaking waves as we crossed onto soundings. With crew in life jackets and ready for anything, we crossed from the ocean depths to the light-blue waters of the island shallows. There was no surf at the line and we quickly motored around to Bennet's Harbor, snugged up to within 100 feet of the beach, dropped two anchors into deep, clear sand and, for the first time in four days, stopped moving.

We spent a day cutting loose the remnants of the genny and dismantling the roller-furling headstay. We had an old number 2 jib that we hanked on the headstay and a storm jib that we used on the baby stay to make a sort of cutter rig.

Hawks Nest Marina was just a day's sail down island but from a quick radio conversation we learned they had no diesel fuel and didn't expect any deliveries until the weekend; not particularly unusual in the Bahamas, last year, the Post Office in Georgetown was out of stamps!

I found that the brand-new fueling depot at New Bight, about halfway to Hawks Nest, was also out of fuel but could get me some pumped by hand. I spent most of the morning ferrying fuel back and forth by dinghy, in five-gallon containers, while Fred, Pat and Lee made their way up to the Hermitage, a miniature monastery and an attraction not to be missed.

Off before dawn, my usual time, trying to make easting in the dying frontal winds. We had little luck because the winds filled in from the northeast putting us right back on a beat.

The No.2 jib we were now using was almost worthless. Not only was it too small to be effective, but it was misshapen and difficult to lead properly, so we wound up motor-sailing the rest of the trip under engine and main. Winds were constantly from the east, when they dropped into the low 30's and high 20's, we thought of them as 'light.'

Easting without a southerly wind component is tough. Most of the time we were basically sailing southeast hoping for the winds to shift from due east to northeast as they had on my last trip in order for us to sail due east which becomes ever more important the farther south one gets.

We stopped for a few hours and had a swim and a good meal in the lee of Crooked Island and the Bird Island Lighthouse, then pressed on in favorable conditions to Cockburn Harbor, South Caicos, where we tied to Cox's fuel dock for the night. It was in South Caicos where we learned that the storm we had experienced had been upgraded to a hurricane.

Finally, after all these days we had some westerly winds. They were light, to be sure, the end of a dying front, but enough to get us past most of the Dominican Republic.

The next gale caught us with vengeful fury right before Cabo Cabron at the eastern end of Hispaniola (Dominican Republic) Forty knots and above from 90 degrees, when you need a course of 110 degrees was simply more than we could handle so before dark we inched our way into Porto Escondido (on the north shore of the Dominican Republic) to seek temporary shelter.

Porto Escondido is simply a dent in the otherwise sheer cliffs tumbling to the sea. If you hold up two fingers in the "Vee" sign, you have the configuration. It is nestled between Cabo Cabron on the west and Cabo Samana on the east. It is, at most, one mile across and one mile deep, completely open to the north and west but absolutely protected from the east. Never, either in the Atlantic or Pacific, have I seen a place with such beauty. It has a magnificent crescent beach ringed with coconut palms, steep valleys and high mountain peaks cascading to the beach. It is totally wild. While there were a few small houses, there was apparently no electricity for we did not see a single light. However, we did see deer walking the beach, whales surfacing almost next to us and sea turtles swimming by.

The following day, not really rested at all because the surge was so great we couldn't sleep, we tried to leave but the winds were still blowing over 40 knots. We returned and anchored for yet another day. Not a chance of getting ashore, the surf looked like the classic 'Pipeline. Next day, with winds reduced to the high 20's, we left for Puerto Rico.

Always the most miserable part of the trip, we were now just slogging to windward into the prevailing easterlies making about 4 knots over the bottom. To add to our already complicated life, the U.S. Coast Guard showed up right in the middle of the Mona passage.

The Coasties probably would have boarded us except for the fact that the seas were running about eight feet and all hands on their boat looked a little green, so they tagged along behind us for an hour or so asking innumerable questions. Actually, I guess, I was kind of glad they were around.

Two days later, we were battling our way into the lee of Puerto Rico. Out of fuel, out of water, out of patience, we closed the sandy shore of the little town of Aguadilla, PR, ran down the beach far enough to achieve a good lee and in about 15 feet of water over clear sand, dropped the hook and went to sleep.

I was the first up the next morning and to my astonishment, there was the Marine Police circling our boat. I waved them over and told them of our need for fuel. They couldn't have been nicer or more cooperative. They led us down the beach to the local launching ramp where they proceeded to organize the entire police department on our behalf. Soon the Chief appeared with a 55 gallon drum and a pump. He whisked me away by truck to the local garage where we filled up, then went back to the ramp where he pumped fuel from drum to assorted containers, while local fishermen ferried them back and forth and the Marine Police helped Fred pour the fuel into the tanks.

Meantime, Pat and Lee were being entertained in the local bar (the sign said "Joe's Coffee House") by enthusiastic residents who kept buying them drinks. Later, locals hosted us all, merriment and camaraderie lasted well into the night. What a great place, what super-nice people. The highlight of the day was the mayor steering the boat for a photo op.

I opted to go around the south shore of Puerto Rico finally having heard a forecast of northeast winds. That way, we could be in the lee of Puerto Rico for a couple of days then beat across to St. Croix, rather than to St. Thomas which was the preferred destination anyway.

Forget it. The north-easterlies never materialized. We slogged directly into easterlies, but behind the southern fringing reefs the entire length of Puerto Rico as far as Roosevelt Roads, where we ran into yet another gale.

It took us two days to beat our way past Punta Tuna, Ensenada Honda, Vieques and Culebra. We were treated to a spectacular display of bombing practice on Vieques as we went by and, as luck would have it, one of the Air Force pilots was none other than Captain Bobby 'D' of the Upper Keys Sailing Club, one of our sailing buddies.

As has happened to me before in this same spot, the wind moderated and the seas flattened on the 30-odd-mile stretch between Puerto Rico and St. Thomas. At 20 minutes past midnight on January 21, we tied to the fuel dock in Yacht Haven, Charlotte Amalie, St. Thomas USVI. We had been gone 18 days, experienced one hurricane, two gales, and stopped in three countries.

It was good to be back. In the USA. The old Cherokee had delivered us safe and sound.

The epilogue to this is, "Don't go to the Virgin Islands from Florida in January!"

Edel Cat 35 - (Standard) Class 4, cruising, racing

Designer: Sylvestre Langevin, Marc Edel

Sailing summary

Edel Cat 35 (standard)

1=Poor 2=Fair 3=Average 4=Good 5=Exceptional

	1	2	3	4	5
Motoring			3		
Docking			3		
Backing			3		
Sailing					5
Weather helm/lee helm				4	
Windward ability				4	
Ease of tacking				4	
Visibility from the helm					5
Tracking				4	
Seakindly/ride			3		
Underdeck Slamming			3		
Convenient Deck Layout				4	
Interior layout				4	
Adequate storage areas			3		

Vital Statistics

Edel Cat 35 (standard)

LOA (length overall)	35' 0"	10.6 m
LWL (waterline length)	33' 0"	10.0 m
BMAX (maximum beam)	20' 0"	6.04 m
SA (sail area)	250 ft^2	65 m^2
Draft: 2' 10"		Keels
Designed displacement:	1,875 lbs	850 kg.
Mast hgt above WL:	43' (est)	16.15 m
Tankage:		
Water:	80 gal.	300 liters
Fuel:	26 gal.	100 liters

Auxiliary power: Outboard or diesel options

Type .... Central cabin
Foredeck ..... Trampoline
Cabins ...... Four
Galley up/down Down
Rudders ..... Skeg mounted
Rig .... Fractional sloop

1. Good choice for a performance oriented cruising boat. Lots of privacy.
2. Good deck layout, easy to get around, good choice with small children, enclosed walkways.
3. Demountable. Can be disassembled for shipment.
4. Rugged construction, skeg mounted rudders.
5. Tiller steering, some wheels have been fitted.
6. Central cabin is small, but well laid out.

7. Some complaints of underdeck slamming with cabin models.

8. Variety of power options. Single or dual outboard, twin diesels. (Last known information, Edel was not in production)

Edel 35 Cabin model (Standard)

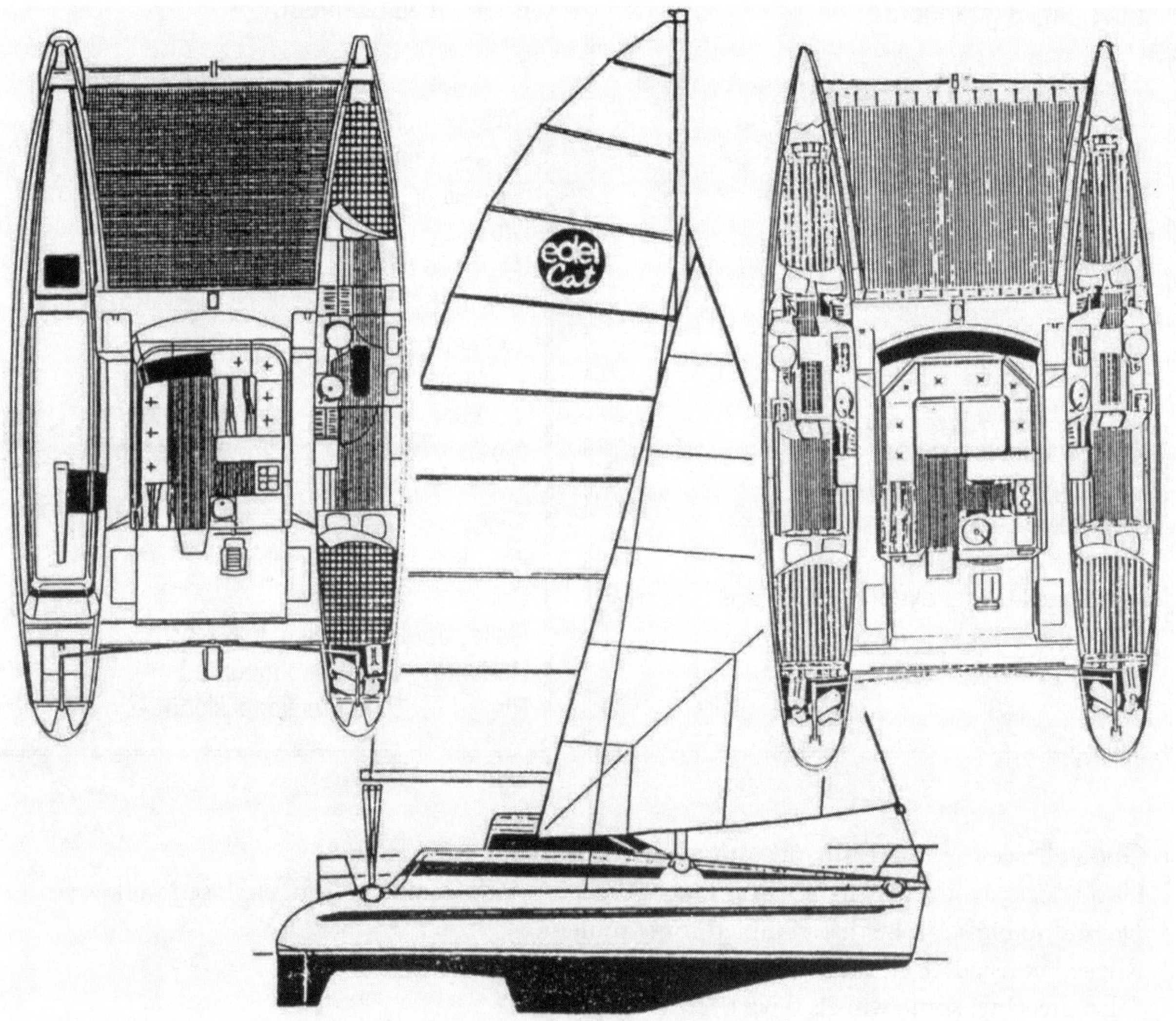

Edel 35 Cabrio

1. This Cabrio version shows a canvas tent over the cockpit for shade. There is no central cabin.

2. Performance is very good. Being freed from the weight and windage of the central cabin makes a big difference on this 35 ft. Vessel.

3. There are no complaints of underdeck slamming.

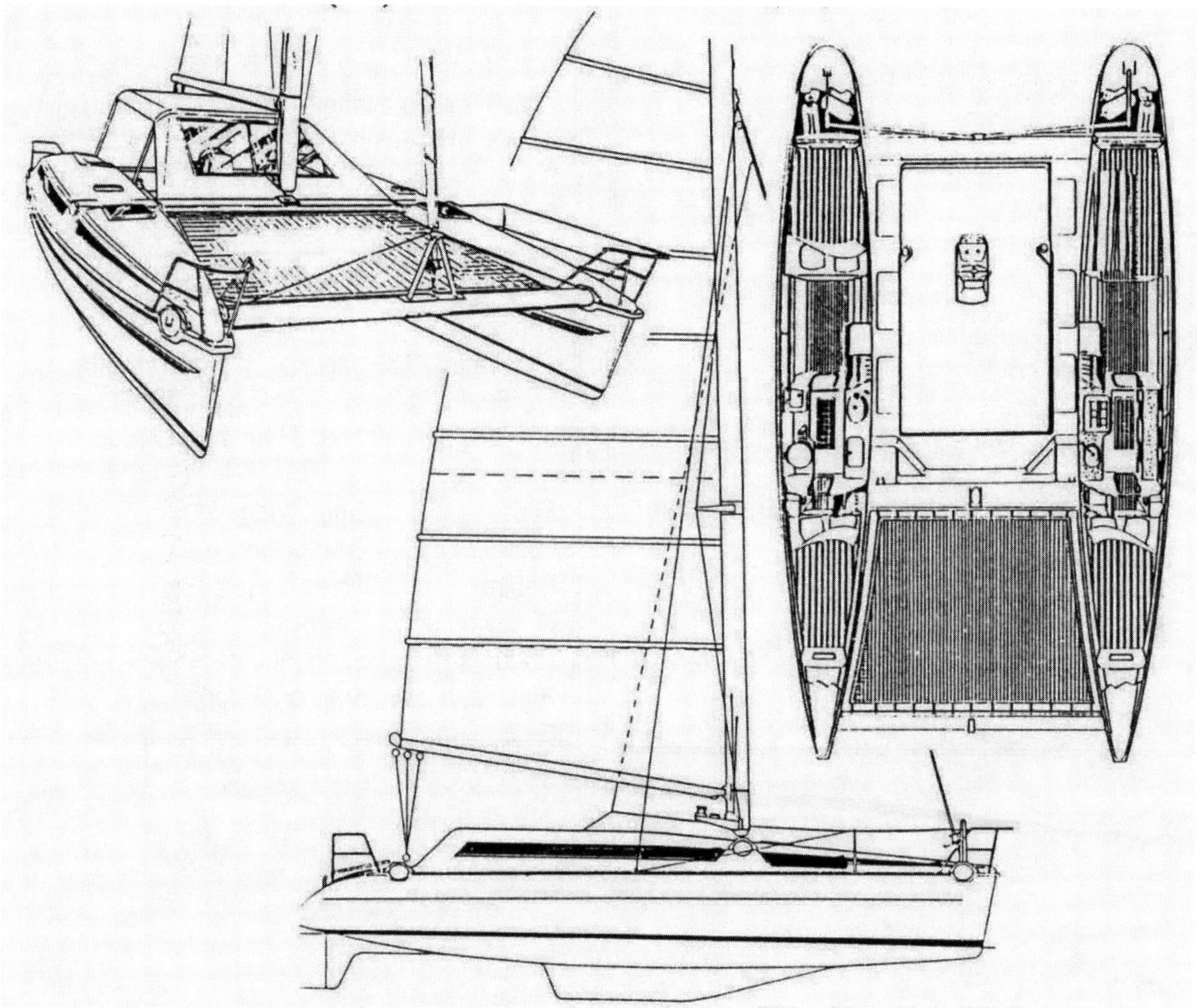

Edel Cat 43 Class 4, cruising/racing

Designer: Marc Edel

Sailing summary

Edel 43

1=Poor 2=Fair 3=Average 4=Good 5=Exceptional

	1	**2**	**3**	**4**	**5**
Motoring				4	
Docking				4	
Backing			3		
Sailing				4	
Weather helm/lee helm				4	
Windward ability				4	
Ease of tacking				4	
Visibility from the helm					5
Tracking				4	
Seakindly/ride				4	
Underdeck Slamming					5
Convenient Deck Layout				4	
Interior layout					5
Adequate storage areas			3		

Vital Statistics

Edel 43

LOA (length overall)	43' 0"	13.3 m
LWL (waterline length)	41' 5"	12.9 m
BMAX (maximum beam)	24' 0"	7.30 m
SA (sail area)	1360 ft^2	414.5 m^2
Draft: 4' 0"		Keels
Designed displacement:	15,432 lbs	7000 kg.
Mast hgt above WL:	61' 5" (est)	18.74 m
Tankage:		
Water:	185 gal.	700 liters
Fuel:	53 gal.	200 liters
Auxiliary power: Twin 28 hp diesel saildrives		

Type .... Open bridgedeck
Foredeck ..... Trampoline
Cabins ...... Three
Galley up/down Down
Rudders ..... Skeg mounted
Rig .... Fractional sloop

Edel 43

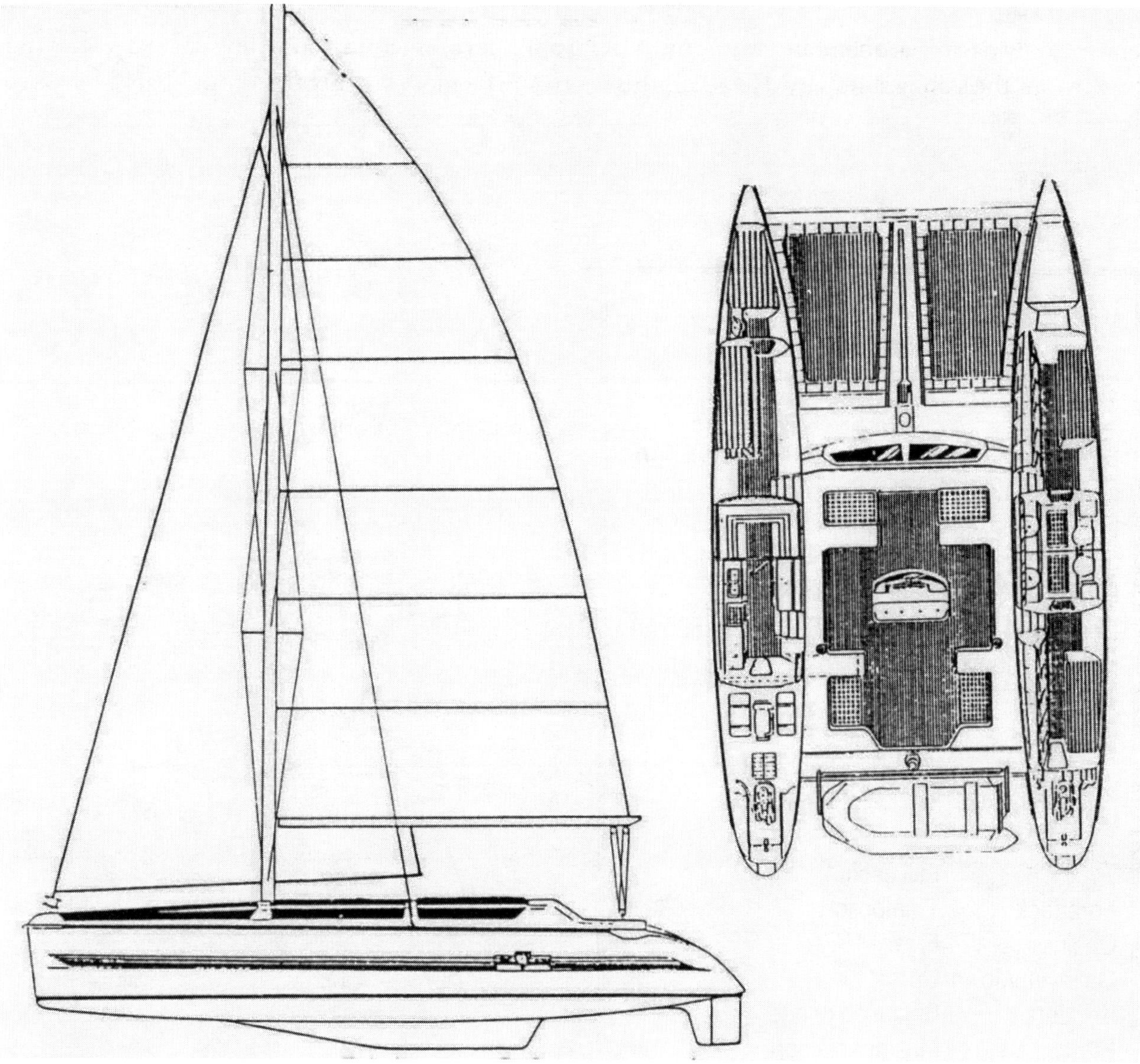

Freebird 50 Class 5, Cruising

Designer: Modular Mouldings, GB

Having only passing acquaintance with this vessel, I am not commenting on sailing characteristics thus leaving out the sailing summary. However, this is one of the most lavishly fitted out boats I have ever seen.

Vital Statistics

Freebird 50

LOA (length overall)	50' 0"	15.24 m
LWL (waterline length)	48' 0"	14.63 m
BMAX (maximum beam)	25' 0"	7.62 m
SA (sail area)	$1250ft^2$	116 m^2

Draft: 3' 3" / 1.0m Keels
Designed displacement: 27,557 lbs 12,500 kg.
Mast hgt above WL: 70' (est) 21 m
Tankage:
Water: gal. liters
Fuel: gal. liters
Auxiliary power: twin 35 hp diesel

Type . . . . Full cabin
Foredeck . . . . . Trampoline
Cabins . . . . . . Five
Galley up/down Up
Rudders . . . . . Balanced spade
Rig . . . . Masthead sloop

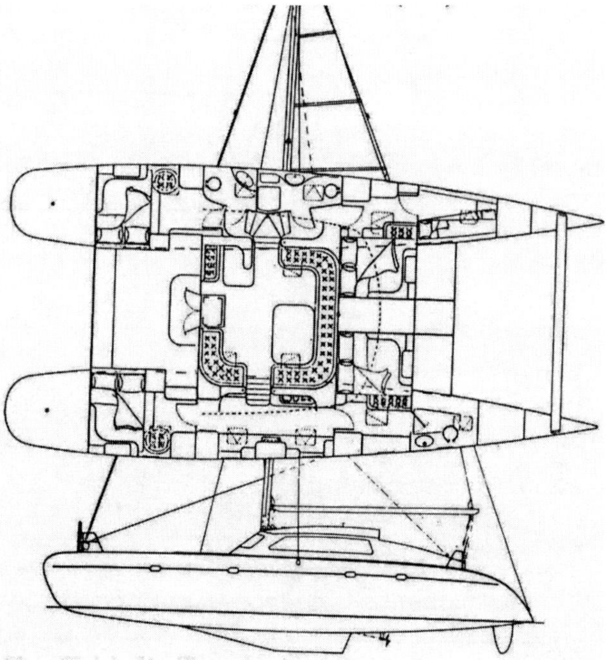

Gougeon 32, Racing, Class 2

Designer: Jan Gougeon

October second was a cold and blustery day on Lake St. Clair, Michigan. The participants in the annual fall racing series were arriving at the public launching ramp and setting up their boats. I had a two o'clock appointment with Jan Gougeon at the ramp for a test sail.

That morning we had driven from Port Stanley, Ontario Canada to Lake St. Clair. Good friends and cruising companions Terry and BarbaraTowner were with us. We had picked them up for this trip in upstate New York. It was a test sail to Corinne and me, but a sales demonstration for them. That meant there would be five (5) full sized adults on this test ride.

My personal testing philosophy is to sail a boat the way it would most likely be sailed in real life. I attempt to avoid the total stripped out ultra-light mode that most demonstrations entail. Except the ultra-committed racers, few boats ever sail with minimum weight. I test-sailed both the *Dragonfly* and the *F-27* with a full complement and wished to do the same thing for the *G-32*. I am estimating the weight of the five of us plus our gear at 180 pounds per person for an approximate total of 900 pounds. That is quite a contrast to the eleven-hundred pounds of designed vessel weight.

Our party had arrived at the launch site early, giving us time to get a feel for the area. We were enjoying lunch at a picnic table when the two trailered G-32 catamarans came through the park entrance. Picking up our gear, we walked over to the boats. Astonishingly, in the time it took us to walk the few yards, Jan Gougeon had his boat rigged. I was flabbergasted by the speed at the ease of raising the mast and rigging the boat. It is accomplished using only the boat's tackle. No trailer winch or special equipment was necessary. Then we watched as the second boat was rigged. That provided a thorough understanding of the uniqueness of this system. Without doubt, it is the simplest and most user-friendly system I have yet seen.

Unlike many other trailerable sailboats, the G-32 has the trailer between the hulls instead of the customary location under the hulls. The greatest advantage to this is the possibility of launching at any ramp or hard beach. Water depth is almost inconsequential. If you want to launch the boat on the ground, which, too, is easily and safely accomplished. This is quite unlike my old swing-keel venture 24, which required a ramp with water deep enough to float it off the trailer. That trailer had

16" wheels. Total minimum required depth of water to launch was about three feet. A tough thing to find.

Once aboard, the roomy cockpit accommodated the five of us with room to spare. Seated visibility was excellent. All the control lines, including those for the ballast tanks and centerboards, are located to be obvious, handy, easy to operate, but not in the way.

The motor started easily and we swiftly moved from the confines of the launch area to open water. We raised sails and sped off. Raising sails is simplicity itself. The jib is roller furling, the main rolls around the boom and the luff is fed into the track by a simple but clever feeder. That subsumes only two lines, the halyard and the roller line for the main. It eliminates the usual rats-nest of reefing lines.

Now came the part we were all anxiously waiting for. How would it sail? Jan steered out into the middle of lake looking for big waves. We eventually found some five-foot waves that we drove right into. This was an astonishing experience. The wave-piercing form of the hulls, quietly and with very little hobby-horsing, slid through the waves with barely a murmur. Hardly a splash of water came onboard. Without doubt, it was an impressive performance.

From there, Jan put the boat through its paces. We tacked, beat reached jibed and ran. Jan went to great lengths showing how the water ballast compartments filled and emptied and gave us demonstrations about the effects of this. I observed that the boat could be sailed with no ballast at all. It sailed a little dicey perhaps, but within the realm of achievement. Ballast was added as the wind increased. With full ballast, hardly any speed was lost. I can see where this boat is going to be fierce competition in light airs. The combinations of boards, ballast and sail trim are almost endless.

We tried all positions of centerboards, conditions of ballast, sailed under main alone, pinched upwind, sailed dead downwind. Always, the boat did well. The helm was always firm, without any tendency to wander or yaw. Windward performance was nothing short of amazing. It seemed as though we were punching directly into the wind. The boat felt almost like an iceboat and accelerated just like one. We did achieve 15 knots of boat speed in ten knots of true wind. This is the first time I could go that much faster than the true wind speed in a boat.

Sailing the boat left us all impressed. Unquestionably, this was a superior sailing machine. However, there is another aspect to this vessel and that is its utility.

We found that it was a difficult vessel to enter. There are no decks as such and there is no easy way of getting forward. The nets are usable while at anchor, but would be difficult to use while underway. The cockpit, however, is large and the seating ample. Seats are both sufficiently wide and deep to provide proper support for adults. The five of us were comfortable all the time and the cockpit long enough to seat four of us on one side. We sat that way occasionally as counterpoise for healing moment. (moveable ballast)

I suggest some type of vestigial deck around the boat to both make easier access to the boat and change the appearance. Adding a deck into the shear-line would enhance the appearance of the boat and diminish the effect of the shape of the cabin top as an eye catcher. It would also make it easier to get forward underway and to board the boat at a launching ramp.

The interior cabin, while making no pretense at anything other than camping, is surprisingly roomy and convenient. There is even a head with a privacy screen and a galley/dining area with full sitting room cleverly arranged by having foot wells into the hulls. Two people could be very comfortable for an extended weekend or two-week cruise. A family also would fit nicely for that time frame. Four adults would be crowding it a bit, but would work out for a race weekend or overnight sojourn. For those looking for offshore adventure, The Bahamas, Cuba, Baja California, the Gougeon 32 provides an inexpensive option. While not usually considered an offshore boat, the long waterline provides an advantage not found in many larger cruising vessels.

Unfortunately, full production was not to be. Only a limited number of boats were produced. The decision not to produce them was a big disappointment to me because I believed the boat fit a truly unfilled niche.

Meanwhile, back at the race course, those few boats actually produced and sailing are making a name for themselves and anyone able to buy one on the used market will have a real prize.

Gougeon 32

Sailing summary

Gougeon 32

1=Poor 2=Fair 3=Average 4=Good 5=Exceptional

	1	**2**	**3**	**4**	**5**
Motoring				4	
Docking			3		
Backing			3		
Sailing					5
Weather helm/lee helm					5
Windward ability					5
Ease of tacking					5
Visibility from the helm				4	
Tracking				4	
Seakindly/ride					5
Underdeck Slamming				4	
Convenient Deck Layout		2			
Interior layout			3		
Adequate storage areas			3		

Vital Statistics

Gougeon 32

LOA (length overall) 32' 0" 9.75 m
LWL (waterline length) 31' 0" 9.45 m
BMAX (maximum beam) 8 ' 4" 2.53 m
SA (sail area) 240 ft^2 22.3 m^2
Draft: 9" / .23 m bd up 4' 3" / 1.29 m bd dn
Designed displacement: 1,875 lbs 850 kg.
Mast hgt above WL: 40' (est) 12.1 m
Tankage:
Water: gal. liters
Fuel: 6 gal. 22 liters
Auxiliary power: one 9.9 outboard

Type .... Partial open bridgedeck
Foredeck ..... Trampoline
Cabins ...... None
Galley up/down None
Rudders ..... Balanced spade
Rig .... Transom hung, kick-up

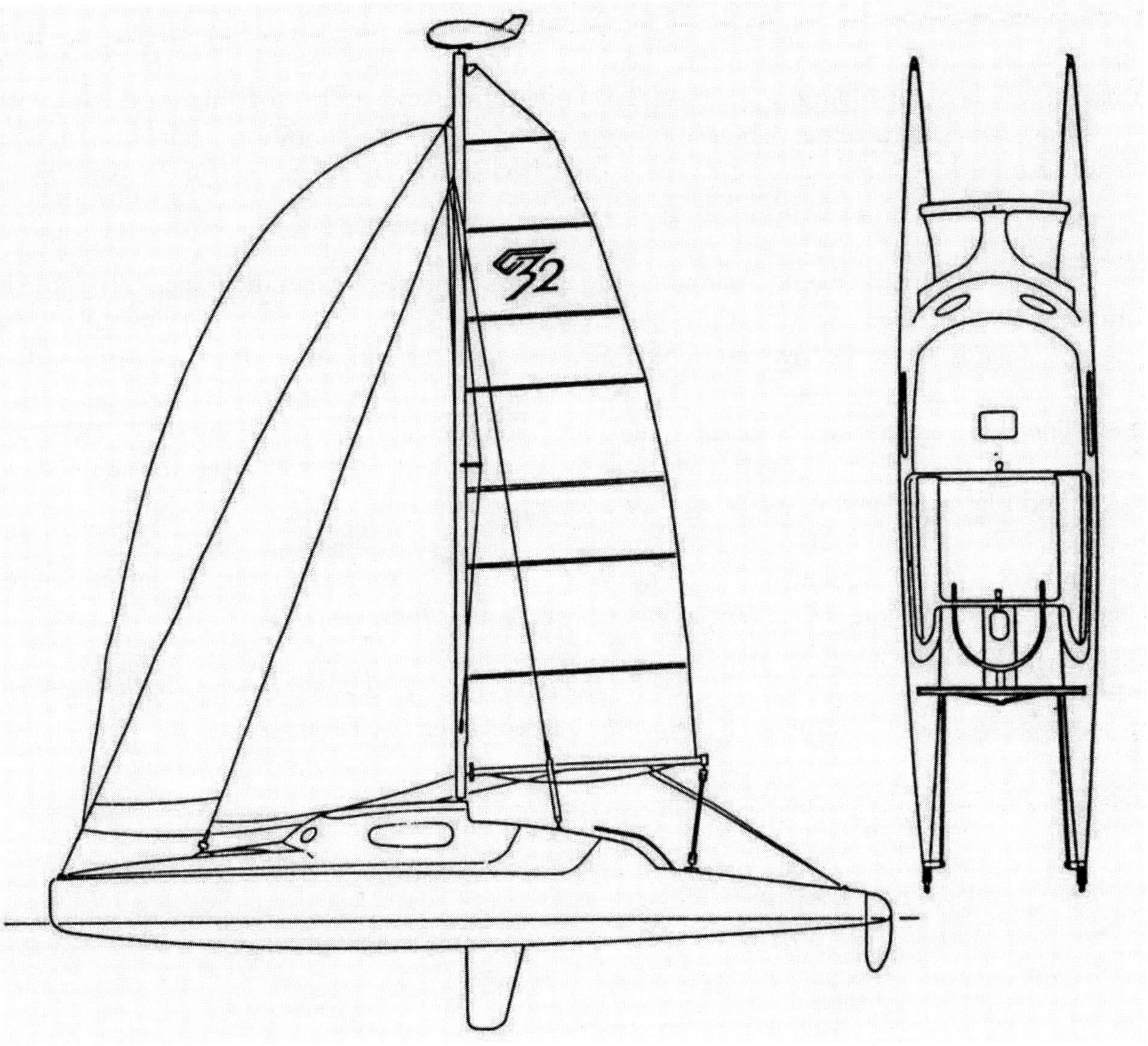
32

Heavenly Twins 26

Heavenly Twins 26 Class 5, Cruising

Designer: Pat Patterson

This is one slick boat. I have to hand it to the designer, Pat Patterson, there is an incredible amount of usable well laid out room on this boat. The compromise and difficulty comes in the outboard installation. Being in the center nacelle, it is difficult to maintain, constantly getting drowned and lets salt water into the cockpit. Two of the boats that I have inspected were built in England by the original builder and were strong and properly built.

There were some built in Puerto Rico in the early eighties. I surveyed one of them that had twin 8 hp Yanmar diesels installed. Actually, they were cleverly installed near the centers of the hulls. After some years hiatus, the twins have resurfaced as a 27 and are currently in production in England.

This is one of the very few center cockpit catamarans. The cockpit is rather narrow with a bench seat but extends all the way across the full width of the boat. It lends itself very nicely to the full width dodger and Bimini top and is a good candidate for a permanent hard top.

Behind the cockpit are two double size cabins that really are large for a 26 foot boat. Some folks have taken the center partition out creating a huge over king-size adult play pen.

Kevin Jeffery, my co-author for SAILOR'S MULTIHULL GUIDE sailed one for several years and wrote a few magazine articles about his experience titled: *Twins On Twins.*

Sailing summary

Heavenly Twins 26

1=Poor 2=Fair 3=Average 4=Good 5=Exceptional

	1	**2**	**3**	**4**	**5**
Motoring			3		
Docking			3		
Backing		2			
Sailing			3		
Weather helm/lee helm			3		
Windward ability		2			
Ease of tacking			3		
Visibility from the helm				4	
Tracking			3		
Seakindly/ride				4	
Underdeck Slamming				4	
Convenient Deck Layout			3		
Interior layout					5
Adequate storage areas				4	

Vital Statistics

Heavenly Twins 26

LOA (length overall)	27' 0"	8.20 m
LWL (waterline length)	21' 6"	6.60 m
BMAX (maximum beam)	13' 9"	1.90 m
SA (sail area)	380 ft^2	35 m^2
Draft: 2' 3"		Keels
Designed displacement:	1,875 lbs	850 kg.
Mast hgt above WL:	35' (est)	10.7 m
Tankage:		
Water:	65 gal.	234 liters
Fuel:	25 gal.	90 liters
Auxiliary power:	one	9.9 outboard

Type . . . . Full cabin
Foredeck . . . . . Trampoline
Cabins . . . . . . Two
Galley up/down Down
Rudders . . . . . Transom hung
Rig . . . . Masthead sloop

Heavenly Twins 26

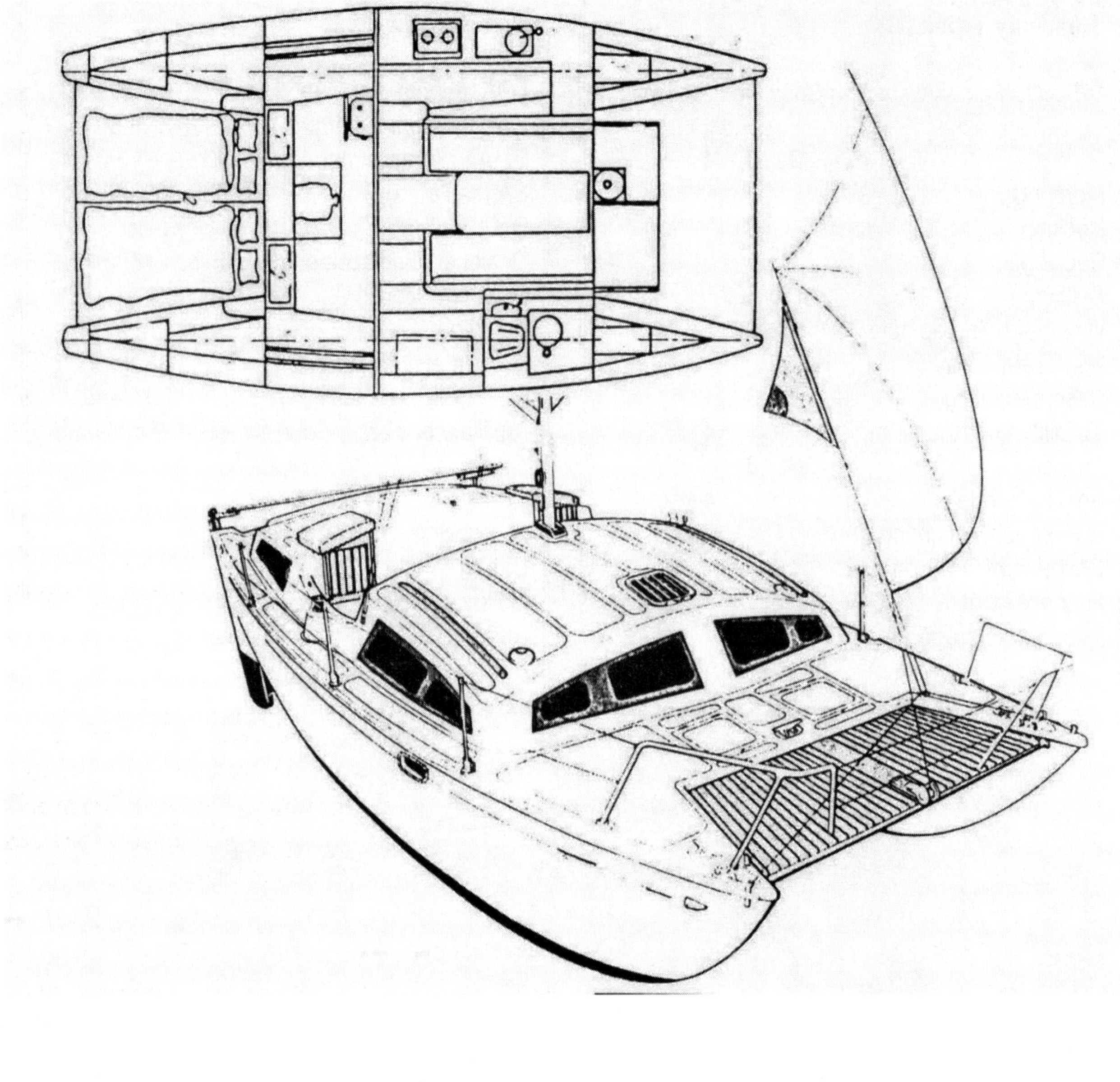

Iroquois, Class 3, Racing/Cruising

Designer: Rod Macalpine-Downey

The Irrepressible Iroquois is a handsome, good sailing catamaran, those are the reasons it has held its value and continues to sell at top prices. I was introduced to the Iroquois the year I bought my Hirondelle, 1971. Occasionally I crewed with Bill Symons in races on the south shore of Long Island, New York. Then I did my first delivery. Bill sold many Iroquois because of all his club racing.

Since those early days I have delivered five, surveyed seven others and managed one at my dock in Maryland for a year. It is difficult to imagine a handier vessel in that size range. She is fast, weatherly, seakindly and tacks assuredly. She has a feather light helm, a useful sail plan, a tabernacle mast and an enthusiastic owners' association. I delivered several from the Annapolis Sailboat show, to New England during late October. These trips included much bashing to windward during really nasty North Atlantic weather. No question, she can take it. I also delivered two, single handed from Annapolis to Florida. I believe I have at one time or other, been aboard every Iroquois in the United States.

Her disadvantage lies in her size. Perhaps perfect for the contemporary English family of the late sixties, she is just a little small for extended cruising, as practiced by contemporary Americans of 2002. However, she is perfect for weekending, club racing or short vacations.

Classification and/or design intent
Many people tend to classify this boat a cruiser. However, the boat design is, in fact, a true racer/cruiser. The design compromises clearly show that proclivity. Her kick up rudders, hung from the transom and her pivoting centerboards make her a delight to maintain, sail, anchor, cruise or race. Several have crossed the Atlantic and you can find Iroquois most any place in the world intrepid cruisers gather. The designer, Rod Macalpine-Downey, was a famous yacht designer.

Thumbnail history
Originally manufactured by Sailcraft, Brightlingsea, UK circa 1968, an indeterminate number of craft were produced. After the demise of the company in the 1980's, the molds were sold to an outfit in Texas that attempted to "Americanize" the vessel. They raised the shear line, enlarged the cabin, cut up the interior layout and named it *The Chieftain*. (See MKIIA)

Iroquois

Galley

This is a galley down layout with a molded single basin sink and drainboard combination. It is surprising just how large the galley really is. For the size of the boat, there is adequate counter space. The originals had a meager icebox in the cockpit. I guess you don't need much cold in England and besides, they drink their beer warm! Symons Sailing, the importer, tried several different arrangements for refrigeration including small LPG powered units as well as extensive cold plate units under the forward starboard bunk.

Saloon

The saloon itself really looks bigger than it is. Severely sloping forward window ate up most of the useable headroom in the semicircular saloon bench. These windows were also chronic leakers and let in a great deal of sunlight that bleached the handsome teak trim in back of the settee.

The mast step is right through the center of the table. It is an exceptionally strong and durable setup if it has been kept from leaking. The mild steel beams and chain plate extensions must be kept from rusting.

The saloon also has a very well located hanging wet locker port side just inside the entrance doorway and a good sized chart table with storage bin type chair under on the opposite side. The chart table flips up for storage to allow more saloon room. This layout really bespeaks its northern racing heritage.

Cabins

The twin aft cabins are a privacy delight. They have a small hanging locker in each and have good headroom. However, the bunks are in the ends and are tight for big people. The forward two bunks are usable for children or occasional guests but usually wind up filled with gear.

Head

The head compartment located on the port side opposite the galley is surprisingly roomy and sports good standing headroom, a hatch, ventilator and a window. There is a lavatory basin and shower. The original vessels were not equipped with holding tanks and there is precious little space to install one.

Deck Layout

Iroquois has a substantial rub-rail a molded in toe-rail, perimeter lifelines and bow pulpits. The original rub-rails were teak, many of the replacements are synthetic products. It has a trampoline foredeck and a metal cross beam with a wire truss arrangement. The standard boat has a single outboard but the flat transom lends itself to twin outboards if desired. The 32' extended version, the Mark IIA, had provision for twin installed engines. In those days, the Vire 2-cycle was quite popular because it was extremely light, simple and inexpensive. Some have installed the 1GM10 Yanmar diesels but that is pushing the weight a bit. The Mark IIA model also has enough buoyancy aft to support davits otherwise, the standard Mark I or Mark II boat is a little small to support stern davits but I have seen them installed. Most keep their dingy on the foredeck.

There is a mainsheet traveler. No Genoa tracks were standard, but may have been owner installed. A cautionary note, careful inspection of track installation is required to insure there is no leakage under as this is a balsa cored deck and improper installation of deck hardware is the number one cause of deck rot and delamination.

Anchoring facilities
There are roller fittings and very substantial cleats on each bow with a double door central anchor locker on the deck just behind the trampoline. Many people when using two anchors, simply secure an anchor to each bow rather than attempt a bridle.

My own experience with anchoring an Iroquois is quite extensive. On one single-handed trip from Annapolis Maryland to Palm Bay Florida, via the Intra-Coastal Waterway, I anchored about twenty times. Each time I anchored with a single anchor from one bow using a 12H Danforth and 1/2" nylon rode. I raised both the centerboards and the rudders. The boat lay head to wind perfectly without any tendency whatever to anchor-sail. It was the ultimate in simplicity. Anchored in three feet of water, with boards and rudders raised, secured from the cleat on one hull, despite howling 40 knots gusts, she lay as docile as if tied to a dock!

Rig
The original Iroquois Mark I had a fractional rig. The Mark II sports a masthead rig with single lower shrouds, a baby stay and a large mast section. Some of the earlier models had those giant "flying saucer" flotation domes on top. It took quite a few decades to rid us of that inappropriate piece of gear. The masthead rig sported a lapper and a Genoa as standard head sails.

Iroquois

Performance

Iroquois goes to windward quite well. Perhaps not as tight as a full blown race boat with six or so feet of draft and a bendy mast with hydraulic backstay, nevertheless, I usually arrives at the windward mark in respectable time. Because of its centerboards and kick up rudders, you have many options available to day sailor cats. Off the wind she goes the way you would expect, occasional double digit speeds are not uncommon and cruising average speeds average higher than similar waterline length monohulls.

Common problems

The centerboard pivot pins often cause leakage. The centerboards and trunks occasionally have problems and the centerboard control lines which run back to the cockpit are attached to the boards with metal shackles. This is a real toe-stubber and I suggest replacing the shackles with splices. The forward crossbeam has a dissimilar metal situation that often results in serious galvanic corrosion. I have not yet inspected one with osmotic blisters but this is not a suggestion that they are immune.

Price history

When I first sailed an Iroquois back in 1971, you could buy one new for about US$15,000.00 Today they sell for as much as US$38,000.00 depending upon condition and equipment.

Using centerboards

Most pundits agree that you should use only the leeward board and change boards during or just prior to a tack. The concepts involved are that only one at a time is required and the leeward board will be deeper in the water and thus be less likely to ventilate. Theory holds that as the board angles, it begins to develop lift just like any other foil, thus with an Iroquois, because the boards tilt inwards, it already has a slight degree of lift on the leeward board, which is great because it is lifting the leeward hull up. However, lifting the windward hull may not be so great, especially in severe conditions. On the other hand, as the boat heels, the less the angle on the windward board becomes thus lessening the lift. Therefore, I suggest it is worthwhile to experiment with each individual Iroquois to discover which configuration works best.

Sailing summary

Iroquois 30

1=Poor 2=Fair 3=Average 4=Good 5=Exceptional

	1	2	3	4	5
Motoring			3		
Docking			3		
Backing			3		
Sailing				4	
Weather helm/lee helm				4	
Windward ability					5
Ease of tacking				4	
Visibility from the helm				4	
Tracking				4	
Seakindly/ride				4	
Underdeck Slamming					5
Convenient Deck Layout		2			
Interior layout			3		
Adequate storage areas			3		

Vital Statistics

Iroquois 30

LOA (length overall)	30' 0"	9.14 m
LWL (waterline length)	27' 0"	8.22 m
BMAX (maximum beam)	13' 6"	04.1 m
SA (sail area)	345 ft^2	32 m^2
Draft: 2' 0"		Keels
Designed displacement:	6,560 lbs	2975kg.
Mast hgt above WL:	4' (est)	13.11 m
Tankage:		
Water:	40 gal.	151 liters
Fuel:	6 gal.	22 liters
Auxiliary power:	one	9.9 outboard

Type . . . . Full Cabin
Foredeck . . . . . Trampoline
Cabins . . . . . . Four
Galley up/down Down
Rudders . . . . . Transom hung, kick-up
Rig . . . . Masthead sloop

Iroquois

Iroquois 32 MKIIA Class 3, Racing/Cruising

Designer: Rod Macalpine-Downey

This vessel was the extended, 32 ft. version of the standard 30 foot model. The design change accommodated a twin diesel engine installation and wheel steering. Except the two foot longer sterns, the rest of the vessel specifications remained the same.

The particular vessel with which I am most familiar, was a Robin's Egg blue one, shown at Annapolis Boat Show. That one, though designed to carry diesel engines, retained the outboard engine and tiller steering. From there, I sailed it single handed, to its new owner in Palm Bay, Florida. It sailed even better than the 30 footer.

Sailing better than the standard model really is expected. The one I sailed, tracked, beautifully, responded to the helm and tacked assuredly. I could self-steer adequately with a rudimentary shock cord to tiller arrangement.

This model became very popular, as is, without the twin diesels. There are quite a few around and I have managed to survey or cruise with three other similar units.

In the late 1980s the molds sold to a prospective builder in Florida. The three units produced, called: *Iroquois Chieftain.* There were some modifications to the original design. Other major extensive renovations include large engines in each hull. One model displayed twin 90 horsepower outboards, could go 25 knots and pull water skiers. Failing to immediately enthral the American public, they then sold the molds to Carlson Marine an outfit in Indiana that further modified them under the name of Ultimate Concept. They are no longer in production. I do not know the current whereabouts of the molds.

The kick up rudders on the original Iroquois were made of wood with wooden rudder casings. Well maintained, they still exist. However, most have been replaced at some point in time and need to be carefully inspected.

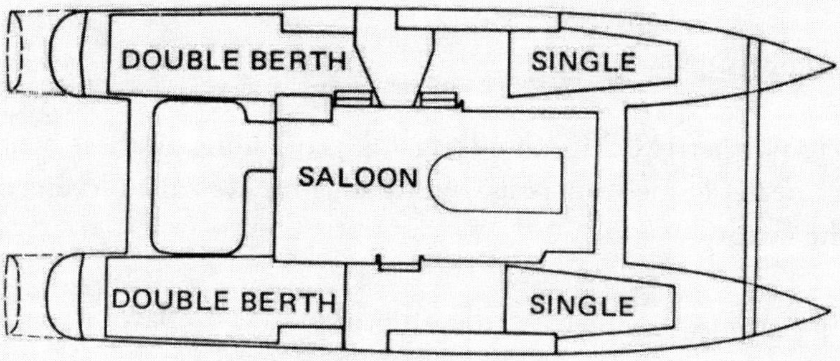

Iroquois MK II

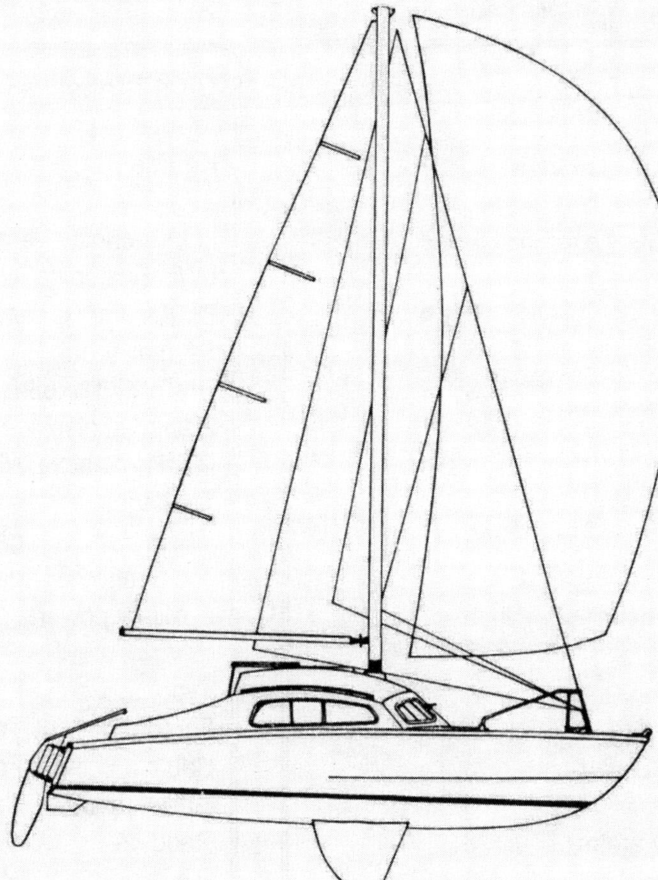

Iroquois MK IIA

Outlines on transoms

Note: protective skeg ahead of rudder

Kaulua 31 class 3, racing/cruising

Designer: Harold Stilson

I flew up to northern Michigan for several days of sailing on these boats for the purpose of evaluating them for a dealer in the Annapolis, Maryland, area. We sailed several different boats and had a grand tour of the factory.

My impressions of the boat included the facts that the layout was excellent, the sailing good, the handling and tacking, fair and the aesthetics, poor.

Sailing summary

Kaulua 31

1=Poor 2=Fair 3=Average 4=Good 5=Exceptional

	1	**2**	**3**	**4**	**5**
Motoring			3		
Docking			3		
Backing			3		
Sailing				4	
Weather helm/lee helm				4	
Windward ability			3		
Ease of tacking		2			
Visibility from the helm			3		
Tracking			3		
Seakindly/ride				4	
Underdeck Slamming				4	
Convenient Deck Layout			3		
Interior layout			3		
Adequate storage areas			3		

Vital Statistics

Kaulua 31

LOA (length overall)	31' 0"	9.45 m
LWL (waterline length)	29' 0"	8.84 m
BMAX (maximum beam)	14' 0"	4.27 m
SA (sail area)	412 ft^2	38.28m^2
Draft: 1' 8" / .53m bdup	4' 0" / 1.22m bddn	
Designed displacement:	6,050 lbs	2744 kg.
Mast hgt above WL:	45' (est)	12.4 m

Tankage:
- Water: gal. liters
- Fuel: 6 gal. 22 liters

Auxiliary power: one 9.9 outboard

Type .... Full Cabin
Foredeck ..... Trampoline
Cabins ...... Four
Galley up/down Up
Rudders ..... Skeg mounted
Rig .... Masthead sloop

Kaulua 31

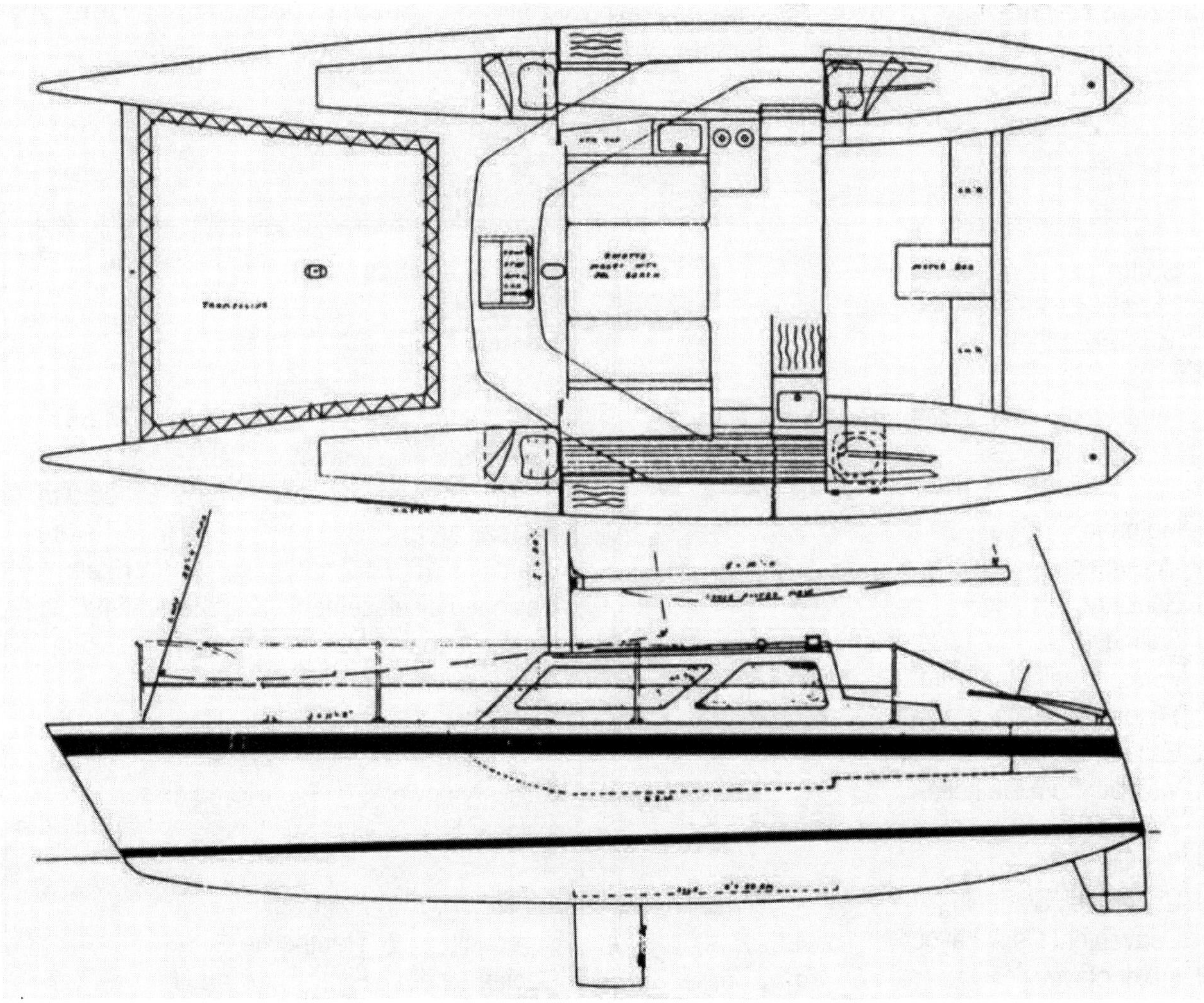

Kennex 380

Kennex 380 Class 4, Cruising/racing

Designer: Pro Kennex / Groupe Graal

In 1996 I flew to Guadaloupe to survey a Kennex 380 for clients. As it turned out, they did not purchase the one they went to see, but another that was on the same dock. This gave us good opportunity to view several boats of the same vintage and get a good idea of its strengths.

Summer, 2000, I surveyed the same boat again, giving me an in depth view of its aging process.

Sailing summary

Kennex 380

1=Poor 2=Fair 3=Average 4=Good 5=Exceptional

	1	2	3	4	5
Motoring			3		
Docking			3		
Backing			3		
Sailing				4	
Weather helm/lee helm				4	
Windward ability				4	
Ease of tacking			3		
Visibility from the helm				4	
Tracking				4	
Seakindly/ride				4	
Underdeck Slamming				4	
Convenient Deck Layout			3		
Interior layout			3		
Adequate storage areas			3		

Vital Statistics

Kennex 380

LOA (length overall)	37' 9"	11.55 m
LWL (waterline length)	37' 6"	11.45 m
BMAX (maximum beam)	19' 9"	06.0 m
SA (sail area)	844 ft^2	78.4 m^2
Draft: 3' 10" / 1.2 m		Keels
Designed displacement:	12,122 lbs	5498 kg.
Mast hgt above WL:	57' 5" (est)	17.5 m

Tankage:
- Water: 184 gal. 700 liters
- Fuel: 58 gal. 220 liters

Auxiliary power: Twin 18 hp diesel saildrives

Type . . . . Full Cabin
Foredeck . . . . . Trampoline
Cabins . . . . . . Five
Galley up/down Up
Rudders . . . . . Balanced spade

Kennex 380 Destiny

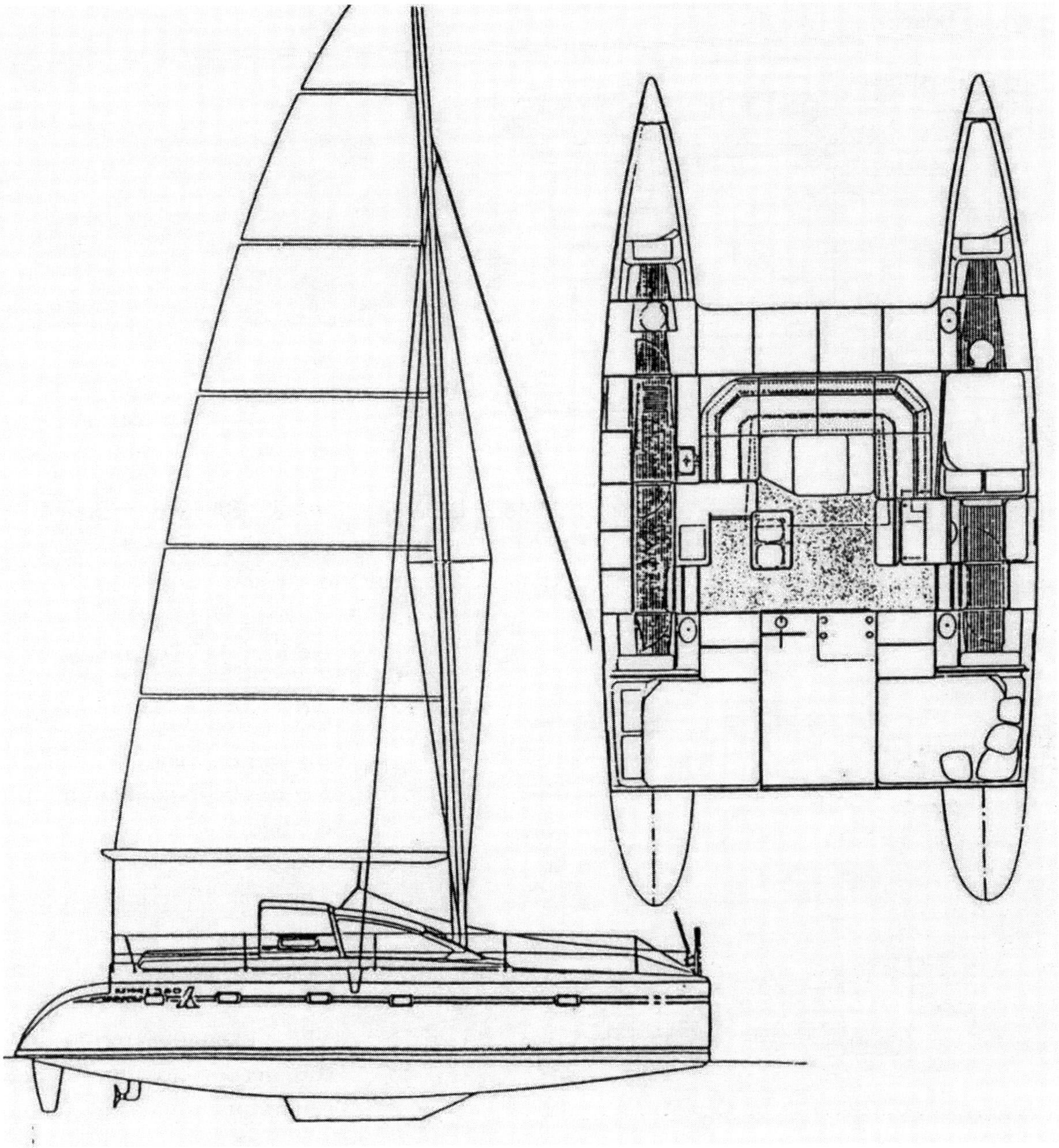

Kennex 380, 420

Kennex 420 Silver Destiny, LOA 42', LWL 40', BMAX 24.5', Draft3.75', SA 1764 ft^2

I surveyed two of these and except for the goofy front windows they are really well built, well laid out, good sailing boats.

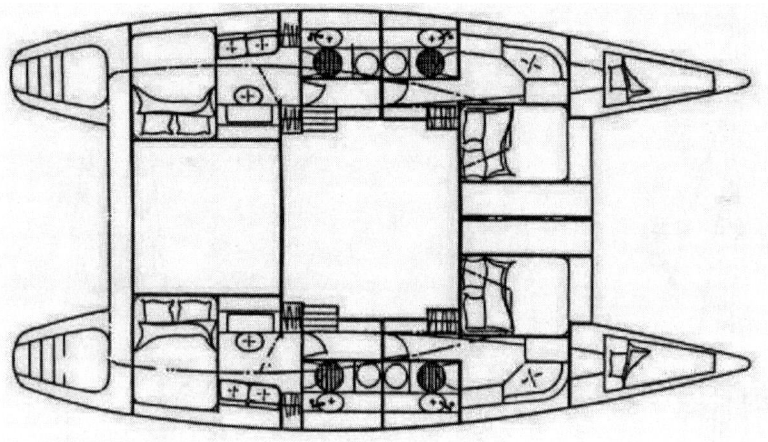

Kennex 480 Golden Destiny, LOA 48.2', LWL, 43', BMAX 28', SA 1356 ft^2, Draft 4'

I spent several days in thorough survey and sea trials of this boat and can say it is among the better boats to be imported from France.

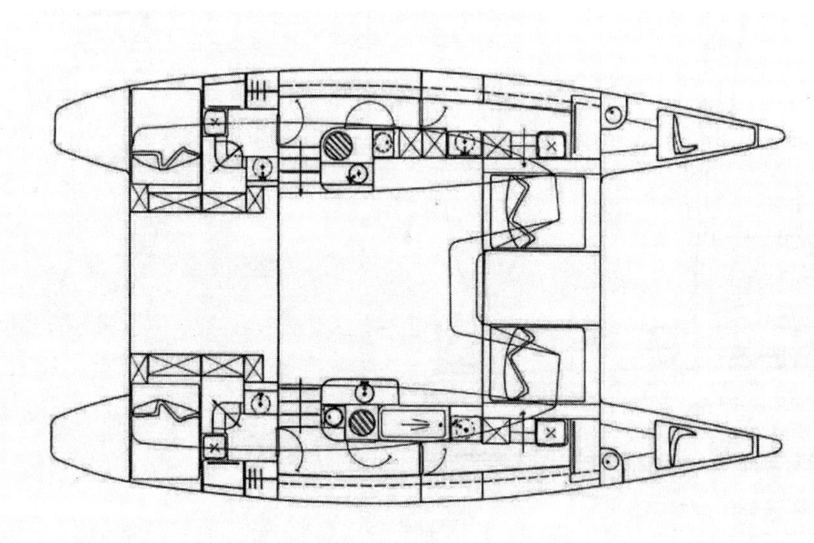

Kennex 445 Legendary

Other Kennex boats include the 445 Legendary. All the Kennex boats come in multiple configurations.

They are presently out of business. The location of the molds in uncertain at this time.

These diagrams are mostly for the reader to use as idea generators.

La Digue 35 Class 4, Cruising/racing

Designer: Alexander Simonis

Corinne and I drove to Stuart, Florida, to review and test sail of the very first La Digue 370 built and imported into this country. The La Digue is a 37 foot catamaran built by Voyage Yachts, Capetown, South Africa. It will compete with the Moorings 3800 and the Fountaine Pajot Athena 38. Stuart is about 110 miles north of Fort Lauderdale and at the confluence of the Indian River and the St. Lucie River. The St. Lucie River headwaters are Lake Okeechobee. The Indian River is famous for its citrus and the Intracoastal Waterway. Having sailed that area many times, it was like old home week.

The La Digue is a new entry into the yacht business by Voyage Yachts. Their other catamarans are the Norseman 430 and 470 and the Mayotte 500. This first boat is actually built of epoxy strip plank composite since it is the plug boat for the mold and it has paint rather than gel-coat. Future boats will be built of fiberglass from the molds made of this prototype. Even though it may be slightly heavier than subsequent yachts, it sat nicely on its lines with its transoms well above the waterline.

User Friendly?

The very first thing I do when I visit a new boat designed as a cruising boat is to judge its user friendliness. That means reasonably flat, accessible decks, lots of hand grips, good toe rails, sturdy rub rail, good visibility from the helm, safe and secure foredeck or trampoline inside adequate lifelines and a host of other important items. This boat will meet any reasonable expectations of user friendliness. All decks are accessible but there are limited handholds for entering the cockpit. However, this problem will solve itself with the installation of a sturdy Bimini top with it's concomitant supports.

Exterior

The non-skid pattern was acceptable, The multi-folding cabin entry doors were a little awkward but my understanding is that this will be changed on future models. The helm has a unique two level step up. Visibility was adequate for me at 5' 9" but would be much better for someone taller. The wheel was in a comfortable location and all instruments were easily visible. There was full instrumentation and single lever controls for the twin engines.

Trampolines and foredeck construction is definitely superior. First, the foredeck has a center support strut which is necessary to prevent the forward crossbeam from bending inward under headstay tension. It also nicely divided the trampoline into two sections. The trampoline itself is made from heavy duty belting, not some piece of fishnet. While I am critical of using sail slides for attaching trampolines, at least these sail slides are black plastic with some sun block.

Then there is the cross beam. So many are cheaply put together with open ends and ragged corners. This beam had many nice touches including sealing the ends.

Between the transoms was a broad deck suitable for an outside activity or for holding a dingy. There were no davits or arch installed as yet. The transoms themselves had steps leading up to the deck. There were no boarding ladders installed. Aesthetics were quite pleasing even though the boat has not yet been "accessorized" with name, graphics or arch and davits.

Interior

The interior and the galley. This galley up model has a nice ergonomic laid out "U" shaped galley with the stove at the center of the "U". Corinne was quick to point out that it had but a single basin sink, and a round one at that. Our preference would be a double sink and square in shape may not be so stylish but gives considerably more room in the same basic dimensions. I also suggest that since most marine stoves are diminutive, a two burner stove is usually adequate and has the additional advantage of being able to accommodate full size pots properly centered over the burners.

The interior layout is not unfamiliar. It is a typical "U" shaped saloon with master table and surround bench seating. The galley up arrangement helps somewhat but I always marvel that all that beautiful saloon space is so unimaginatively used. Comfortable chairs, recliners and a host of other options are available for such a luxurious craft. This arrangement is a throwback to the charter days when all boats were designed for that purpose. But for an owner who intends to take long cruises or live aboard, there are far more comfortable arrangements.

There is a nice navigation station/computer area/radio room area facing aft just opposite the dinette. Behind it is the main electrical panel and access to the steering and other operational systems. It is both convenient and well done.

The head. I always suggest to people who are in boat buying mode that they actually go into the head compartment, close the door, sit down and simulate all functions you would normally do in that room. The reason is simple. Most heads are considerably smaller than home size bathrooms and you need to make sure that you can actually accomplish all essential body movements in that area. This compartment was very adequate for me at average stature.

The engine rooms are a mixture. In the three cabin arrangement that this prototype is, one is beautifully located port side behind the head with access through a panel. In that engine room there is almost stand up head room and lots of very accessible space to install and service accessories such as a watermaker. It is also an ideal location for an above waterline KISS system MSD. The other engine is located under the starboard bunk. Under bunk installations bring several problems with them such as poor service access, latent heat and noise. In the two-cabin arrangement, both engine rooms are accessible through the twin heads.

Engines are Yanmar sail drives turned backwards. This is a clever way of getting space forward of the engine. Nicely done, as in this boat, it does not add too much extra effort to service the front of the engine. The standard engines are Yanmar 18 horse power.

How does it sail?

We motored away from the dock and out a very narrow creek into the St. Lucie River and from there into the Indian Creek or Intracoastal Waterway. Since this boat is brand new and everybody is new to it, it took longer than expected to raise the sails. I handled the yacht during this period. With both engines just idling in forward, I was easily able to keep the boat arrow straight into the wind until the main was raised. Once the main was raised, I shut the engines and sailed under main alone until the jib was raised.

The boat handled like a good dingy under main alone. Sailing in that area of the ICW is very tricky. Being just opposite St. Lucie Inlet, it is full of shoal water and strong cross currents. At no time did I feel apprehensive about sailing this boat.

The speedometer transducers were not in their scabbards so we did not have those instruments to use but the GPS gave us readings over the bottom, plus or minus the current. The boat sailed better than I expected but not faster than I expected. We generally did half the wind-speed which was 8-12 knots. Frankly, this is good performance for this class of yacht and about the same as most of the other yachts of similar design intention. Do not expect to sail cruising catamarans of this type at warp speed past all the monohulls. You will be disappointed, at least in light air. As the wind picks up, so does your speed advantage. You will sail faster in many conditions but not all and not against all monohulls.

This vessel has good bridgedeck clearance with no appendages hanging down in the tunnel. Our test sail could not possibly test for under-deck slamming since we sailed inshore. The La Digue handles astonishingly well for a vessel with skeg rudders, just reinforcing my point that you do not have to have the disadvantages of spade rudders to achieve good handling. (Spade rudders are more prone to damage, especially if you want to beach your boat and much more prone to snagging various underwater lines from traps, nets, anchors, etc.)

The La Digue tacked just like a good dingy being positive and not needing to backwind the jib. It tacked just as well under main alone. Having a boat that will sail well under main alone is a distinct advantage. Gibing posed no problems but then again it was light air. I would not expect to have any problems jibing as this vessel has a nicely laid out traveler.

The basic configuration and equipage of the La Digue is rather straight forward and well tested. There are no radical departures or experimental installations on this boat which gives a new boat buyer a comforting feeling. My only concern is the enormous forward windows. While the manufacturer assured me they were as strong or stronger than the hull itself, I still question the advisability since it will be difficult to install curtains, very expensive to repair or replace and because of the differences in coefficients of expansion possibly create a chronic leaking problem, especially in the tropics.

Sailing summary

La Digue 35

1=Poor 2=Fair 3=Average 4=Good 5=Exceptional

	1	**2**	**3**	**4**	**5**
Motoring				4	
Docking			3		
Backing			3		
Sailing				4	
Weather helm/lee helm				4	
Windward ability			3		
Ease of tacking				4	
Visibility from the helm			3		
Tracking			3		
Seakindly/ride			3*		
Underdeck Slamming			3*		
Convenient Deck Layout				4	
Interior layout				4	
Adequate storage areas			3		

*minimal expose due to light air& calm seas

Vital Statistics

La Digue 35

LOA (length overall)	35' 9"	10.9 m
LWL (waterline length)	33' 3"	10.1 m
BMAX (maximum beam)	20' 6"	6.27 m
SA (sail area)	915 ft^2	85 m^2
Draft: 3' 2" / .98 m		Keels
Designed displacement: 13,200 lbs		6000 kg.
Mast hgt above WL: 56' (est)		17.3 m
Tankage:		
Water: 68 gal.		250 liters
Fuel: 39.6 gal.		150 liters

Auxiliary power: Twin 18 hp Diesel saildrives

Type . . . . Full cabin
Foredeck . . . . . Trampoline
Cabins . . . . . . Three
Galley up/down Up
Rudders . . . . . Balanced spade

La Digue

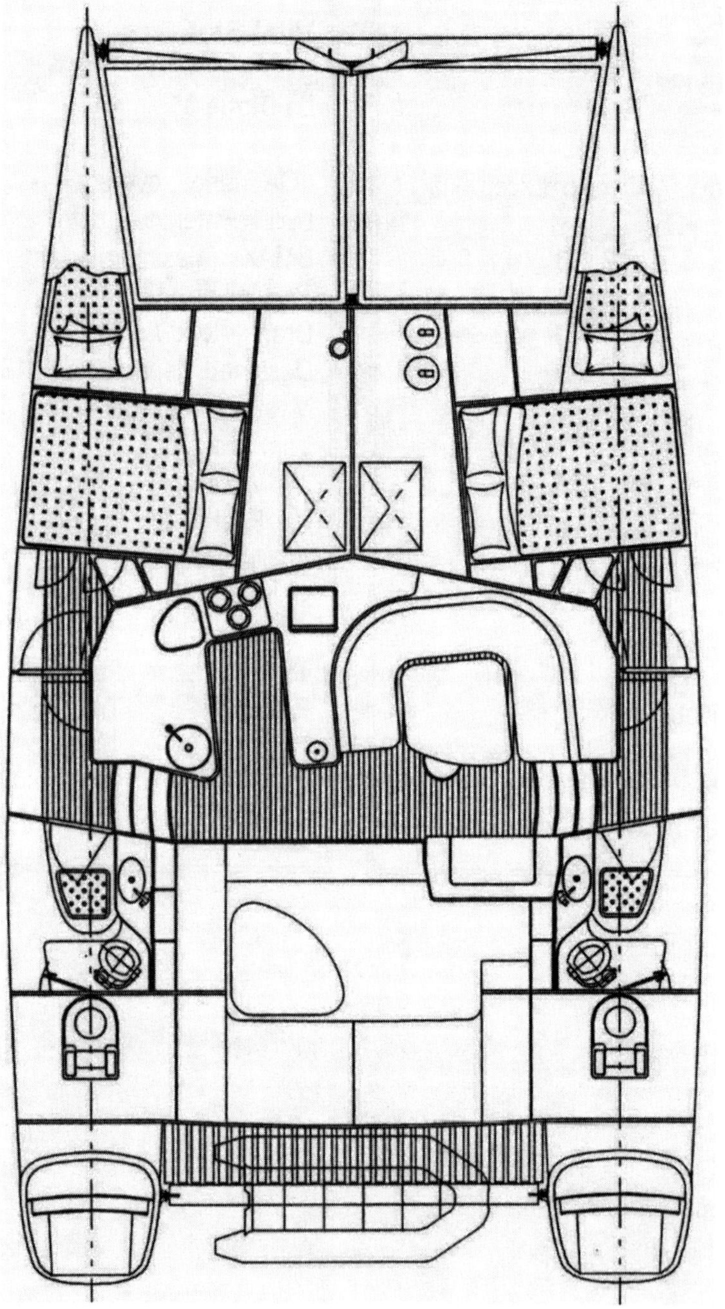

Lagoon 42 Class 5, Cruising

Designer: Marc Van Peteghem, Vincent Lauriot-Prevost

Way back in 1991, at the Annapolis boat show, the Lagoon 42 was introduced. No question, it made a big hit and was the surprise darling upstart of the show. I climbed all over it, poked around, peered into the dark recesses of bilges and virtually inaccessible compartments and could find only one major drawback, the boat had no rub-rails. Now you may think I have a rub-rail fetish, but to prove my point about their necessity, this boat had big scratches on the sides that may not have happened had there been proper rub-rails. Imagine, the very first of the breed ever shown, in all her beauty, had big scratches down the sides the most likely could have been avoided with proper rub rails

As the years ticked by, I watched several of these boats under sail in various charter operations. I was not impressed by their performance. However, I had to give them the benefit of the doubt because, after all, they were operated by charter people. I did insurance surveys on a couple, but that did not really let me get to know the boats in depth. Beginning November of last year, I was engaged to do a pre-purchase survey and sea-trails on one. Finally, I had my opportunity to really get into the nitty gritty. As it turned out, I surveyed that boat several additional times and gave it extensive sea trials in order to use-test the added equipment.

Plunging out in the Atlantic Ocean through Port Everglades inlet (Ft. Lauderdale) into six to eight foot seas and thirty knot winds is not for the faint hearted. My intrepid crew, consisting of the owners and a couple of 'wannabes' made the trip. It was my second sea-trial on this particular vessel. They had just completed extensive refit and equipment installation and sea trials were definitely in order, the tougher the better.

This is one sea-kindly boat. No number of tacks, jibes, rough conditions or deliberate attempts could make this boat slam its bridgedeck. A single reef in the main and a full jib gave us good speed and perfect control. No question about it, this is a boat you can feel safe, secure and comfortable anywhere in the world. We never missed a tack and found that the boat would tack with main alone, even with a reef in it. Despite the fact that the boat has keels rather than daggerboards, it points quite nicely. We had no trouble tacking through ninety degrees and holding thirty-five degrees apparent even in those rough and tumble conditions.

Lagoon 42

What I do not like about the boat is a complaint I have with several of the French boats, that is too much is sacrificed for style alone. The extremely sloped cabin face turns the saloon into a hot house. Outside window covers are necessary to keep the sun out. When you cover the windows you lose one of the greatest catamaran attributes: panoramic visibility. The extreme slope also makes it difficult to get up on the top deck and causes the loss of considerable interior space. (I note the new Lagoon series has a completely different "vertical" look)

Jeanneau Lagoon 42, built by Tillotson-Pearson in Rhode Island. It is a blue-water boat with twin diesel auxiliary and a fractional sloop rig. It is a beautiful vessel with exceptionally fair lines. It has excellent deck access and a secure trampoline area. The forward crossbeam appears to be well made. The dinghy davits are massively strong. It has a four-cabin two-head arrangement with ample room.

It has a modest but adequate sized galley-down layout. Both the galley and the typical; horseshoe shaped saloon are smallish by contemporary standards. Visibility from the interior is good without the exterior curtains.

One of the nicest things about this boat is the engine room access from the stern steps. This puts the engines completely outside the accommodation. Accessories such as watermakers and generators fit securely in the huge, airy, accessible engine rooms.

These are not the easiest boats to anchor but are not bad. Anchors can be stored in the bow rollers, the windlass in back by the mast. This arrangement works well in benign weather. However, when the weather gets rough and/or you are in a marginal anchorage it is not a simple task to fix a bridle or a snubber and Bahamian moor is particularly difficult.

The fractional sloop rig with a moderate roach mainsail has both advantages and drawbacks. Mainsails of this type are heavy and hard to raise. Having a full roach with battens pressing inward makes things even more difficult. But from the sailing and performance point of view, it is usually worth the effort.

The price history of these vessels is interesting and those looking for pre-owned boats have an expanding market. Boats are selling for as low as fifty percent of their original cost. They started out in 1991 in the US$300,000 range with sails and engines and now you can see them offered at competitive prices. Caution, many of the low price ones are ex-charter boats needing major refit.

One big advantage to these boats is that even though they are French designed boats, they were built in the United States by Tillitson/Pearson thus they are American boats and not subject to the restrictions on foreign boats covered by the infamous Jones act, thus for those with chartering on their minds, here is an advantage.

Sailing summary

Lagoon 42

1=Poor 2=Fair 3=Average 4=Good 5=Exceptional

	1	**2**	**3**	**4**	**5**
Motoring			3		
Docking				4	
Backing				4	
Sailing				4	
Weather helm/lee helm				4	
Windward ability			3		
Ease of tacking				4	
Visibility from the helm			3		
Tracking				4	
Seakindly/ride					5
Underdeck Slamming					5
Convenient Deck Layout				4	
Interior layout				4	
Adequate storage areas			3		

Vital Statistics

Lagoon 42

LOA (length overall)	42' 6"	12.95m
LWL (waterline length)	38' 9"	11.81m
BMAX (maximum beam)	22' 8"	6.918m
SA (sail area)	1119ft^2	104 m^2
Draft: 2' 0"		Keels
Designed displacement: 16,500lbs		7484kg.
Mast hgt above WL: 64' (est)		19.5 m
Tankage:		
Water: 160 gal.		6002 liters
Fuel: 160 gal.		6002 liters
Auxiliary power:	Twin diesel "vee" drives	

Type .... Full cabin
Foredeck ..... Trampoline
Cabins ...... Four
Galley up/down Down
Rudders ..... Balanced spade
Rig .... Fractional sloop

Lagoon 42

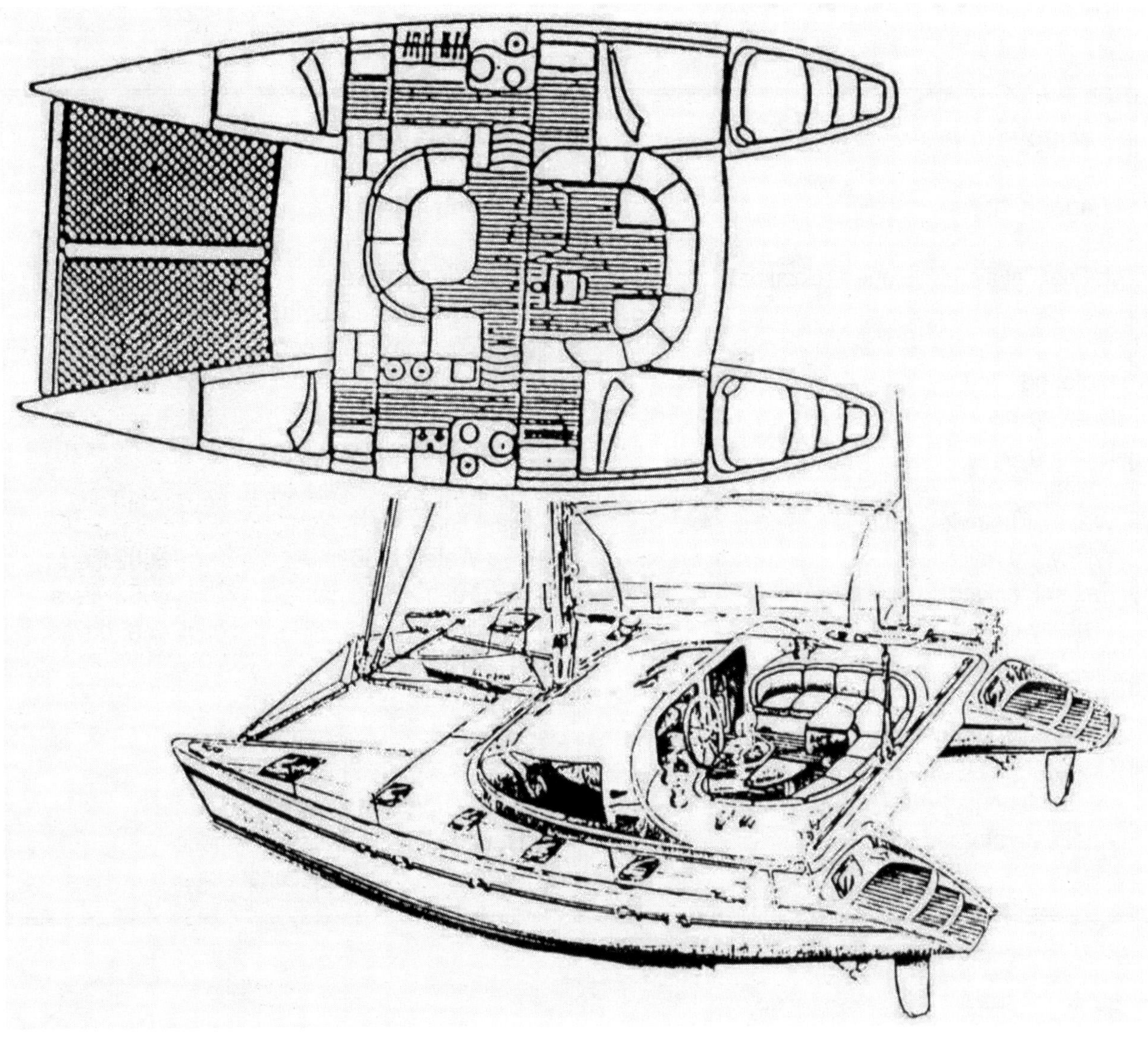

Louisiane 37 Class 3, Racing/Cruising

Designers: Michel Joubert/Bernard Nivelt, Built by Fountaine Pajot

I sailed this boat a few times in Annapolis harbor. It certainly is a handsome boat that sails reasonably well. In my opinion, it never sold well in the United States because you must go through the cockpit to get from one cabin to another, it has outboard power and tiller steering. Even though I consider outboard power and tiller steering advantages, most customers for boats in that size range do not see it that way especially circa 1980 when it was introduced into the United States.

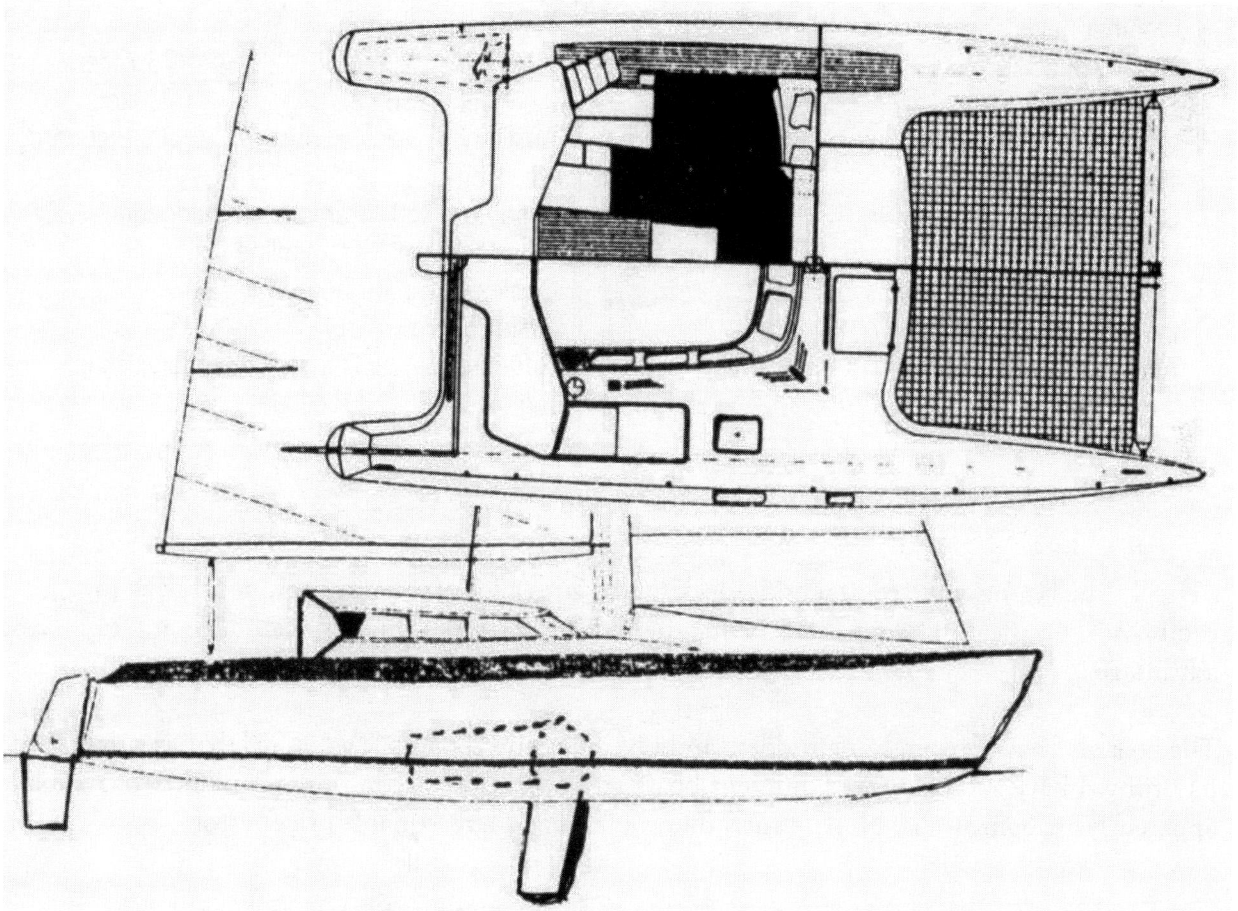

Louisiane 37

Sailing summary

Louisiane 37

1=Poor 2=Fair 3=Average 4=Good 5=Exceptional

	1	2	3	4	5
Motoring			3		
Docking			3		
Backing			3		
Sailing				4	
Weather helm/lee helm				4	
Windward ability				4	
Ease of tacking				4	
Visibility from the helm					5
Tracking				4	
Seakindly/ride				4	
Underdeck Slamming				4	
Convenient Deck Layout		2			
Interior layout			3		
Adequate storage areas			3		

Vital Statistics

Louisiane 37

LOA (length overall)	37' 0"	11.28 m
LWL (waterline length)	36' 0"	10.97 m
BMAX (maximum beam)	19' 7"	06.0 m
SA (sail area)	768 ft^2	71.4 m^2
Draft: 1' 4" / .41 m bdup	5' 4" / 1.65 m bddn	
Designed displacement:	6,000 lbs	2721kg.
Mast hgt above WL:	50' (est)	15.24 m
Tankage:		
Water:	20 gal.	75.7 liters
Fuel:	6 gal.	22 liters
Auxiliary power:	one	9.9 outboard

Type . . . . Partial open bridgedeck
Foredeck . . . . . Trampoline
Cabins . . . . . . Four
Galley up/down Up
Rudders . . . . . Transom mounted kickup
Rig . . . . Fractional sloop

Actually, the layout of the central bridgedeck cabins gives a lot of interior room for a sleek low profile catamaran. Its nearest competitor being the Edel Cat but having considerable interior layout advantage.

The boat also has transom hung rudders and pivoting centerboards. Two items that greatly enhance its utility. I have surveyed three different boats and one boat a second time for a new owner. They apparently are aging well with no structural problems and a reasonable price history.

MacGregor 36 Class 2, Racing

Designer: Roger MacGregor

This is strictly a class 2, racing catamaran. Grudgingly, you might want to let it slide into class three because it does have a head and a galley and people do cruise them extensively. However, it also is a full trampoline boat. In my classification system, it is not sensible to categorize an all trampoline boat as a cruising boat.

I feel badly about this boat going out of production, it was the perfect speedophile's toy. It is fast, handsome and most of all, it had exceptional value. It is one of the few boats to come with a really good Owners Manual that actually had some usable information in it. Too bad recognition of the virtues of this boat went into speedophile criticism. We could have for the first time, had some really good level racing in fast boats that most of us could afford, but it was not to be.

MacGregor did some experimenting with daggerboards on theses boats. Some had only one daggerboard in the port hull. MacGregor claimed that as long as the board was vertical and you sailed the boat flat, you could not tell the difference. There is a great deal of merit to that claim (see chapter on lateral plane devices) Other models had articulating daggerboards in a mildly wedge shaped trunks so you could angle the board to windward to induce lift. This was an utter failure. Surprisingly, I would have suspected that he knew better than to do that because of all the monohulls and America's Cup boats that had tried similar things and also flopped.

There is a great deal of art to using daggerboards properly for most efficiency. However, if you think real hard about it, you can understand why an articulating board just causes a lot of extra drag! If its centerline is out of line with the hull centerline it must cause drag. Aside from assembling about a half dozen and surveying seven, I delivered one from Amityville, New York to Norfolk, Virginia. An offshore distance of about 350 miles.

As usual, we had the prevailing winds. Southwest at 10 to 15 knots. This meant that the wind was mostly forward of the beam. It was a wonderful trip all the way to Cape Charles, Virginia, where we snapped our daggerboard right off at the hull line. After that, we did not get to windward so well. We had no instrumentation, making it a little difficult to record accurately. However, we made this 350 mile trip in only one night offshore sailing, which means we averaged somewhere around ten knots for the trip. Not bad for a delivery with an amateur crew and no instruments.

MacGregor 36

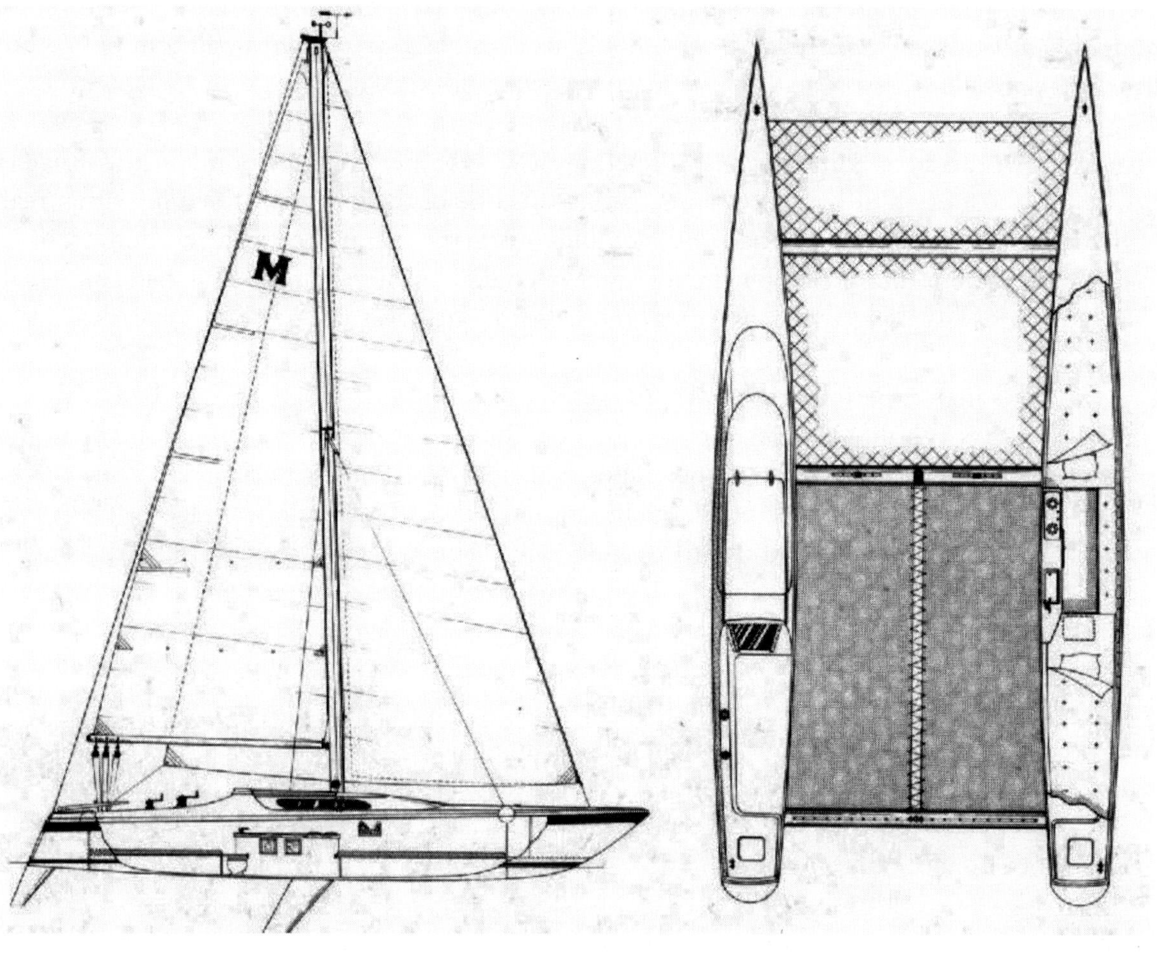

Sailing summary

MacGregor 36

1=Poor 2=Fair 3=Average 4=Good 5=Exceptional

	1	2	3	4	5
Motoring	1 (cavitate& drown)				
Docking		2			
Backing		2			
Sailing					5
Weather helm/lee helm				4	
Windward ability				4	
Ease of tacking				4	
Visibility from the helm					5
Tracking				4	
Seakindly/ride				4	
Underdeck Slamming					5
Convenient Deck Layout			3		
Interior layout			3		
Adequate storage areas			3		

Vital Statistics

MacGregor 36

LOA (length overall)	35' 6"	10.8 m
LWL (waterline length)	32' 0"	9.75 m
BMAX (maximum beam)	18' 0"	5.49 m
SA (sail area)	834 ft^2	77.5 m^2
Draft: 24" / .61m bddn	5' / 1.52m bdup	
Designed displacement: 3,800 lbs		1723kg.
Mast hgt above WL: 46' 6" (est)		14.17 m
Tankage:		
Water: gal.	liters	
Fuel: 6 gal.	22 liters	
Auxiliary power:	one/two	9.9 outboard

Type .... Open bridgedeck
Foredeck ..... Trampoline
Cabins ...... Three
Galley up/down Down
Rudders ..... Balanced spade

Maldives 32

Maldives 32 Class 4, Cruising/racing

Designer: Michel Joubert / Bernard Nivelt

An early Fountaine Pajot boat that had many interesting and novel features. The galley was against the rear bulkhead and the top deck lifted to provide headroom. This allowed the low sleek look to prevail yet give headroom at rest.

This was one of the first of their line to be taken from production and was ostensibly replaced by the Tabago 35 which itself was taken out of production in the late 90's. When Fountaine Pajot began concentrating on ever larger catamarans.

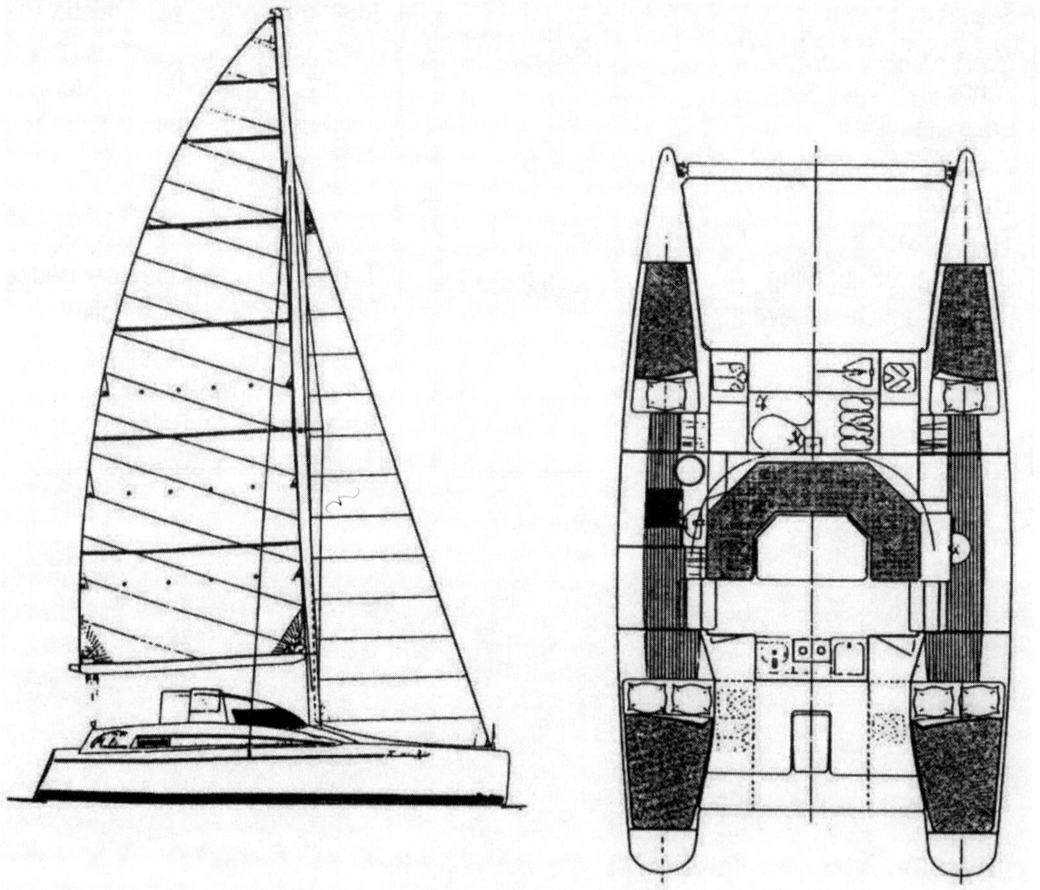

Sailing summary

Maldives 32

1=Poor 2=Fair 3=Average 4=Good 5=Exceptional

	1	**2**	**3**	**4**	**5**
Motoring			3		
Docking			3		
Backing			3		
Sailing				4	
Weather helm/lee helm				4	
Windward ability			3		
Ease of tacking				4	
Visibility from the helm			3		
Tracking				4	
Seakindly/ride				4	
Underdeck Slamming				4	
Convenient Deck Layout			3		
Interior layout				4	
Adequate storage areas			3		

Vital Statistics

Maldives 32

LOA (length overall)	32' 0"	9.90 m
LWL (waterline length)	NA	
BMAX (maximum beam)	17' 5"	5.30 m
SA (sail area)	592 ft^2	55 m^2
Draft: 3' 0"		Keels
Designed displacement:	6,612 lbs	2999 kg.
Mast hgt above WL:	43' (est)	13 m
Tankage:		
Water:	26 gal.	98 liters
Fuel:	6 gal.	22 liters
Auxiliary power:	one	9.9 outboard

Type .... Full cabin
Foredeck ..... Trampoline
Cabins ...... Three
Galley up/down Up
Rudders ..... Balanced spade
Rig .... Fractional sloop

This is a nice boat for those believing that "small is beautiful". It is cleverly laid out with four cabins in the ends and a single outboard linked to the rudders for power steering. It is a good sailing boat with a large rig. There were no escape hatches as the French Government considered the boat too small for offshore.

Norseman 400

Norseman 400, Class 4, Cruising/Racing

Designer: Alexander Simonis

The original Norseman 400 was manufactured by, Norseman, South Africa. After only four vessels, the company was sold to Voyage Yachts, SA. The bulk of my experience and the basis of this critique comes from the original Norseman 400. It is my understanding from Voyage that improvements have been made. However, I may differ from Norseman on our concepts of what needed to be improved.

Voyage yachts has discontinued the Norseman 400 and replaced it with the Voyage 430.

Very many of the South African built yachts suffer from the same problems. Those problems are: insufficient bridgedeck clearance, too fine bow entry and excessive beam. All three of those items are interrelated thus leading to some undesirable handling and sailing traits. The Norseman 400 I am most familiar with by virtue of survey and sea trials exhibits all three of those traits. The Norseman Classification and/or design intent is blue water. Much of the design shows this with the curious anomaly of not only low bridgedeck clearance but vertical forward facing flat panels on the underdeck.

Aesthetics
This is one handsome boat with nice lines.

Cabins
This four cabin two head layout really works well. The heads are large and spacious for a boat of this size and each head is located between two cabins. There are four sets of stairs going to the four cabins thus providing exceptional privacy. This puts the cabins as far away from each other as possible with no common bulkheads.

Galley up/down
This lavish galley-up layout could be the envy of all boats in that size range. From the perspective of galley and saloon layout, you can really see where that enormous beam makes its best use. The galley features a very comfortable three burner stove with oven. By comfortable I mean easy to use, well placed controls and a sufficiently large surface to actually accommodate three burners so that normal size pots will fit on them.

The cavernous refrigerator and separate freezer are located on the opposite side of the "U" shape thus keeping those items away from the stove. The double sink is also located on that side. The "U" shape is interrupted somewhat by the entrance to the stair leading to the port side forward cabin.

Saloon layout/visibility from interior

The saloon is laid out with the usual central table and surround settee. This setup evidently works well on charter boats due to the charter lifestyle. Frankly, with all that floor space I would prefer more comfortable arrangements for owners who are going to be on board for long periods of time. The entire interior of the vessel is gel-coat. This serves a very practical function but many people complain that it is too antiseptic looking. Frankly, after you are living aboard for a while and have the vessel suitably decorated you will never even notice the bulkhead cosmetic material.

Keels/Rudders

The boat has low aspect ratio fin keels. The keel is sectioned off into tanks and appears to create a useful safety factor in the event of a catastrophic grounding. The two tanks are physically well separated, one being just empty for buoyancy and the other is the ship's holding tank. A clever arrangement. This is one of the very few boats where the bilge pumps are mounted in recessed sumps so you can actually pump the bilges dry if required. The boat also has skeg mounted rudders which I much prefer over the more common spade rudder, especially for a cruising boat.

Cockpit

The cockpit has an unusual shape due to the aft cabins. It is also a big step down into the cockpit which is difficult for the less agile. The main bulkhead is quite high and a novel approach to visibility, seating and storage is in the form of two mounted Igloo coolers on the cabin tops to be used for the mentioned purposes.

Sheet winches are accessible as are other sail and engine controls. The base of the mast is reasonably accessible.

The vessel I sailed handled reasonably well. It tracked nicely and responded to the helm with reasonable assurance. Maneuvering under power was excellent. Visibility from helm was satisfactory, there was sufficient storage in deck lockers, deck layout worked well, but the cockpit is really unusual.

The engine rooms give above average engine access even though they are entered through the aft cabins. It is not necessary to get under bunks to service engines which is a big help.

The rig was simple and easy to control. The full batten main raised easily on a double purchase main halyard. So many mainsails are hard to raise because of inward pressure on the battens. The Norseman has considerable aft-rake to the mast and I believe this alleviates much of the pressure. (It does on my boat)

Common problems

The forward windows are much too large. It appears almost impossible to keep them from leaking. They are simply too large for the panel they are being supported by. There is always a certain low level amount of torque in any flat panel on any boat. It is my opinion that because the window is so large in relation to the panel it is mounted in and that panel has insufficient stiffening for to support it that the imperceptible slight torque is sufficient to break the seal between the window and the frame. The results of differing coefficients of expansion between the plastic windows and the FRP cabin top exacerbate the situation. The brutal tropical sun can destroy anything and given an opportunity like this it is almost a foregone conclusion.

There are no metal fasteners in the windows. I believe that if there were, it would make the situation even worse. Perhaps replacing polycarbonate or acrylic windows with auto safety glass might be a viable solution. In addition to the leakage problem, the huge windows leave little room for curtains. There is simply insufficient edging to put proper curtain rods. Because the windows are radically sloped and have no overlapping roof line to shade them, they require extensive exterior sun covers which negates the beauty of the huge windows.

Underdeck slamming

Strangely, the boat has almost vertical panels on the underdeck. On one boat I surveyed, the slamming was obviously so severe it tore the gel-coat right off, exposing the bare fiberglass. There were also cracks beginning at the junction of the underdeck and the main hull. Underdeck slamming and nose diving are complaints I hear from owners. The extremely fine bows leave insufficient buoyancy forward to prevent nose diving. This tendency is exacerbated by the extra five feet of beam.

Anchoring

Perhaps a more important aspect than the sailing performance of a cruising boat is the ease and reliability of its anchoring system and setup. In recent years, marketing types have tried to disguise the dirty old anchor in a number of ways and Norseman is no exception. Let me begin by reminding people purchasing cruising boats that you will spend over ninety percent of your time at anchor or at a marina. Your anchor and your ability to use it safely and properly is of utmost concern. Any tricky little ploy to hide it, get it out of the way or keep you from safely getting your hands on the chain, rode or windlass may cost you some body parts that are dear to you.

Unfortunately, Norseman is no exception. At least the anchor is on the bow, where it belongs, but they run the chain through a tube back to the chain locker. The controls are the pneumatic types that commonly are stepped upon for operation. However, they are down inside the locker. A good place to keep them from harms way but a terrible place for safe utilization and visibility . The clearance between the windlass and the chain tube exit is much too small for safety. The windlass itself is below deck level thus negating any other use for it and requiring your hands below decks for service or to clear a simple jam.

Having a good windlass up on deck where you can use for all sorts of other purposes such as kedging off, raising the mainsail, hauling a second anchor, etc. is an important item. Having the chain captive in a small tube is simply inviting disaster. Just think of the ramifications if you are raising the anchor and the chain drags something into the tube and jams it? Or, the person operating the throttles runs ahead of windlass retrieval and the chain jams between the windlass and the tube? There are simply too many potential disaster scenarios with this setup to let it go by unchallenged.

The big problem here is the truly savvy and experienced sailors subliminally take note of those problems and avoid most catastrophes simply by experience, thus scoff at these criticisms. Yet in the real world, the overwhelming majority of their customers are rank novices and really need better safety engineering than that.

Trampolines
The trampolines on this boat are divided into two mirror image sections with a promenade through the center. (This fixture contains the previously discussed anchor tube.)

Rub rail and toe rail
There is a marginal rub rail. There is no toe rail. Both of these items are important for a cruiser.

Norseman 400

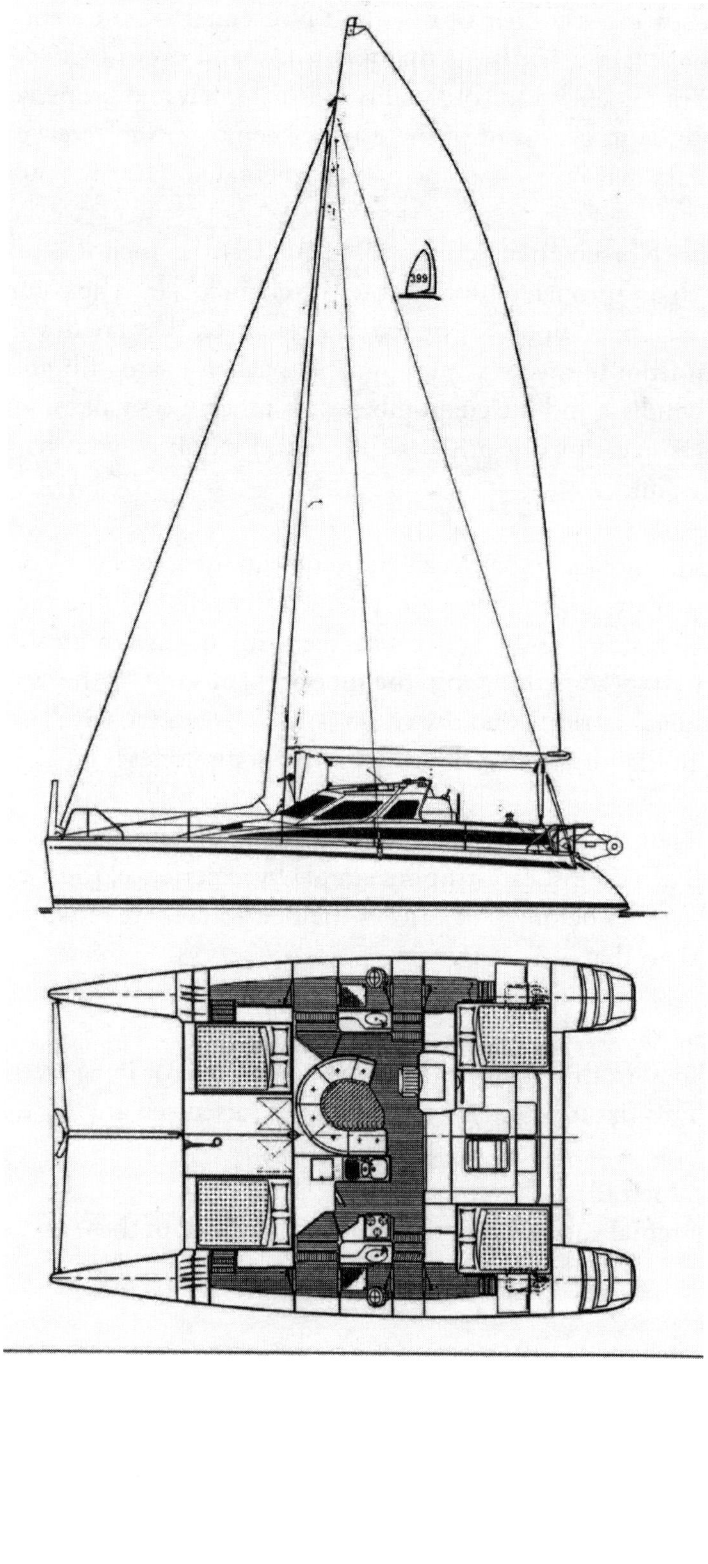

Sailing summary

Norseman 400

1=Poor 2=Fair 3=Average 4=Good 5=Exceptional

	1	**2**	**3**	**4**	**5**
Motoring			3		
Docking			3		
Backing			3		
Sailing				4	
Weather helm/lee helm				4	
Windward ability			3		
Ease of tacking			3		
Visibility from the helm			3		
Tracking				4	
Seakindly/ride		2			
Underdeck Slamming		2			
Convenient Deck Layout			3		
Interior layout				4	
Adequate storage areas			3		

Vital Statistics

Norseman 400

LOA (length overall)	40' 0"	12.19 m
LWL (waterline length)	38' 2"	11.64 m
BMAX (maximum beam)	25' 0"	7.60 m
SA (sail area)	1130ft^2	105 m^2
Draft: 3' 3"		Keels
Designed displacement:	11,089 lbs	5029 kg.
Mast hgt above WL:	65' (est)	19.8 m
Tankage:		
Water:	450 gal.	1700 liters
Fuel:	119 gal.	450 liters
Auxiliary power: Twin Yanmar diesels		

Type . . . . Full cabin
Foredeck . . . . . Trampoline
Cabins . . . . . . Four
Galley up/down Up
Rudders . . . . . Balanced spade
Rig . . . . Fractional sloop

Ocean Winds 33

Ocean Winds 33 Class 5, Cruising

Designer: Pat Patterson

I surveyed the one that Pat Patterson sailed up the Amazon River. It was quite a boat. Heavy, rugged, and well equipped. However, it had huge bulbs added to the bows and I suspect it was done to quiet a pitching problem. If you are considering purchasing one, take this into account, check it out.

I sailed this boat in the North fork of The St. Lucie River, Florida. The air was quite light, I got the boat to tack, jibe beat reach and run modestly well. Considering its size and equipment it did well.

Even though I have surveyed two of these, boats I do not have a floor plan in my files. However, the accommodation plan is virtually identical to its little sister the Heavenly Twins 27.

Sailing summary

Ocean Winds 33

1=Poor 2=Fair 3=Average 4=Good 5=Exceptional

	1	2	3	4	5
Motoring			3		
Docking			3		
Backing			3		
Sailing		2			
Weather helm/lee helm			3		
Windward ability		2			
Ease of tacking			3		
Visibility from the helm				4	
Tracking			3		
Seakindly/ride			3		
Underdeck Slamming			3		
Convenient Deck Layout			3		
Interior layout			3		
Adequate storage areas			3		

Vital Statistics

Ocean Winds 33

LOA (length overall)	33' 0"	10.6 m
LWL (waterline length)	26' 0"	7.92 m
BMAX (maximum beam)	16' 0"	4.88 m
SA (sail area)	594 ft^2	55.1 m^2
Draft: 3' 3"		Keels
Designed displacement:	11,000lbs	4990kg.
Mast hgt above WL:	42' (est)	12.8 m
Tankage:		
Water:	40 gal.	180 liters
Fuel:	40 gal.	180 liters
Auxiliary power:	Twin 13 hp Volvo diesels	

Type .... Full cabin
Foredeck ..... Trampoline
Cabins ...... Three
Galley up/down Down
Rudders ..... Transom hung
Rig .... Masthead sloop

Packet Cat 35 Class 5, Cruising

Designer: Bob Johnson

No question about it, the Packet Cat 35 is a new and different concept for a cruising catamaran. Knowing that in any field of endeavor, breakthroughs often come from outside that field, I squirmed with impatience waiting for my first sail on the Packet Cat. The production cruising catamaran itself, first introduced in the late 1950's and early 1960's, was a breakthrough from outside the traditional field of multihull sailboats.

A multihull orthodoxy has grown up which has as its central theme, “In a multihull, if you haven't got speed you haven't got anything.” The designs and the numbers of sales in recent years of production cruising catamarans deny that premise. Many people are taking advantage of the other sterling qualities of cruising catamarans such as shallow draft, non-heeling level sailing, seakindliness, large deck area and interior volume. These cruisers are willing to sacrifice the possibility of high speed (and I stress possibility) for creature comforts. The concern of this multihull orthodoxy, shared by me, was that the bow design of the Packet Cat would simply turn the boat into a huge blundering barge in serious weather.

November 27, the rain and West wind whipped the Manatee River into a turbulent froth. The River channel extended almost due west, through a pass and out into Tampa Bay, Florida. I had mixed emotions about a sail in that type of weather, yet I wanted it rough as a cob to obtain a realistic idea of the boat's seakindliness.

Two 27 horsepower diesels spread wide apart by the hulls of this catamaran made short easy work of getting away from the dock, thus passing the first test of a user friendly cruising boat. Under those same conditions, a single-screw boat, would have had a difficult time getting out and away from being pinned against the slip by the winds. Once away from the dock, with the wind directly on the nose, we motored out the inlet into the teeth of a nasty chop, low tide exacerbating both the period and height of the waves. At times, the waves were three to five feet and very close together. That was exactly the situation I wanted. I had to answer the question; would the Packet Cat slam under these conditions? Would she just push a huge frontal wave like a barge?

In all fairness, it must be said that all boats of this size range (powerboats, keel boats and catamarans) will slam under these conditions. It is simply a matter of comparison of how badly they will slam. Starting from the worst case, there is the kidney-jarring, confidence-shaking, smashing and banging, scenario usually associated with cathedral hull powerboats or IOR style monohulls. Then, working up, there is the mild bump when lunging forward off the occasional odd-direction wave and smashing headlong into the next oncoming wave, condition. If we construct a one through five comparative measure, with one (1) being terrible and five (5) being superb, the Packet Cat, without doubt, ran a good solid four (4). There was no doubt in my mind that the Delta-Pod-Bow achieves its objective of preventing the bows from plunging headlong, deep into the oncoming waves. I spent more than an hour lying on my stomach with my head over the bow watching the wave action between the hulls. I was watching for what some believed would be the inevitable result of this hull shape. It never occurred. Green water never came aboard any place whilst boats alongside of us were being swept from stem to stern.

For those just beginning their consciousness raising about cruising catamarans, I must inject that bridgedeck clearance has long been a major drawback to using catamaran technology for cruising. The bridgedeck is the bottom of the supporting structure between the hulls. A catamaran of less than forty feet, built with the bridgedeck low enough to have reasonable headroom in a central cabin, may often slam unmercifully, especially when overloaded. If it is built high enough not to interfere with the boat's sailing qualities, on a smaller boat it almost precludes having a central cabin unless a person is willing to make substantial sacrifices to aesthetics. Several recent catamaran designs have unusually high freeboard to control this problem. Bob Johnson, the Packet Cat designer, chose another approach, The Delta Pod Bow® This bow configuration absorbs the tendency to plunge thus lessens the slamming. The technology is not without precedent. Lock Crowther, the popular Australian designer, has a wave piercing ferry design using a similar bow concept. More recently, a rash of vessels with that design have cropped up.

Packet Cat tacked, jibed, beat, reached and ran with reasonable level of performance. I estimated our speeds to be higher than a comparable monohull cruising sailboat under the same conditions. We easily sailed seven to eight knots, tacked with reasonable aplomb with no need to backwind the jib, and handled all points of sail without any particular bad habit showing up, although I would have liked a little firmer helm. The balanced spade rudders, which are shaped like a DC3 tail section, create the lack of even a little weather helm. This gave me a lack of feeling at the helm when we were reaching. We did this with eight adults in the cockpit trying to keep warm and dry.

That much weight so far aft, subsumes the boat being out of trim, yet the boat shrugged it off performing within reasonable parameters. The boat had little tendency to pitch. Pitching, or hobby-horsing, can be the bane of some catamarans, especially at that size range. Long narrow hulls with fine ends tend to oscillate fore and aft. Its called hobby-horsing. The asymmetrical water plane (footprint) of a monohull in that size range dampens the tendency. It did, however, heel more than many wider cats. It heeled as much as five degrees when hard pressed.

The mainsail handling system was a delight, requiring only two lines to perform all functions. The sail rolls up around the boom and there is a halyard to raise it and a furling line to retrieve it. Operation was simple, secure, uncomplicated and effective. The jib was a roller reefing jib that worked well. All lines led to the cockpit and did not interfere with the usage nor leave tails strewn helter-skelter around the deck, a very important point. Later models have used in-mast furling rather than in-boom furling for a variety of reasons.

For a catamaran, this boat has many unique features. It is the most monohull-like cat I have ever seen. From the configuration of the cockpit, location of the helm, and interior layout, it is more like a typical monohull than a typical catamaran. The engine rooms are a pleasant surprise to most. Unlike so many modern boats, instead of having equipment buried in esoteric hard to service parts of the vessel, the engine rooms contain all the operating systems. They are clear, uncluttered, well laid out and exceptionally large with full standing headroom. However, the engine noise level is high and a good sound insulation job is called for.

This first sail, whet my appetite for more extensive sailing. While definitely not in the catamaran speed-demon class, its performance is better and more seakindly than many comparable cruising monohulls, that is motor sailors, not pure sailboats. Although I admit, it gets awfully hard to tell the difference nowadays. Speeds were definitely better, between 5 and 15 percent faster, depending upon the point of sail. Without doubt, this is a first-class vessel.

Workmanship, layout, equipage, innovation and material all spell top quality, typical of Island Packet Yachts. The truly unique features of this vessel really number three major items and a host of secondary ones.

First, is the interior layout. There are two, mirror image "apartments." Apartment really is the correct term rather than stateroom or cabin because each apartment is totally private and complete with its own queen size bunk, hanging locker, dressing area and full service bathroom with separate shower stall. Again, the term "head" just doesn't quite make it in this context. Sliding doors access the apartments from the main saloon and galley area. Engine rooms are accessed through doors in the shower compartments. Engine rooms have full standing headroom. Doing anything in the engine room is a breeze.

The galley area is ergonomic in its horseshoe layout. I do have two criticisms of it. First, there is only a single basin sink and second, it still has a top loading refrigerator much like it's monohull ancestors. (In a monohull, you always find the galley sink about the center of the boat and the galley stove mounted on the fore & aft axis, that is because the boat heels. In a catamaran you lose those restrictions. See chapter on galleys in book 2)

Second, is the cockpit and deck layout. It is four steps up from the saloon to the cockpit just like a monohull. This gives you a monohull-like cockpit with a wheel mounted on a central binnacle and best of all, full forward visibility for those sitting in the cockpit. Forward visibility for anyone but the helmsman is very rare in catamarans.

The sail controls are utter simplicity as are the placement of sheet leads and spaghetti boxes. The solid foredeck has a unique to the field, full lounge area, complete with cushions. If you do not want to lounge, you can sit on the forward facing bench seats.

Third, is the immensely practical engine and drive setup. For any of you out there thinking about selling your sailboat and buying a trawler, STOP! Go look at this boat. It can outperform just about any comparable trawler in the same size range, especially in maneuverability. It has all the accouterments of your favorite sailboat, it does not have the rolling associated with trawlers and you won't need flopper-stoppers at anchor. It is far more seakindly and you will find yourself voyaging when others are still at anchor because you perceive conditions to be better, just like sailing catamarans. But best of all, when all is said and done, it still sails quite well and is a proper sailboat.

When the Packet Cat was first introduced there were howls of protest and screams of righteous indignation that someone other than an orthodox multihull designer had the temerity to introduce a boat with features not approved by the speedophile crowd. Specifically was the Delta-pod bow configuration. In fact, Bob Johnson, a world class highly respected designer simply used an accepted technology from another part of the marine world to solve a very real, small catamaran problem, pitching (hobby horsing) Always the bane of catamarans and some monohulls in that size range, the Delta-pod bow cleared it up. There was none and there was no burying the bows in waves as some of the skinny bow cats are prone to do.

I often have the thought that this will eventually become a sought after premium vessel with a vigorous following.

For those who understand the requirement to barter potential maximum performance in exchange for creature comfort, utility, and heavy construction, Packet Cat has many attributes.

Sailing summary

Packet Cat 35

1=Poor 2=Fair 3=Average 4=Good 5=Exceptional

	1	2	3	4	5
Motoring			3		
Docking			3		
Backing			3		
Sailing				4	
Weather helm/lee helm				4	
Windward ability				4	
Ease of tacking				4	
Visibility from the helm				4	
Tracking				4	
Seakindly/ride				4	
Underdeck Slamming				4	
Convenient Deck Layout			3		
Interior layout			3		
Adequate storage areas			3		

Vital Statistics

Packet Cat 35

LOA (length overall)	35' 0"	10.6 m
LWL (waterline length)	31' 0"	09.4 m
BMAX (maximum beam)	15' 0"	04.5 m
SA (sail area)	600 ft^2	5.49 m^2
Draft: 2' 6"		Keels
Designed displacement:	11,000 lbs	5,000 kg.
Mast hgt above WL:	50' (est)	15.1 m
Tankage:		
Water:	100 gal.	378 liters
Fuel:	50 gal.	189 liters
Auxiliary power:	Twin 27 hp diesel	

Type . . . . Full cabin
Foredeck . . . . . Full solid foredeck
Cabins . . . . . . Two
Galley up/down Up
Rudders . . . . . Balanced spade
Rig . . . . Fractional sloop

Packet Cat 35

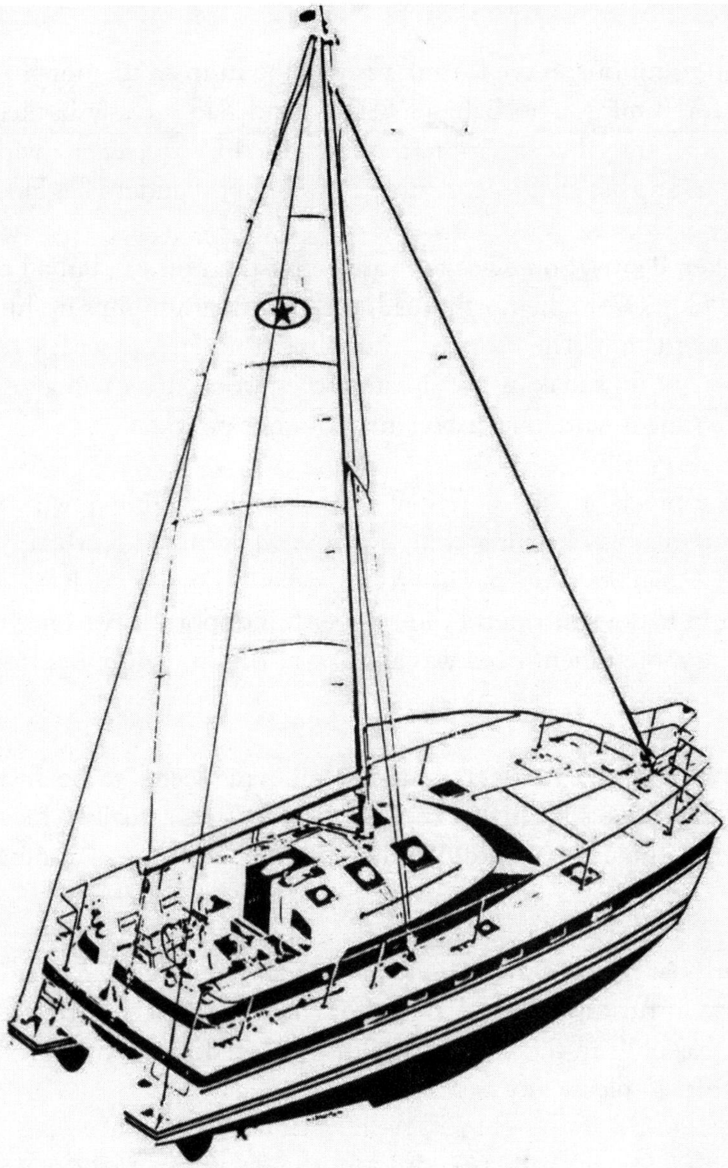

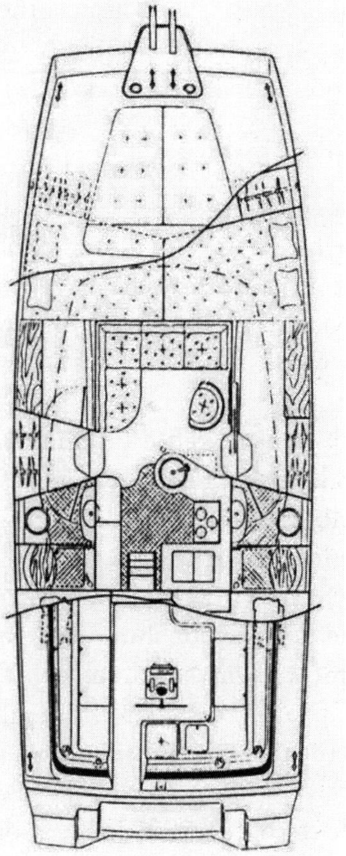

Privilege 43

Privilege 43 Class 5, Cruising

Designer: Marc Lombard

April 26, 1992. The weather report: ominous. A cold front was moving in from the north as we left Port Everglades, Florida, bound for Bimini, The Bahamas. The wind had already backed to the northeast and was gusting to 15 knots. Into that situation, the Privilege 43, *Expectations*, with owner and wife aboard, with me as skipper, and assisted by Corinne, departed Ft. Lauderdale, Florida.

Crossing the Gulf Stream is not taken lightly. Some passages are better than others, but all crossings get proper respect. When the wind blows from the north quadrant, the stream humps up huge steep seas. A classic, wind against the tide situation. The vectored course line from Ft. Lauderdale to Bimini has a considerable southerly component thus making the situation even worse. If a catamaran is going to pound its bridgedeck, this is the time it will do it. *Expectations* did not pound.

Following seas prevented us from appreciating the potential sailing ability. Sailing downwind is no test of any vessel except those particular sea keeping abilities. *Expectations* handled the downwind sleigh ride in good fashion. She never had any tendency to broach, never pounded the bottom, never felt insecure at the leeward bow and had a soft ride. I did not get full opportunity to put the boat through its sailing paces until later; a more urgent need was arriving at Bimini before the front came through.

The steering felt a little mushy. I believe the rudders are too small. This seems to be an industry tendency because designers design the rudders for higher speeds than the users actually achieve. From a close quarters handling perspective this is not a problem because maneuvering is done with the engines.

The deck layout shows the charter boat concept. Its design is for a maximum of passenger creature comfort, with a minimum of crew intrusion. Much design thought went into making passengers happy. There are large lounging areas both in the cockpit and on the foredeck trampolines. Below, there are sufficient space and amenity to please any passenger.

The stern quarter helm location is ideal for a charter boat. It keeps the operation of the vessel clearly separated from the passengers. The passengers can come and go with little worry about tripping over lines or interfering with the helmsman. On the other hand, basic forward visibility is so poor we almost ran into a freighter leaving Port Everglades inlet! (Yes! We almost hit him!)

Being a charter boat, it has cabins everyplace. Five double cabins and five heads. The cabin in the central nacelle is designed so the crew are completely separated from passengers, an ideal setup with maximum privacy for all. (This is a prime example of the use of extra-wide beam)

There is a three cabin, master suite, private owner option. Since this vessel, Privilege (Jeantot Marine) has come out with a multitude of other models.

Sailing summary

Privilege 43

1=Poor 2=Fair 3=Average 4=Good 5=Exceptional

	1	**2**	**3**	**4**	**5**
Motoring				4	
Docking				4	
Backing				4	
Sailing			3		
Weather helm/lee helm		2			
Windward ability		2			
Ease of tacking			3		
Visibility from the helm	1				
Tracking			3		
Seakindly/ride				4	
Underdeck Slamming				4	
Convenient Deck Layout					5
Interior layout				4	
Adequate storage areas			3		
Dual outside helm stations					

Vital Statistics

Privilege 43

LOA (length overall)	42' 8"	13.0 m
LWL (waterline length)	41' 4"	12.6 m
BMAX (maximum beam)	24' 0"	7.30 m
SA (sail area)	1292 ft^2	120 m^2
Draft: 3' 11"		Keels
Designed displacement:	18,700 lbs	8482 kg.
Mast hgt above WL:	63' (est)	19.2 m
Tankage:		
Water:	148 gal.	560 liters
Fuel:	106 gal.	40 liters
Auxiliary power:	Twin 27 hp diesel	

Type .... Full cabin
Foredeck ..... Trampoline
Cabins ...... Three/five
Galley up/down Down
Rudders ..... Balanced spade
Rig .... Fractional sloop

Privilege 43

Privilege 43

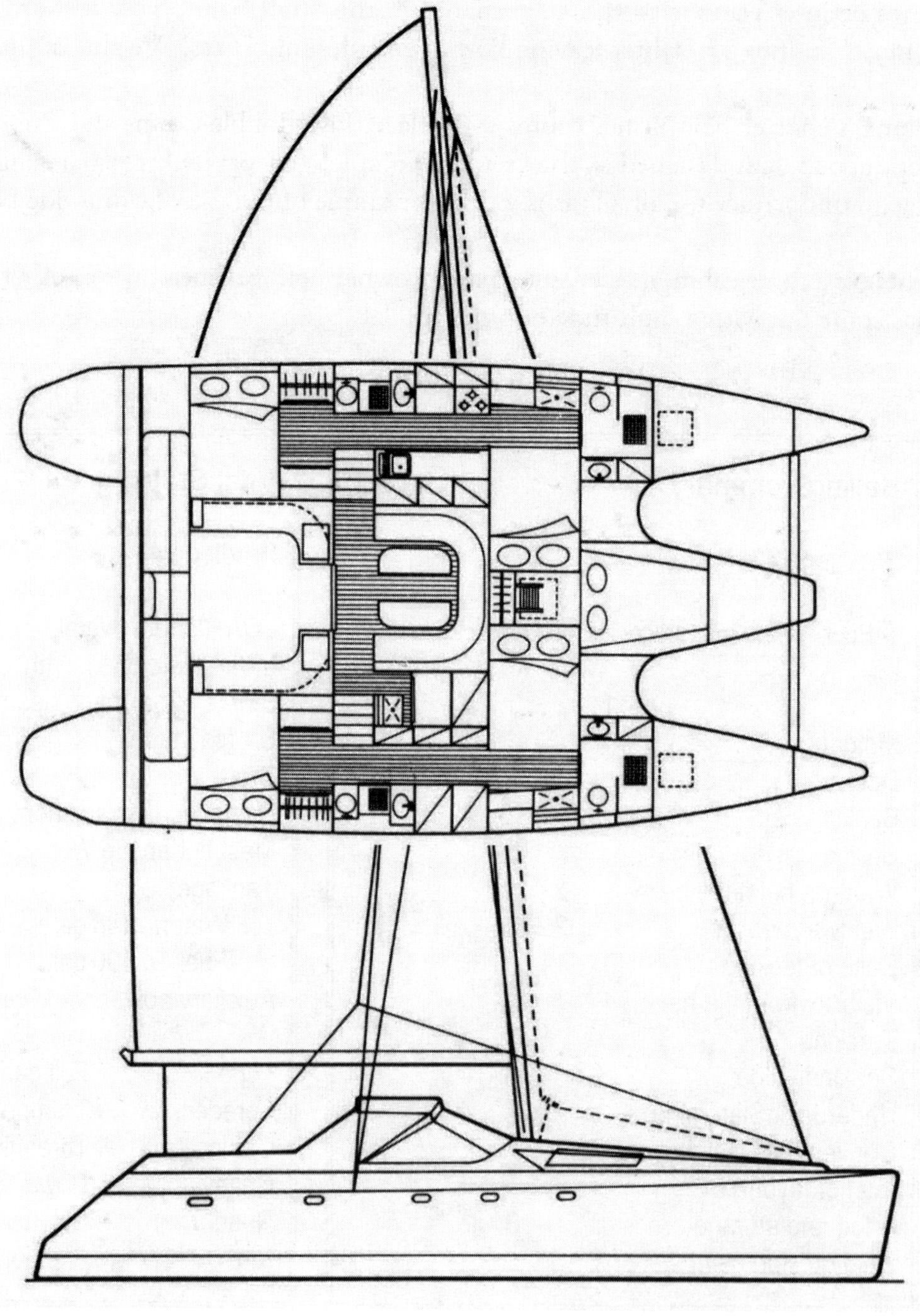

Sable 36, Class 5, Cruising

Designer: Ron Carlson

April 1992 saw me leaping from a marina dock in Ft. Lauderdale, Florida to the decks of a brand new *Sable.* The twin screws kept the bows right into the strong current. That is their job and they do it well.

After the usual travail of getting through Port Everglades inlet, we had to face the fact that this was a true shakedown cruise. The boat was not completely assembled, nor had its reefing lines been installed. A fact with which we would soon come to grips.

It was a brisk South Florida day. Northeast winds gusting to thirty knots with seas running from three to five feet were our lot. Maybe it was a little rough for the timid, but just right for testing a new catamaran.

We put *Sable* through her paces. We tacked, jibed, reached and beat. We reefed and shook out reefs. The newness of the boat hampered us, but it was a worthwhile exercise. The search for bad habits knows no bounds.

This is a powerful boat. It was a pleasant surprise to see how well it carried full sail under the conditions. As the wind built, we reefed. *Sable* always sailed flat, with a neutral helm, no matter what point of sail. We did not have the opportunity to see just how high it would point because of conditions. We did, however, beat back into the inlet under reefed main and reefed jib. It was a satisfying performance. I have not been aboard a true class 5 cruising boat that sailed any better than this *Sable.*

You get a good feel of a boat hanging around waiting for draw-bridges, crossing powerboat wakes and dodging jet-skis. A busy narrow channel really shows up a boat's bad habits and clumsy traits. The *Sable* has none. During the entire sail, through the inlet across the sportfisherman wakes, out in the ocean in high winds and all the miscellaneous possibilities, we never had a wave pound the bridgedeck.

This boat is hull number one. Therefore, it is premature to criticize appearance and finish. In both cases, I would have been happier if it were better. The interior layout sports a conventional layout.

Sable 36

Finish is adequate, but not luxurious. Machinery and plumbing are above average, a pleasant change from others I have sailed. What makes the boat appear like a looming colossus, is the high cabin. The builder assured me that profile would be much reduced on the next boat. The rakish shear stripe does help the lines.

The original Loch Crowther design had daggerboards. The builder replaced them with fixed, low aspect ratio keels. Having sailed the original and this model, I could see the difference, but I wonder if that difference bares any relationship to a class 5 cruising boat.

Sailing summary

Sable 36

1=Poor 2=Fair 3=Average 4=Good 5=Exceptional

	1	**2**	**3**	**4**	**5**
Motoring			3		
Docking			3		
Backing			3		
Sailing			3		
Windward ability			3		
Ease of tacking			3		
Visibility from the helm			3		
Tracking				3	
Seakindly/ride				4	
Underdeck Slamming				4	
Convenient Deck Layout				4	
Interior layout			3		
Adequate storage areas			3		

Vital Statistics

Sable 36

LOA (length overall)	36' 0"	10.97m
LWL (waterline length)	31' 7"	9.66 m
BMAX (maximum beam)	15' 6"	04.7 m
SA (sail area)	585 ft^2	178 m^2
Draft: 2' 11"		Keels
Designed displacement:	13,500lbs	6123kg.
Mast hgt above WL:	49' (est)	14.94 m
Tankage:		
Water:	290 gal.	110 liters
Fuel:	48 gal.	182 liters
Auxiliary power:	one	9.9 outboard

Type	Full cabin
Foredeck	Trampoline
Cabins	Three
Galley up/down	Down
Rudders	Balanced spade
Rig	Fractional sloop

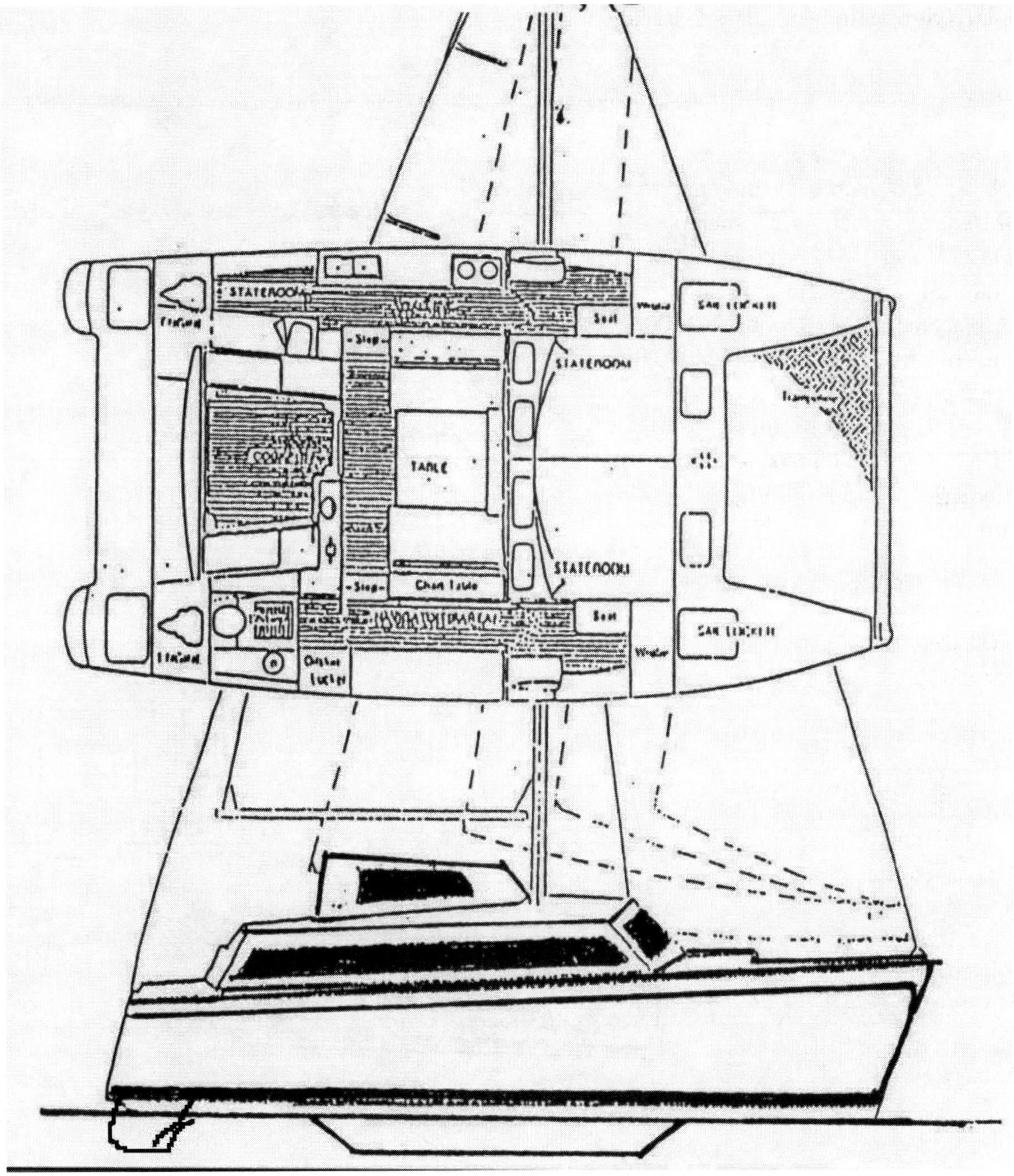
STATEROOM
STATEROOM
TABLE
STATEROOM

Solaris 24, 32, 36, 40

Solaris Sunbeam 24 Class 3, Racing/Cruising

Designer: Derek Kelsall

LOA: 24'0" feet, 7.3 meters
LWL: 22'0" feet, 6.71 meters
BMAX: 14'0" feet, 4.27 meters
SA: 295 $feet^2$ 27.40 $meters^2$
Draft: 2'1" / 0.64m Keels
Displacement: 4410 pounds, 2000 kg.
Mast height above WL: 36' / 10.97m
Tankage:
Water: 55 gal. 20.18 liters
Fuel: 6 gal. 23 liters
Auxiliary power: 9.9hp outboard

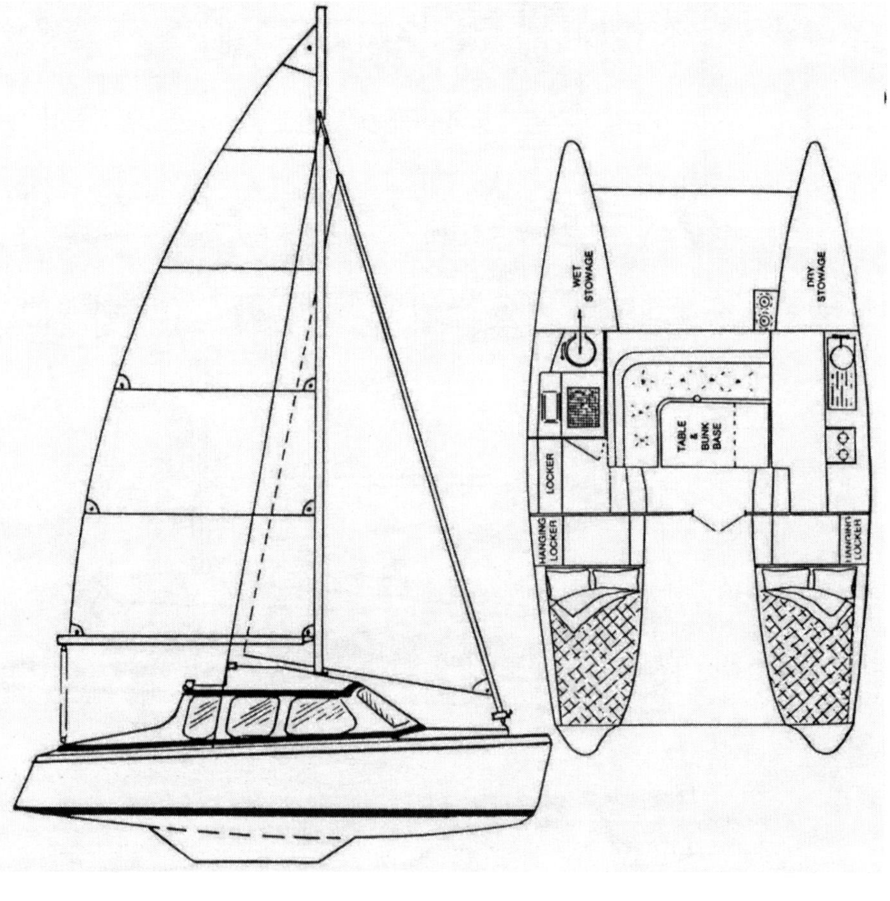

Solaris Sunstar 32 Class 4, Cruising/Racing

LOA: 31'10" feet, 9.69 meters
LWL: 31'5" feet 9.57 meters
BMAX 17'3" feet, 5.26 meters
SA: 540 $feet^2$, 50.17 $meters^2$
Draft: 3'0" / .91m Keels
Designed displacement: 8,379 lbs, 3800 kg.
Mast height above WL: 48' / 14.63m
Tankage:
Water: 63 gal. 238.46 liters
Fuel: 23.8 gal. 90.08 liters
Auxiliary power: Diesel option

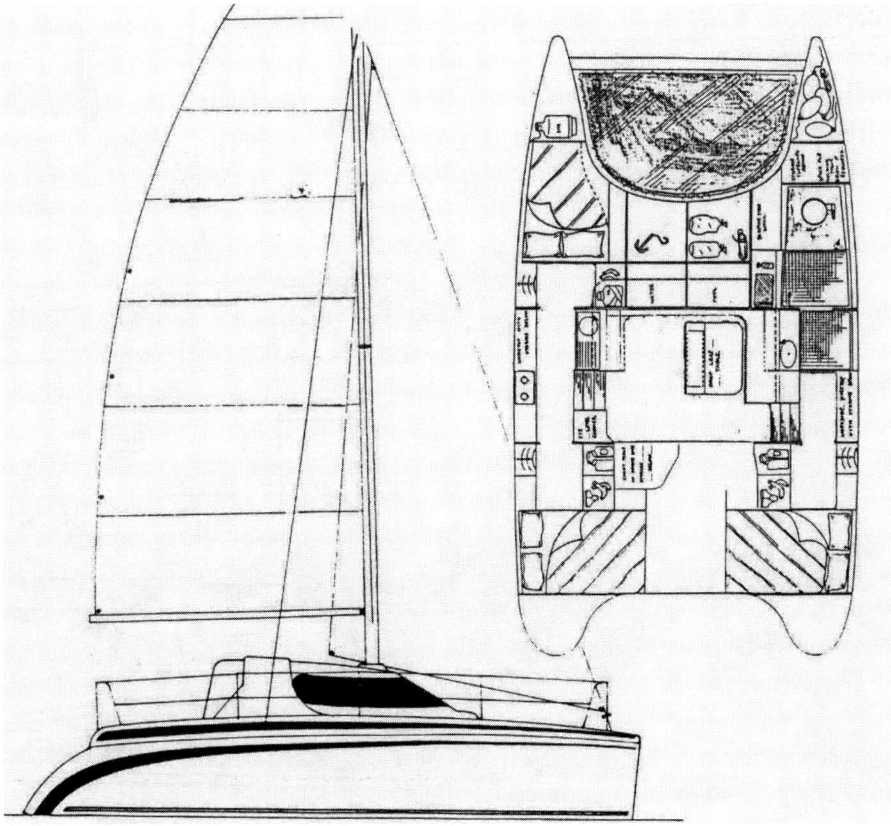

Solaris 24, 32, 36, 40

Solaris Sunrise 36 Class 5, Cruising

LOA: 36'0" feet, 10.97 meters
LWL: 43'0" feet, 13.11 meters
BMAX: 15'10" feet, 4.80 meters
SA: 565 feet^2 52.49 meters^2
Draft: 2'10" / .84 m
Displacement: 9,520 pounds, 4318.27 kg.
Mast height above WL: 48' / 14.63m
Tankage:
Water: 84 gal. 317.94 liters
Fuel: 30 gal. 113.55 liters
Auxiliary power: Options, 27hp Yanmar diesel w/Sonic drive, or, 2 Yanmar 9hp saildrive

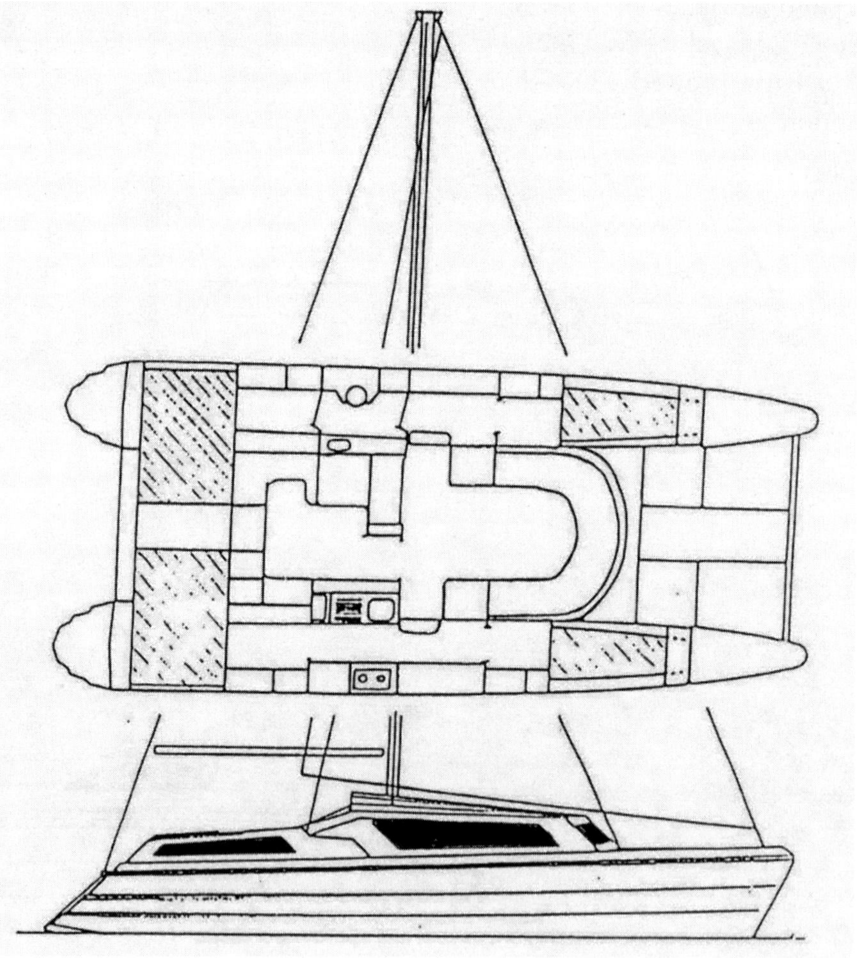

Solaris Sunstream 40 Class 5, Cruising

LOA: 39'10" feet, 12.10 meters
LWL: 37'3" feet, 11.35 meters
BMAX: 16'6" feet, 5.03 meters
SA: 663 $feet^2$, 202.13 $meters^2$
Draft: 3'6" / 1.10m
Displacement: 14,560 pounds, 6604 kg.
Mast height above WL: 53' 0" / 16.15m
Tankage: Water: 120 gal. 454.2 liters
Fuel: 65 gal. 246.03 liters
Auxiliary power: Twin 18hp Yanmar saildrive

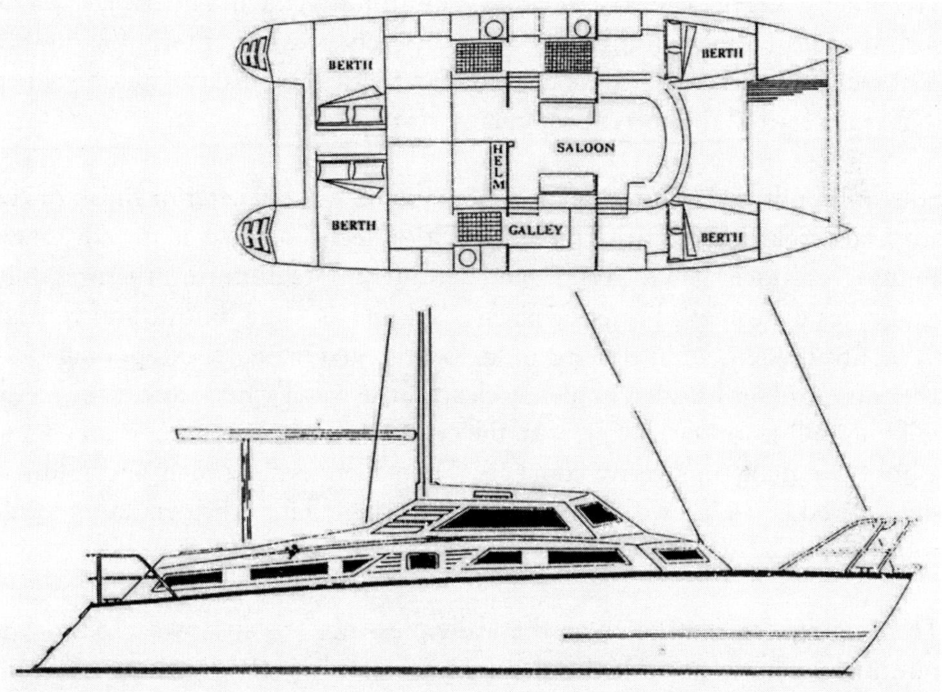

Solaris was a maufacturer of very high quality yachts. I have surveyed several and found them rugged and heavily built. It was Solaris that built Jacques Cousteau's famous Moulin A Vent.
Solaris is no longer in production.

Stiletto

Stiletto 27 Class 2, Racing

Recalling a trendsetting speedster: Stiletto 27

Designers: John V. Cloud, Bill Higgins

Out my back door, bobbing gracefully at the dock is *Bahama Hunter*, a beautifully refurbished Stiletto 27. It brings memories of a trend setting innovative boat.

The Stiletto was the first true yacht class multihull made in America. It was a production vessel built with the latest aircraft technology and met every possible criteria for superb workmanship, modern high-tech materials and exotic beautiful aesthetics and exciting performance. It is one of the vessels that really started the current catamaran revolution.

Stiletto is built from Nomex prepreg honeycomb with an exterior finish of AwlGrip. It is baked in an autoclave to kick the resin. This construction technique has many advantages some of which are an ability to produce lighter, stronger and definitely more uniform structures. Eliminating the gelcoat also eliminates the osmotic blistering plague. *Stiletto* has some unique and very advanced features such as the control system for its rotating mast/loose-footed main, an exceptional kick-up rudder system which keeps the rudder blades completely clear of the water when raised, the solid bridgedeck rather than full trampoline for the cockpit area, the center daggerboard rather than having boards in the hulls and a host of equally innovative concepts. Then there is the trailer and folding mechanism. The vessel slides together on its own extendable trailer bringing it to the maximum legal without special permits width of 8' 6".

The original *Stiletto* though aesthetically pleasing, was strictly a racer/day sailor and had the most minimal interiors possible. Demand grew and a "Special Edition" was built which had a beautiful custom interior and portlights integrated into extraordinary graphics. Nevertheless, it was still too small for any serious cruising. Some years later, the Stiletto 30 was introduced which attempted to address the internal space problems by adding wide bunks on the bridgedeck, trimaran style. This concept never caught on.

The *Stiletto* is a handy boat to manage. It has broad flat decks and even wide flat deck areas in the stern, making it simple to dock or raft up. With the mast and all controls in the solid cockpit area it is dry to sail. The intriguing flip forward fighter pilot hoods give good access to the hulls and also make a quick solution the "open hatch in-the-way problem." Cleats at the ends of the bows eliminate the need for chocks.

Stiletto could be either class two, Racing or class three, racing/cruising depending upon how the boat is fitted out. Stiletto Marketeers put much emphasis upon the performance aspects of the boat and not enough emphasis upon its other superb qualities. Thus they lost a terrific market and eventually went out of business, but not before they sold several hundred boats.

I used a Stiletto 27 for the Multihull Sailing School. The school at the time operated by Denis Blaise and myself. I took out boatloads of students from Yachthaven, Spa Creek Annapolis Maryland. We sailed out into the Chesapeake Bay, then into Back Creek and proceeded to short tack all the way up that very narrow and congested waterway. A good deal of the time, we would do it under the mainsail alone. Onlookers were incredulous. They would just stare at this catamaran, zooming in at the docks at high speed and at the last possible moment deftly tacking away without a murmur. In other words, doing what multihulls were allegedly unable to do. If there ever was a sweetheart of a boat to sail, this is it. The Stiletto 27 is a sailboat! We sailed both into and out of our slip each time.

Another experience took place in Baltimore, Maryland when I was engaged to teach new Stiletto owners how to sail it. The brand new outboard failed to start. The owners were nonplused, we simply pushed the boat out of the slip in the three-hundred slip marina and we tacked out into the Chesapeake Bay. On return, we simply sailed right into the slip on main alone. No problem.

Jim Butterworth, then with Annapolis Sailing School, was so completely taken with his demo ride on the boat that he convinced the school to buy one for his St. Croix, USVI operation. They did and he proceeded to win the Around St. Croix Race the following year sailing the Stiletto.

I also sailed in the 1979 Stiletto Nationals held at Pier 1, Kent Island Maryland. With Charlie Reddick's boat, Charlie, Tod Knowles and myself finished in a respectable fifth place. I could go on and on, but I think you understand my profound respect for Stiletto sailing ability.

I have surveyed twelve, including one severely damaged by capsize. There are a couple of improvements I would like to see made to the existing boats as they upgrade or transfer to new owners. Most important, is the addition of watertight flotation compartments fore and aft. While the Stiletto will not sink, it does not float high enough to prevent capsizing if flooded, nor to allow people inside the hull whilst flooded. It really helps if there enough buoyancy in a flooded hull for a person with a bucket to stand inside and bail. The situation now is, if you climb inside a flooded hull, your weight may be sufficient to force the hull below the level of the coamings. If that happens, more water will pour in. The boat should have enough encapsulated flotation to prevent that from happening.

The second item is a relocation of the outboard fuel tank. The original design, has the hose crossing the traveler and the tank itself is loose in the cockpit. I had once suggested incorporating the tank into the motor mount as a solution. I believe, that, or other ideas could solve this annoyance and safety problem.

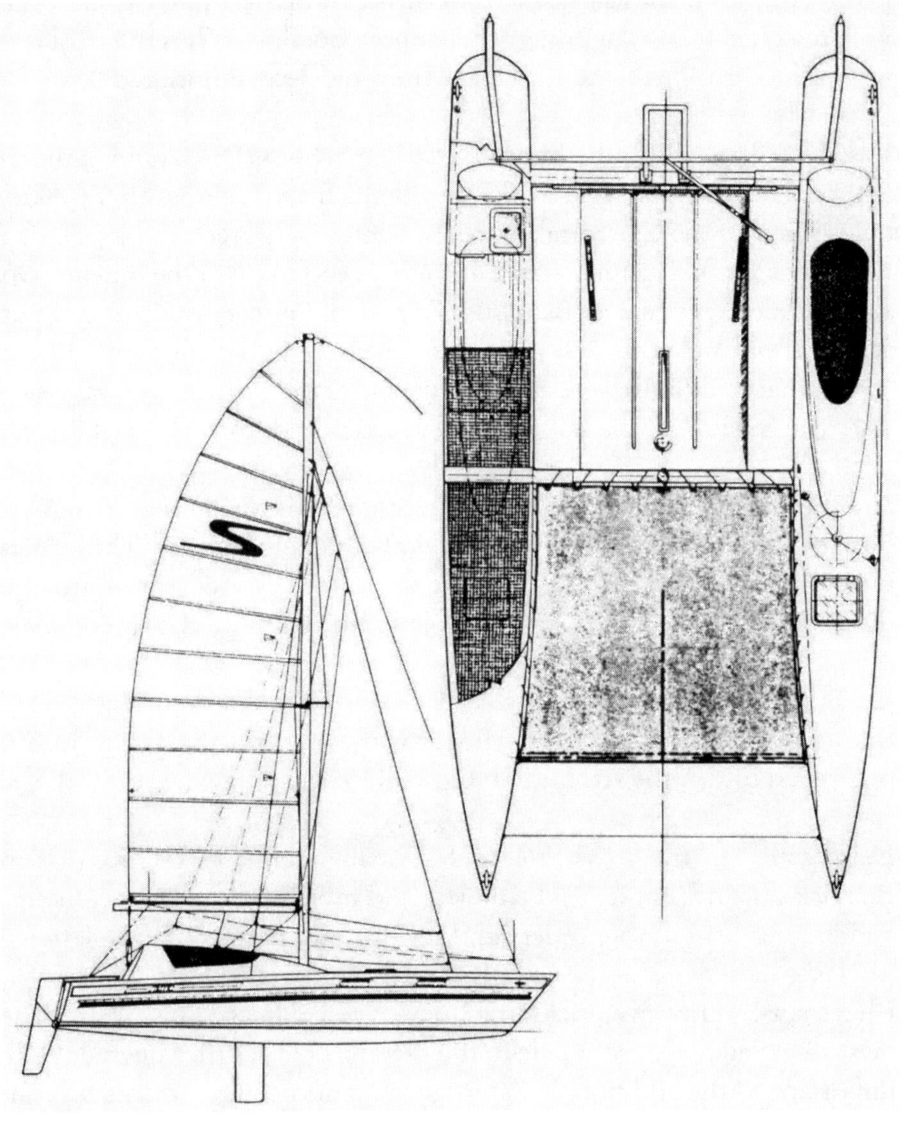

Sailing summary

Stiletto 27

1=Poor 2=Fair 3=Average 4=Good 5=Exceptional

	1	2	3	4	5
Motoring				4	
Docking				4	
Backing			3		
Sailing					5
Weather helm/lee helm				4	
Windward ability				4	
Ease of tacking				4	
Visibility from the helm				4	
Tracking				4	
Seakindly/ride				4	
Underdeck Slamming					5
Convenient Deck Layout			3		
Interior layout			3		
Adequate storage areas			3		

Vital Statistics

Stiletto 27

LOA (length overall)	26' 10"	8.17 m
LWL (waterline length)	24' 4"	7.42 m
BMAX (maximum beam)	13' 10"	4.23 m
SA (sail area)	352 ft^2	107 m^2
Draft: 10" / .25 bdup 4' / 1.22 bddn		
Designed displacement: 1,150 lbs		522 kg.
Mast hgt above WL: 36' (est)		11 m
Tankage:		
Water: 10 gal.		38 liters
Fuel: 6 gal.		22 liters
Auxiliary power: one		9.9 outboard

Type .... Open bridgedeck
Foredeck ..... Trampoline
Cabins ...... Three
Galley up/down Down
Rudders ..... Transom hung kick up
Rig .... Fractional sloop

Are there faster boats? Sure there are. Are there more weatherly boats? Sure there are. But none of those miracle boats instantly telescoped onto a specially designed trailer and ramp launched. None easily sailed in and out of an ordinary boat slip. Few were as handsome or of such superb quality and most important of all, none ever made it with the public. (It is not difficult for an individual to build a boat that will out perform a production design. A little more on the WL, a few feet of sail area, a couple of pounds less and you have it. What happens is you kill the sales for perfectly good sensible designs thus hurting the entire industry. That is a lesson for *speedophiles*.

Stiletto

Stiletto 30 Class 2, Racing

Designer: Peter Wormwood

LOA: 29' 04" / 8.95 m
LWL: 26' 10" / 8.17 m
BMAX: 17' 10" / 5.43 m
SA: 415 ft^2 / 38.55 m^2
Draft: 10' 5"/1.24 m bd up
4' 0" / 1.22 n bd dn

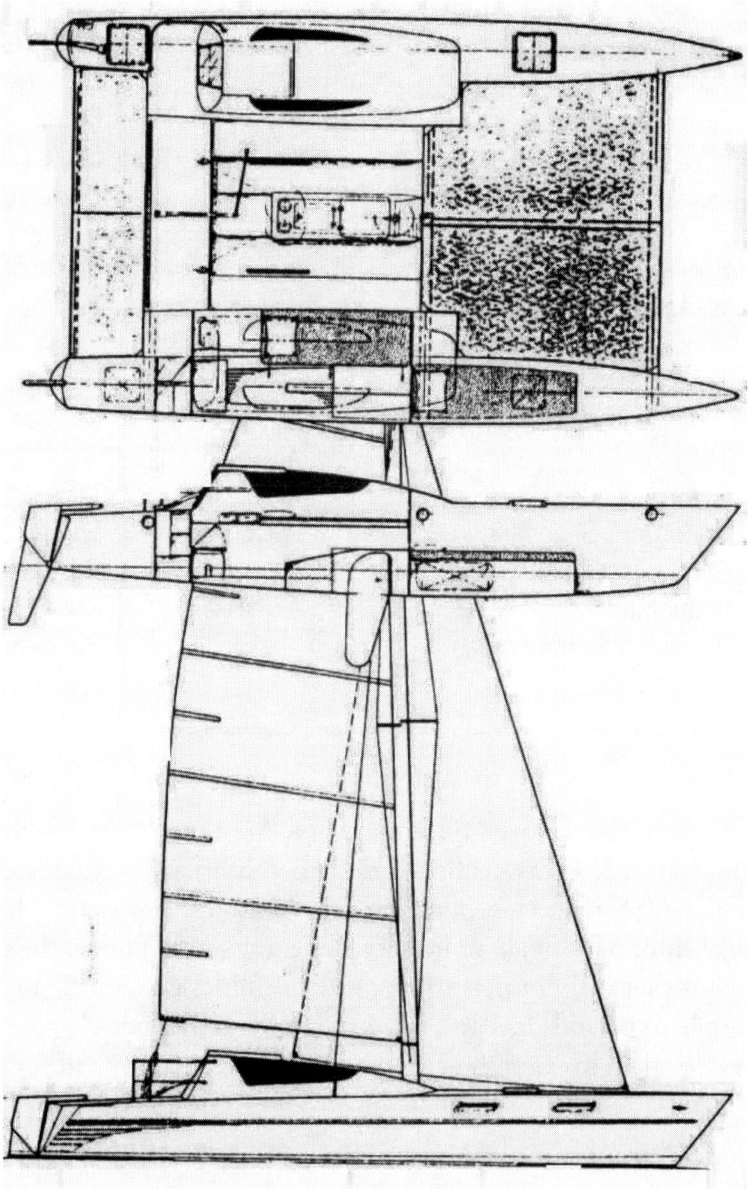

There are a few Stiletto 27 that have been modified to 30'. These vessels do not have the extended side bunks.

There is a factory in Florida that specializes in rebuilding and repairing Stiletto.

The original factory is no longer producing.

The Stiletto 30 has taken line honors in many races.

Ultimate Concept 32 Class 5, Cruising

Designer: Ron Carlson

As a standard equipment, there are twin 9.9 hp outboards. You can purchase options for something as powerful as a pair of 90's. These engines are cleverly fitted into the stern lockers. Special retracting rudders are set off-center of the hulls to allow clear access to the centerline engine installation.

Carlson is a pioneer in a really new field, that of a true cross between a powerboat and a sailboat, trying for a yet-to-be- filled niche in the market. This boat will motor at 22 knots and pull water skiers with the twin-90 installation!

If you are unaware of what is happening outside the field of multihull sailboats, let me point out that much of the world is turning to the catamaran configuration for its powerboats. The reasons for this are seakindliness, stability, lack of wake, and low drag. There are units under construction everywhere, from offshore racers, and customs patrol boats to huge passenger ferries.

The Ultimate Concept is an awesome powerboat. Even with the rig in place, which has to add not only drag but pitching moment, it is still a seakindly delight to power. According to the instrument panel, I was motoring at 20+knots, through the usual powerboat slop outside Annapolis harbor. We experienced no pounding. The boat planed easily, stayed level, and did all the things expected of a good powerboat. Those who have experienced the incredible maneuverability of twin screws on a monohull will delight in the further increased handling ability of a twin-screw catamaran. This is due to the greater distance between the engines.

So much for power, how does it sail? The boat was equipped with a Doyle Stack-Pac full-batten main and a luff-groove roller-furling genoa. According to the yachts' instruments, we sailed to wind speed on a tight close reach. That is quite good! For a few moments I could hold the boat at 8 knots with the true wind at 8 knots about 12-14 apparent. Close-hauled, we were easily able to move at half the wind speed. With the apparent wind indicator on 30, we could hold between 6 and 7 knots with the wind-speed indicator between 12 and 14 knots. The true wind speed was between 5 and 10, gusting to around 12. Downwind, or broad reaching, we sailed at 3/4 wind speed with both main and Genoa drawing; that is, with the wind coming over the stern quarter, but not far enough aft to blanket the Genoa.

Ultimate Concept 32

The Ultimate Concept has a very good reefable roller-furling Genoa of about 150%. Reaching and running, we used full sail; beating, we rolled in about 25% of the Genoa area without any noticeable difference in speed, but the motion was more comfortable, as a full sail was a little too much to use upwind in high winds.

One aspect I did not care for was the turbulence caused by the motor well openings, in the bottom of the hulls, and the off-center rudder configuration. The rudders are far enough off center to become surface-piercing foils. I also believe this hampered our ability to tack, for we managed to mess up nearly every time we did. The original English boats tacked with predictable efficiency. However, the turbulence did not seem to affect our sailing performance but, then we were lightly loaded. Perhaps I was part of the problem, I am used to tiller steering and this boat was steered hydraulically, with a wheel, a system notorious for its lack of feel. All told, I believe the boat retained the ability to sail very well, while also being a high-performance powerboat.

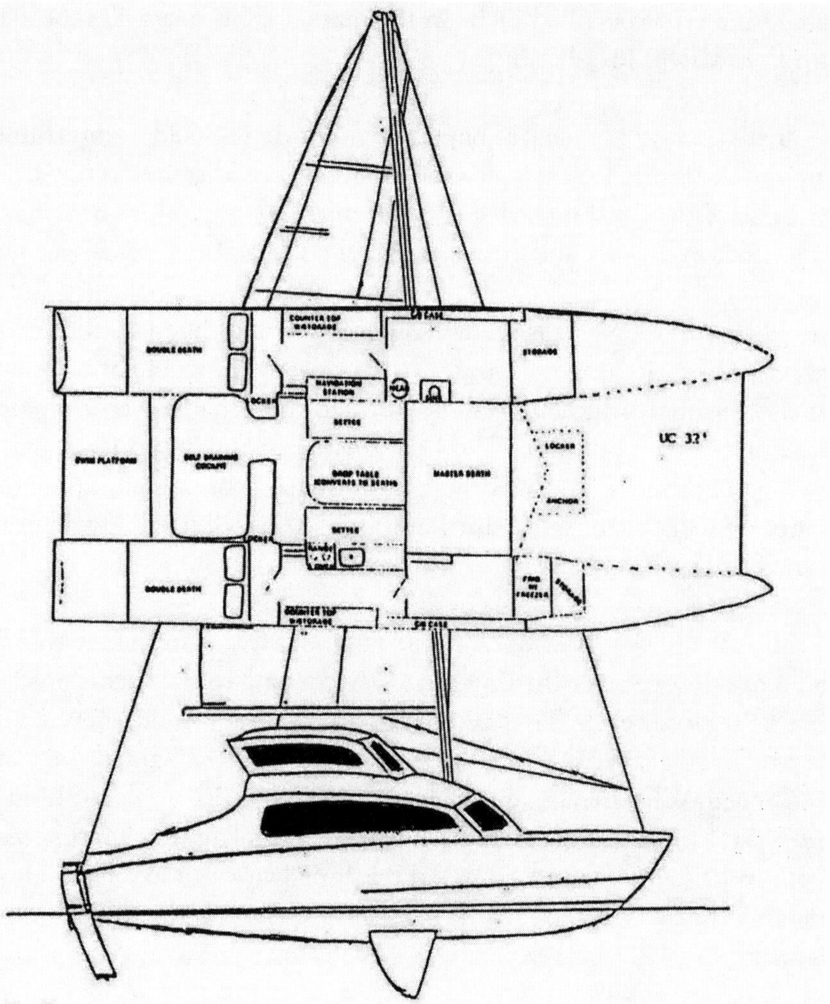

These boats have an interesting history having started out from the original Macalpine-downey Iroquois molds which went to Texas and were modified into the Chieftain before they went to Carlson.

Sailing summary

Ultimate Concept 32

1=Poor 2=Fair 3=Average 4=Good 5=Exceptional

	1	**2**	**3**	**4**	**5**
Motoring					5
Docking			3		
Backing			3		
Sailing				4	
Weather helm/lee helm				4	
Windward ability				4	
Ease of tacking				4	
Visibility from the helm			3		
Tracking				4	
Seakindly/ride				4	
Underdeck Slamming				4	
Convenient Deck Layout			3		
Interior layout		2			
Adequate storage areas		2			

Vital Statistics

Ultimate Concept 32

LOA (length overall)	32' 0"	9.75 m
LWL (waterline length)	28' 0"	8.53 m
BMAX (maximum beam)	13' 4"	4.03 m
SA (sail area)	513 ft^2	156 m^2
Draft: 1' 6" / .457 m bd dn		5' / 1.52 m
Designed displacement:	6,250 lbs	2835 kg.
Mast hgt above WL:	41' (est)	12.19 m
Tankage:		
Water:	50 gal.	189 liters
Fuel:	48 gal.	182 liters

Auxiliary power: Optional hp twin outboards

Type .... Full cabin
Foredeck ..... Trampoline
Cabins ...... Four
Galley up/down Down
Rudders ..... Transom hung

These vessels are no longer in production. No figures for how many actually have been produced are available. My survey experience shows wide variation in both quality of manufacture and finish.

Venezia 42 Class 5, Cruising

Designer: Michel Joubert / Bernard Nivelt

LOA: 41.34' feet 12.60 meters
BMAX 22.80' feet 6.95 meters
SA: 968.76 feet^2 90 meters^2
Draft: 3.95' 1.2m Keels
Designed displacement: 14,987.2 lbs 6798.19 kg.
Mast height above WL: 62' 18.91 m
Tankage:
Water: 132 gal. 500 liters
Fuel: 92 gal. 350 liters
Auxiliary power:Twin 28 hp diesel
sail drive

This is a typical Fountaine Pajot vessel of which I have surveyed well over fifty different ones.

Much depends upon how the vessels was maintained during its life.

Older vessels are noted for chronic window leakage, winches pulling from mountings and bent crossbeams.

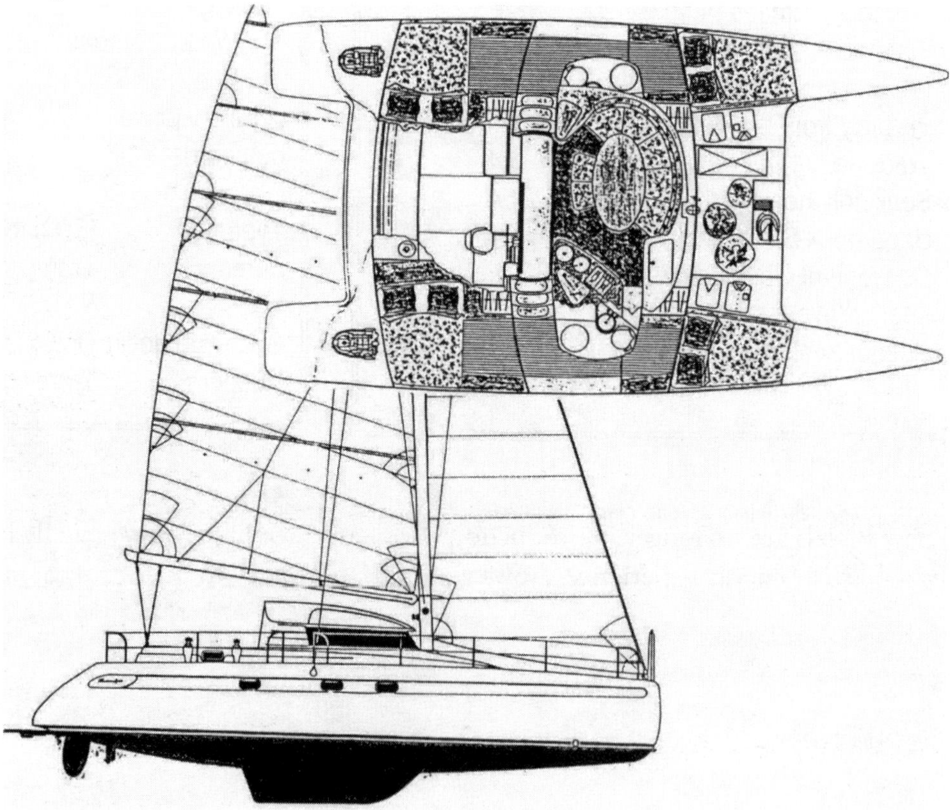

Witness 34 (36) Class 4, Cruising /Racing

Designer: Koogle/Loch Crowther

I have surveyed one, have a long time friend and ex business partner living permanently on one and first hand knowledge of two others.

In this boat, I can clearly visualize the basic industry faults. The vessel, is essentially the Loch Crowther designed "Spindrift 36." As built by the original builder, the boat left a lot to be desired.

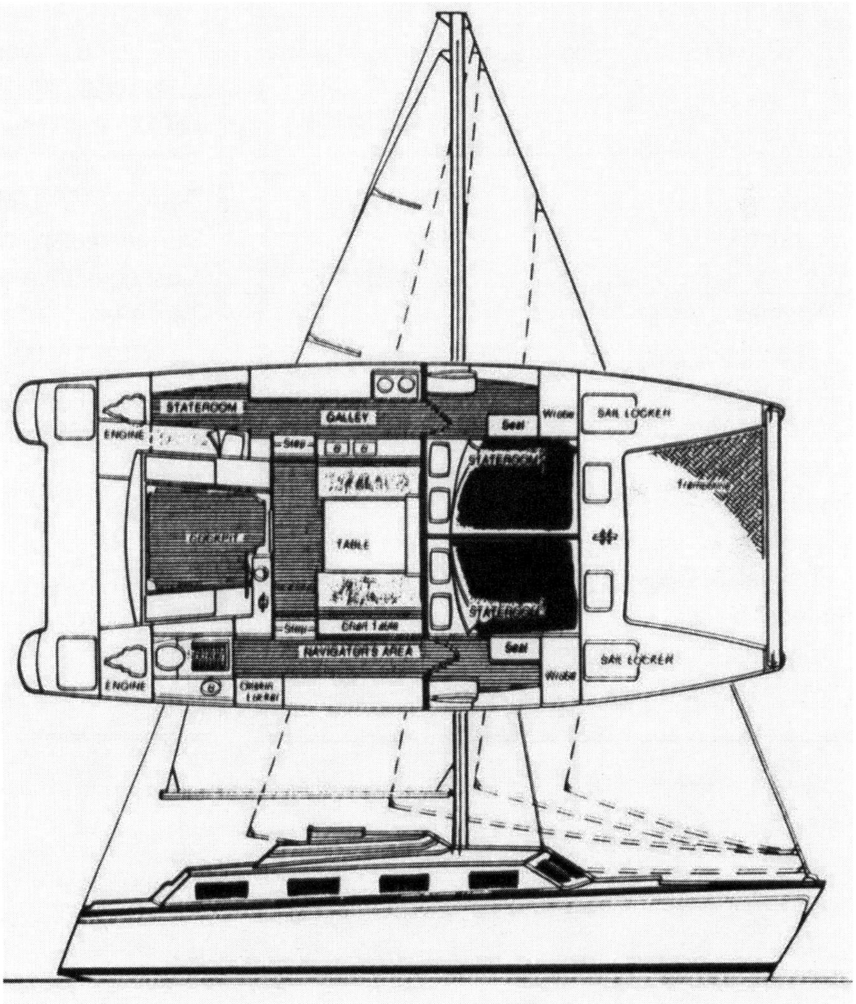

On the *Witness* I found bow NO FLOTATION chambers or crash boxes (unforgivable, but easily correctable) Unsupported panels in the bows, poor hull-deck joints, a poor forward crossbeam, marginal rigging and a host of other irritants not atypical of so many small boat builders.

The engines were accessed in typical monohull style, under the bunks. I could go on and on but it would not be appropriate. Suffice it to say that all of these boats were eventually stretched in the sterns to accommodate the extra weight and resulting stern drag of the engines.

Sailing summary

Witness 34 (36)

1=Poor 2=Fair 3=Average 4=Good 5=Exceptional

	1	**2**	**3**	**4**	**5**
Motoring			3		
Docking			3		
Backing			3		
Sailing				4	
Weather helm/lee helm				4	
Windward ability				4	
Ease of tacking				4	
Visibility from the helm				4	
Tracking				4	
Seakindly/ride				4	
Underdeck Slamming				4	
Convenient Deck Layout			3		
Interior layout			3		
Adequate storage areas			3		

Vital Statistics

Witness 34 (36)

LOA (length overall)	34' 8"	10.59 m
LWL (waterline length)	30' 3"	9.22 m
BMAX (maximum beam)	17' 0"	5.18 m
SA (sail area)	625 ft^2	58.06 m^2
Draft: 2' 2" / .66 m bdup	5' / 1.52 m bddn	
Designed displacement:	11,500 lbs	5216kg.
Mast hgt above WL:	49' (est)	15 m
Tankage:		
Water:	100 gal.	379 liters
Fuel:	18 gal.	68 liters
Auxiliary power:	Several diesel options	

Type .... Full cabin
Foredeck ..... Trampoline
Cabins ...... Three
Galley up/down Down
Rudders ..... Balanced spade
Rig .... Masthead sloop

My Relationships with Trimarans

It was the Miami Boat Show. Vendors were doing demo sails for customers. Kevin Jeffery and I were toggling back and forth among boats taking enthusiasts for rides and at the same time doing both still and video footage. The day was magnificent, the wind light, the water smooth when I boarded the F 31 *Matador*, belonging to Pablo Aguilera. The contrast between the cruising catamarans I had been sailing and this feather-light sailing machine was profound. In that light air, the tri had every advantage. We out sped, out pointed and out tacked everybody, and did it with ease. I came away with the feeling this was some awesome machine. It was so light and agile that you felt like you were not even touching the water. Since that time, I have been on three other F 31's which confirmed my opinion.

The F 27 I was about to survey was suspended in the air on davits. Neat, I thought, now that's the way to dry-sail a boat! Being a practically new boat, I was curious as to why the owner was selling. His explanation was simple, direct and to the point, he was buying an F 31! I poked around finishing my survey confirming the details I picked up through exposure at boat shows, demo rides, sales pitches and just being around them for so many years, they are fine boats.

My first real exposure to the F 27 came in Newport, Rhode Island, 1985. The boat had only been on the market a short time and Ian Farrier, the designer, was there to help promote it. It was a typical blustery New England autumn day when Ian, with a gaggle of prospects in tow, invited me to go along for the ride. Naturally, I jumped at the chance. Six of us boarded the boat which was tied at the end of the show dock. Ian and I sat on the windward trampoline and just observed. The dealer, using the prospects as crew, got underway with little fuss and we tacked our way through very crowded Newport Harbor. The boat tacked, handled and responded with the sureness and ease I expected from any good boat, no matter how many hulls it had. It was obvious to everybody, onlookers and participants alike, that we were sailing at close to twice the speed of the other boats in the harbor.

Hide-bound, hair-shirt-sailing-mentality, Newport-variety sailors would try to head up to show us that they could go to windward better than we could as soon as we would begin to close on their boat. It was futile for them, fun for Ian and me, and mind-blowing for the prospects. They did not stand a chance. We blitzed through the harbor, out the straights to Brenton Reef lighthouse, tacked and reached back through the harbor at blistering speed. There was no competition. We had eaten their lunch!

Back at the show dock, the prospects were awe-struck. Two people put down deposits immediately. Ian smiled, it had been a good sail.

During that same period, as a sailing coach at the United States Naval Academy, I had ample opportunity to sail *Moxie*, Phil Weld's record breaking Newick designed, Gougeon built 60' trimaran. Phil had donated it to the Naval academy and CCMA (Chesapeake Cruising Multihull Association was called upon to supply coaches.

Other trimarans things were happening. Phil Herting was producing the Condor 40 in downtown Annapolis. It was interesting to watch the progress. In later years I had my opportunity to sail and survey them. Phil developed a Condor 30 which many people thought was his undoing. Too bad, the Condor 40 was the perfect class racing boat. It was demountable and fit in a standard shipping container thus had the promise of world-wide competition.

Several years earlier, Corinne and I flew to St. Croix, USVI, to pick up *Tanith*, a 36-foot Kelsall-Thompson trimaran, built in England way back in the early sixties. Howard Beagles raced it hard in the daily St. Croix galaxy of multihull racing stars, but it was now being retired to make way for a more modern vessel. I was to sail it to Aruba, NA and deliver it to the irascible Art Lamour. *Tanith* had no engine.

We picked up one crew off the beach, a young army corporal on leave and looking for adventure. He found it. We blasted across the Caribbean in two and one half days.

This was the third trimaran I had brought to Aruba for Capt. Art Lamour. A couple of years earlier, I sailed a Symons 49 *Trece Nino's*, from Atlantic City New Jersey to Aruba. That boat, at least, had an engine. Nevertheless, we ran out of fuel about latitude 22^0 N because the selling owner gave us erroneous information about fuel capacity. By the time we got to Puerto Rico we not only had no fuel, but had blown out most of the dilapidated old sails. Art's old boat, a Symons 36, was sold to a charter operator on St. Martin, but not before I had lots of time to sail aboard it also.

That same year I delivered a 38 foot Kelsall designed trimaran from Montego Bay, Jamaica WI to Marsh Harbor, Abaco, Bahamas. It was an aging derelict formerly named *Worcestershire Sauceress* built as an entry into the single-handed trans-Atlantic race; it was in rough shape. A full account of that voyage and its aftermath is available in a chapter titled; *Is It Worth the Money?* published in my book *The 13th Trip and Other Sea Stories.*

Several other trimarans have crossed my bow. I sailed *FT*, a singled handed transatlantic racer that had amas longer than its main hull, A rule beater for sure and a forerunner of things to come. I sailed several Brown Searunners and sailed with Norman Cross on his 31' trimaran *Crossfire.* I did a demo sail on a beautiful Dragonfly with Joanne and Karsten Steenberg aboard. I was trying to influence Suzanne Pogell of Womanship to get with the action and teach on these boats that really sail. The demo ride convinced her that they do really sail.

Of course, I have sailed extensively on several of Tony Smith's *Telstar* trailerable trimarans. *TelStar* being an early design folding trimaran that is still advanced technology even by today's standards.

There have been a few others, in the swirl of time I have forgotten their names. One was a Dutch built 30-odd footer that I single handed from somewhere in Massachusetts to Bill Symons place on Long Island, N.Y. It had a little one-lung diesel and practically no clearance under the bridgedeck in an attempt to have all-around visibility like a catamaran.I also sailed a beautiful Harris 46, and Hartley Sparkle.

Then there was that absolute colossus of a Texas Tri. At 48 feet it was a regular sailing condominium. By the time I got to it in Newport, Rhode Island, it was in a semi-derelict state and had been used as a flop-house for itinerant hippies complete with graffiti decorating the interior. I worked for days just trying to establish navigation lights and engine controls. The sail from Newport to Long Island, New York was harrowing to say the least.

The results of this eclectic experience left me with certain indelible impressions. First, the old style, original Piver, Brown and Cross designs were extraordinary vessels. One of the best sails I ever experienced was the voyage from Atlantic City, New Jersey to St. Croix, USVI in the Symons (Piver) 49, *Bon Bini II* (nee *Trece Nino's*). That vessel was dry, never ever pounded a wing deck and maneuvered like a powerboat. We, even, docked that boat in one of the notorious cross-the-current slips in Beaufort, North Carolina. With its four full-fledged apartments, it was spacious, commodious, elegant and it even sailed quite well.

As fast, weatherly sailing machines, the modern trimarans got high marks. As comfortable cruising machines, they got mixed reviews, depending upon beam, percent of buoyancy in the amas and whether the decks are open or closed. The older, high-buoyancy, solid-deck trimarans were satisfactory. The newer, open-deck, wide-beam, lower buoyancy trimarans were wet, uncomfortable and had precious little in the way of cruising accommodations.

The thing I remember most from my trans-Caribbean sail aboard *Tainith* was the windward ama knocking the tops off the waves, turning them into spume which shot back directly into the cockpit. It was like standing in front of a fire-hose for two and one half days.

Then, there is see-saw-syndrome. The racing style trimarans that have one ama out of the water when at rest and are difficult to live with indeed. Any shift in weight from one side to the other toggles the boat across to the opposite hull. The sensation is not pleasant. First, you more or less adapt to being slightly inclined. Then the balance shifts by someone moving across the hull and the boat heels in the opposite direction. The movement starts in a peculiar way, accelerates, then stops with a shuddering crash as the opposite ama hits the water: just like a see-saw when it hits the ground.

There are certain other characteristics that may or may not appeal to you. The more a trimaran is built for performance, the more "monohull-like" it becomes. Because the main hulls are narrow, you wind up sitting sideways in the cockpit. Because the akas must be high out of the water to prevent pounding and drag, they restrict visibility from the cabin windows. When you go below, you are below in your own environment very much like a monohull. In some trimarans, looking out the cabin portholes gives you a clear, unobstructed view of the inside of the amas.

Sailing them, however, usually makes up for any real or imagined shortcomings. As a gross generalization, you will find that comparing similar class trimarans and catamarans that the trimarans are most often the better sailors, have faster response times, tack more surely, are lighter on their feet and easier to maximize your sailing performance.

If you really like high performance and are not absolute driven by maximizing accommodations and conveniences, there are some nice trimarans around. Give yourself a treat. Go take a demo ride on one. I guarantee you will be impressed.

One of my heros is Tony Bullimore. When he was sailing his trimaran *Apricot*, he wistfully gave us his critique of trimaran designers:

"Fascinating to compare trimaran designers;
- Piver, Pragmatist;
- Brown, Realist;
- Cross, Perfectionist;
- Newick, Artist, Minimalist;
- Horstman, Maximalist."

I am only going to give a full blown description of four different trimarans with which I am familiar simply because there really are so few others in the marketplace and to the best of my knowledge, none in production. The excellent Corsair and Farrier boats barely fit into the category of cruising boats so I am including the two with which I am most familiar.

If you really need to know more about any of the others listed below, contact me via email: Surveyor@sailcopress.com

Brown 25, 31, 34, 37, 46
Cross 36, 50
F-27
Horstman Tri Star
Condor 30, 40
Dobler Intercontinental 30+
Firefly 30
Kelsall Tango
Contour 30, 34
Dragonfly 800
Harris 36

The Dragonfly 800 Class 2, Racing

Designer: Borge Quorning

The Miami International Boat Show is a wonderful opportunity to test sail boats. The winds blow, the currents flow, and the shoals lurk just beneath the surface. You better have a boat that does what you want, or you may be in deep trouble.

It was a typical idyllic Miami day. Southeast winds around 15 knots, puffy little cotton ball clouds, temperature around 80. Susan Pogell, of *Womanship* Sailing School, was dying for a ride on a fast boat, so when the opportunity came to test sail the *Dragonfly*,I both took her and Erika Brigham along.

That meant we would have five people on the test sail. Carsten and Joanne, the importers and distributors, were a little nervous about the extra weight, but I assured them I was also interested in observing how the boat handled with many people, since in real life, it probably would often be sailed that way.

The Dragonfly sailing ability borders on the spectacular. Even with the extra weight and lack of familiarity by the crew, we effortlessly tacked, jibed, beat, reached and ran all with feather light control and sports car response. What a vessel! We all came away impressed.

However, undeniable superior performance is only a small part of the advantages of this remarkable little speedster. Its handiness around docks, wharves and piers is, I believe, far superior to any other boat in its size range, including comparable monohulls. The reason for that is the way the boat retracts its floats.

By pulling the floats in parallel with the hull, the boat remains level and accessible its entire length. Unlike a monohull in that size range, which heels significantly when boarded, the Dragonfly does not heel when you step on the retracted float. You can leave the vessel retracted without fear of fouling the topsides, since even retracted, it maintains the normal waterline, thus exposing the bottom paint to the water rather than the sides of the floats as other types of trimaran folding mechanisms do.

Sailing summary

Dragonfly 800

1=Poor 2=Fair 3=Average 4=Good 5=Exceptional

	1	2	3	4	5
Motoring			3		
Docking			3		
Backing			3		
Sailing					5
Weather helm/lee helm				4	
Windward ability					5
Ease of tacking				4	
Visibility from the helm				4	
Tracking				4	
Seakindly/ride				4	
Underdeck Slamming				4	
Convenient Deck Layout			3		
Interior layout			3		
Adequate storage areas			3		

Vital Statistics

Dragonfly 800

LOA (length overall) 26' 3" 7.99 m
Beam trailered: 7' 10" / 2.50 m
BMAX (maximum beam) 19' 7" 6.0 m
SA (sail area) 377 ft^2 35 m^2
Draft: 1' 2" /.35 m bdup 4' 7" / 1.40 m bddn
Designed displacement: 2,315 lbs 1,050 kg.
Mast hgt above WL: 41' (est) 12.5 m
Tankage:
Water: NA gal. liters
Fuel: 6 gal. 22 liters
Auxiliary power: one outboard

Type .... Open bridgedeck
Lateral plane: . Pivoting centerboard
Rudders ..... Transom hung kick up
Rig .... Fractional sloop

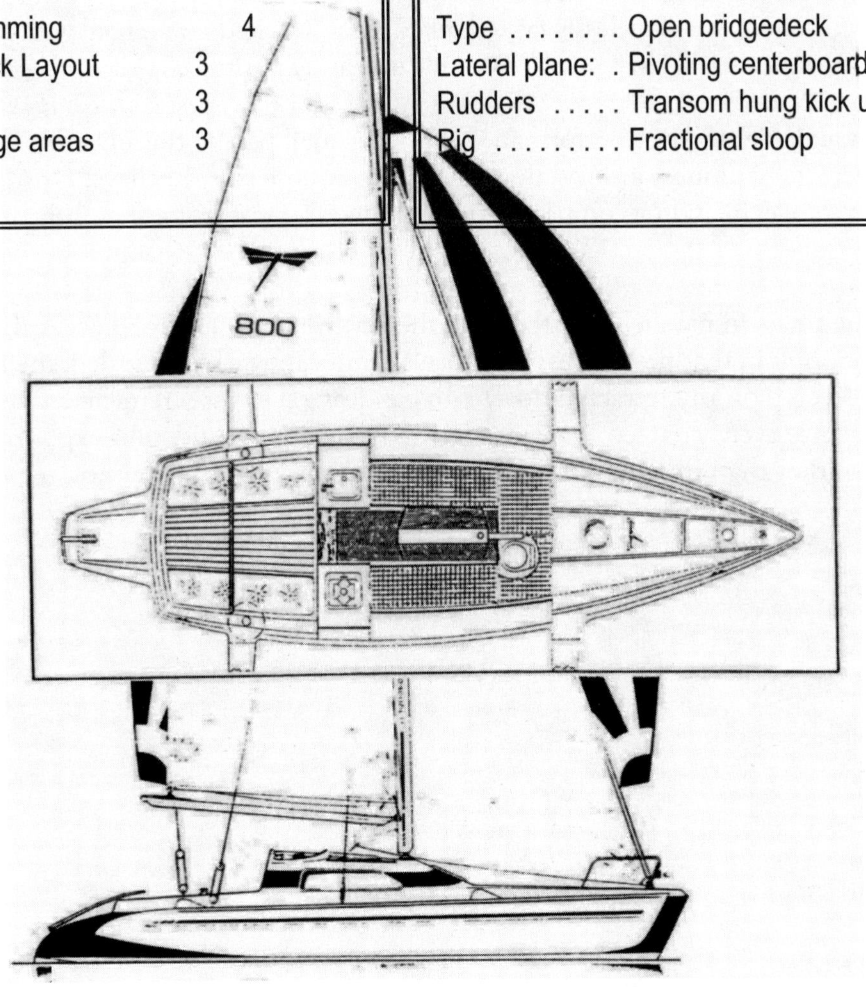

Dragonfly 1000 Swing Wing Class 2, Racing

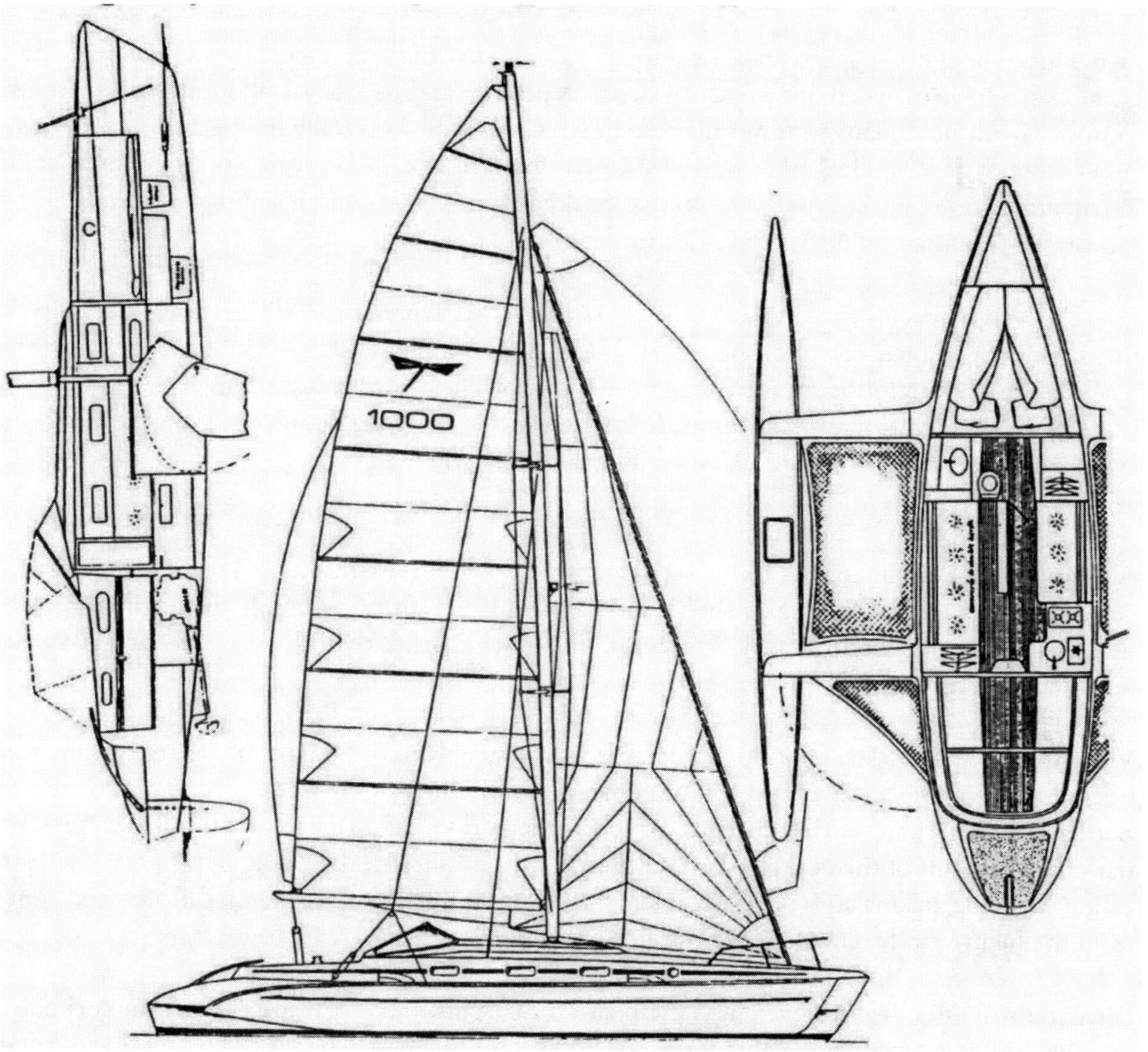

LOA (length overall)	33' 0" feet	10.00 meters
Beam, folded ;	11'10" feet	3.30 meters
MAXB (maximum beam)	11'8" feet	7.30 meters
SA (sail area)	582 $feet^2$	54 $meters^2$
Draft:	1'8" / 0.50 m bdup	5'3" / 1.60 m bddwn
Designed displacement:	3,969 pounds	1800 kg.
Mast height above WL:	50' / 15.24	
Tankage:		
Water:	23 gal. 87.06 liters	
Fuel: 30 gal.	113.55 liters	
Auxiliary power:	18hp diesel	

Brown Searunner 37 Class 5, Cruising

Designer: Jim Brown

I write little about cruising trimarans because there are so few of them, especially on the United States East Coast. However, when one particular design is popular and people ask about it I will share what I know. If you have plans to travel in areas in which the size, especially the beam of your cruising vessel, will not be a handicap, then consider one of these vessels. I have many thousand miles sailing in trimarans under my keel in a variety of vessels. Each is different and has different characteristics. Jim Brown trimarans hold a special place for me.

I have surveyed four (4) Jim Brown *Searunner* 37 and sailed aboard others. I can't help but marvel at the superior ergonomic layout of this vessel and its extraordinary sea-keeping ability. In my opinion, it is the finest layout for ocean voyaging I have yet seen. The center cockpit, which in this execution is more like a "command post," leaves the helmsman within touching distance of everyone and almost every operating system on board. An incredible layout at sea.

Advantages
All the necessary requirements for vessel operation are right there in the cockpit. The helm station is just behind the mast leaving all sheets and halyards where you can handle them without a lot of heroics such as complex turning block systems. The centerboard is right there and reefing takes place in the cockpit by nature.

Visibility is excellent. The cockpit deck is raised considerably in order to create a large engine room below it. You can see all six corners of the vessel from the helm station. Crew contact is also excellent. Berths are just forward of the central cockpit and the saloon, or great cabin as Jim Brown prefers to name it, is just aft of the cockpit. The galley is at the entrance to the cabin. Thus people cooking, eating, laughing, talking are a full cockpit length away from those sleeping, yet in full view and contact from the helm, a truly remarkable layout for sea-keeping.

Weight distribution is excellent. The centerboard trunk splits the central space under the cockpit sole and the resulting structural member forms a solid base for the mast. The raised cockpit sole integrates it into a major structural component and forms two large volume compartments exactly where the weight should be. With the engine installed on one side of the center board trunk and the water tank, fuel tank and batteries on the other side, you have, what is termed, natural ballast. This allows unfettered hull shape design optimization unlike many twin screw catamarans that need extra buoyancy aft to handle the weight of the engines, etc.

Drawbacks

On the other hand, for just plain cruising or living aboard, this arrangement has a few drawbacks. First, putting the cockpit in the center means it is more time consuming and difficult to load and unload your dinghy with provisions. It means that you must go through the cockpit to get from the forward berths and navigation station and in some cases the galley, to get to the main saloon. Essentially, going outside to get back inside. This is the classic objection to a center cockpit.

Even though the engine and batteries are strategically located for weight distribution, there is basically poor access to them. The centerboard trunk splits the area lengthwise and the supporting bulkheads for the cockpit sole and mast intersect across an already narrow beam. The cockpit sole being the top of the compartment and the horizontal support member for the centerboard trunk mitigates against having access hatches on top of the engine. The mirror image compartments that house the batteries hatches are not easily accessed and require removal of floorboards, stairways, etc. for access.

For ocean crossing, the cutter rig is renowned. For inshore work, the inner jib-stay is a headache while tacking. The center cockpit arrangement, also, creates a minor problem arranging a Bimini top without raising the boom to excessive heights. This may cancel out some of the advantages of being in the cockpit.

The original design featured a double chine. Construction was out of plywood sheathed in glass. This is a time proven method of construction, but because of the myriad small pieces there are intricate details and lots of little corners to collect dust and trap water thus creating an above average need for vigilant maintenance.

The design has solid wing decks. To me, this is a distinct advantage as it gives enormous deck area, eliminates high expense and maintenance trampolines and stops the spray, that particular trimaran vice of the windward hull turning the tops of waves into spray that shoots back into the cockpit.

Trimaran bunks on the bridgedeck always have limited headroom sometimes so claustrophobic you cannot even roll over without banging your hips. Normal human activity usually associated with beds is severely restricted.

As in a monohull, having the head forward means walking through the boat past the bunks, a compromise endemic to narrow beam single hull boats.

Years ago, wide beam was the kiss of death for trimaran technology. However, it is becoming less and less of a problem as other boats are expanding their beams thus coming ever closer to the trimaran beam. Actually, at 22'6" the brown 37 is not as wide as some modern catamarans in that size range. I have lifted, for survey, the Brown 37 in an ordinary travel lift without a problem.

Sailing summary

Brown Searunner 37

1=Poor 2=Fair 3=Average 4=Good 5=Exceptional

	1	**2**	**3**	**4**	**5**
Motoring				4	
Docking			3		
Backing			3		
Sailing				4	
Weather helm/lee helm				4	
Windward ability				4	
Ease of tacking				4	
Visibility from the helm					5
Tracking				4	
Seakindly/ride				4	
Underdeck Slamming					5
Convenient Deck Layout				4	
Interior layout					5*
Adequate storage areas				4	

* for seakeeping

Vital Statistics

Brown Searunner 37

LOA (length overall) 37 '4" 3.46 m
BMAX (maximum beam) 22' 3" 2.067 m
SA (sail area) 681 ft^2 37.0 m^2
Draft: 2' 0" Keels
Designed displacement: 11,000lbs 850 kg.
Mast hgt above WL: 42' (est) 3.90 m
Tankage:
Water: gal. liters
Fuel: gal. liters
Auxiliary power: diesel

Design intent. . .Blue water
Type . . . . Closed bridgedeck
Foredeck . . . . . Trampoline
Cabins . . . . . . Three
Lateral plane . . Centerboard
Rudders . . . . . Transom hung
Rig . . . . Cutter

Facilities, layout Summary

- Eight bunks
- Galley and Saloon in stern cabin
- Visibility from interior better than most trimarans
- Adequate rub rails. life lines optional
- Cutter rig with running backstays
- Fair storage not including amas, amas not be used for heavy items.

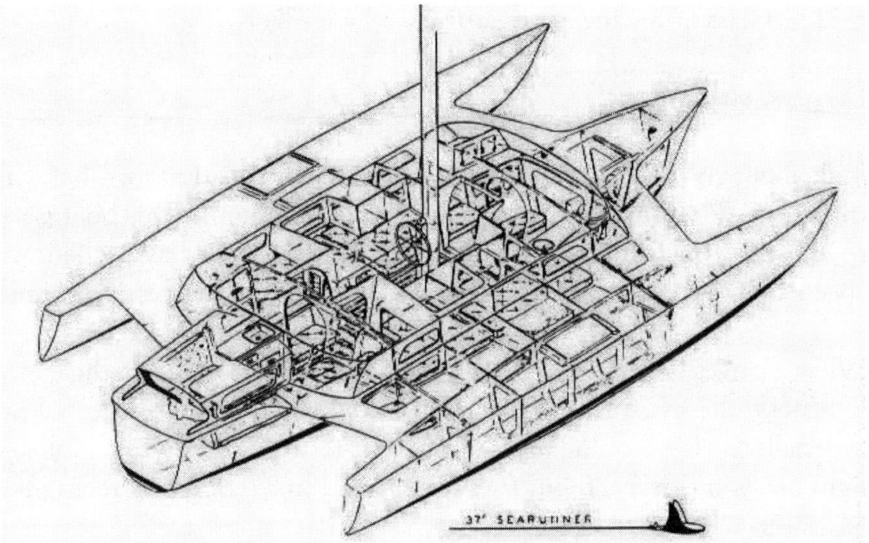

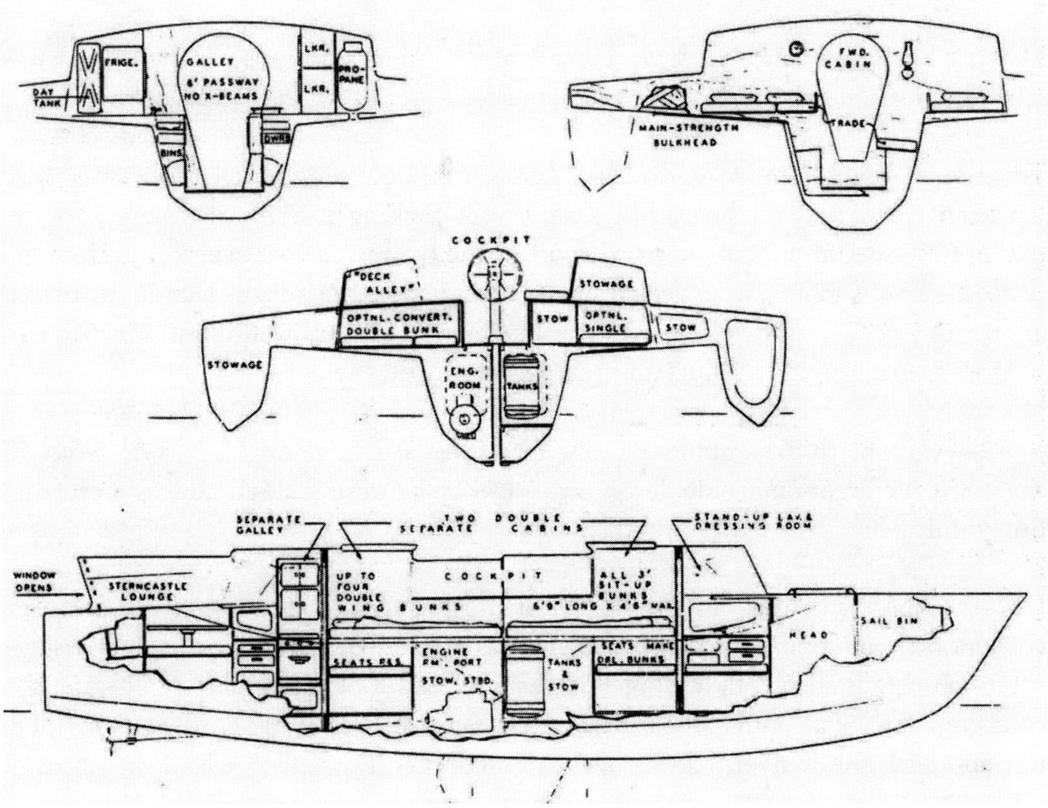

F-31 Class 2, Racing, sport-cruiser

Designer: Ian Farrier

Talk about good sailing, you just do not have appreciation for what really good sailing is until you sail this boat. When most boats are wallowing in light airs, this boat is sailing faster than the true wind speed. Do you think I was impressed? I guess so. I have sailed several Farrier boats previously, specifically the old reliable F 27. They, too, are very impressive sail boats.

What changes between the 27 and the 31 is the magnitude of the experience. On the 31, you feel the lightness, the mass, the sensation of velocity to a greater degree. I have always faulted catamarans, especially the big cruising boats for losing the "rush" of sailing. Not so on the F 31 trimaran. Even in light air, you can feel the horses under you. In a word, as a racing boat, as a sport boat, this boat is *awesome*!

My first F 31 experience took place on Pablo Aguilera's famous *Matador*. That was just prelude to really sailing one since it was only in an enclosed area at Miami Beach during the boat show.

Yesterday, I sailed *Flash* for an extensive period of time during sea trials out on Florida Bay. From the perspective of sailing the boat, things went well. Tacking was virtually the best of monohull like. No matter what we did or how poorly we worked the boat, we never even came close to going into irons or losing a tack. Jibing, as expected, was a breeze since the air was very light. I won't bore you with angles and details, suffice it to say I would challenge *any* monohull sail boat up to almost twice the F 31 waterline length to a duel and put my money on the F 31.

I guess everyone thinks I am going to sell *La Forza* and buy an F31. Not quite yet. The other side of the coin, the ergonomic side, left a great deal to be desired. Actually, as a cruising boat, it will be uncomfortable even for me, a hard-core frugalphile.

I don't expect amenities in the boat and the thought of destroying the performance by adding generators or air conditioners or diesel engines is horrifying. The extremely sparse galley has enough room to upgrade slightly by adding a deeper sink & a foot pump plus a salt water foot pump and we could replace the alcohol stove with the new Kenyon KISS butane stove. (Alcohol just doesn't burn hot enough) The Kenyon KISS stove is named after *The Galley K.I.S.S. Cookbook.* There is enough physical room inside to store a frugalphile's stores and equipment.

The bunks are reasonable, the head is ok, there is room to add my PUR water-maker and my HAM radio, so I guess I'm all set. Not so fast. The things that deny me the use as a comfortable fast cruiser are mostly outside.

First and foremost is visibility. Visibility is nil sitting in the cockpit at the tiller. You simply cannot see over the cabin top from anyplace in the cockpit. Further dampening your ability to see forward is the boom. (Some are boomless, but that makes little difference) No matter where you are standing and in some places sitting, the boom is just at head level. The boom is fixed on the mast so there is no chance of raising it. This also precludes the use of an extremely important cruising accessory, a Bimini top with dodger, mosquito curtains, etc.

Gone is the ability to sit serenely in a cockpit surround, out of sun or rain and motor through a waterway or up the ICW. The cockpit itself is divided into two sections and is basically a smallish monohull cockpit where you always sit sideways. The center divide precludes easy access and perhaps a more comfortable stance. Decks are expanded and round so that except for the trampolines, there is nary a spot anywhere a person can stretch out on a towel.

There is a beautiful anchor locker built into the foredeck, but anchoring with that bowsprit and whisker stay arrangement makes it a bear. Multiple anchors? Maybe the three bears.

Getting back inside, the electric system is completely minimal. In order to add virtually any electric gizmo, you need an entire new electric system, starting with the battery. The single schedule 24 battery is stuffed into a tiny inaccessible compartment that needs to have trim panels unscrewed to get at it. However, there is plenty of room in adjacent lockers to have a proper installation.

On the other hand, there is an enormous amount of unused volume in the stern section of the boat, under the cockpit. Slight modifications would make that volume accessible and useable. As it is now, you could use it to smuggle refugees.

Meanwhile, back on deck, we are struggling with poorly placed winches and jib leads. The factory settings for these things have the sheets chafing badly on the cabin top. The screecher needs entirely different sheet leads. It is not possible to tack the screecher unless you have the jib set.

F 31

The huge roach assures one of getting plenty of exercise rasing the mainsail. The complex mainsheet block system dangles very close to the tiller and is a knuckle knocker when jibing and clunks and clatters against the top of the tiller. The four-part traveler with endless control line adds to the clutter and complexity of that small area. This is a huge, small boat. What I mean by that is that it is treated the same way you would treat any small sport boat: Hobie, NACRA, Tornado, etc. It has few amenities, a minimal cockpit, is rigged with the same types and theory of sheets, leads, tiller and tiller extension as any day-sailing sport boat.

This boat is a classic example of the complete dichotomy between racing boats and cruising boats and is the foundation for my several decades long lament that the multihull industry hurts itself by not naming boats for what they are and calling every boat in existence a cruising boat. This is in no way meant to detract from the sterling qualities of this fine vessel, only to show people looking for cruising boats that boats of this type simply do not fit the job description.

Because designers and manufacturers called their products cruising boats thus completely confusing the public who are basically familiar with descriptive nomenclature, I set up classification standards that would identify multihulls for what they were. This was done in 1992 in my book: *Cruising On More Than One Hull.* I have referred to that classification system ever since in an attempt to bring some sense of marketing order to our vessels, just as has always been with powerboats. That system is fully explained in the first section of this book, Classifying Multihulls. I use automobiles rather than powerboats for my comparison chart because I believe it is easier for most to grasp the concepts since many people are not at all familiar with powerboat classification.

F-31 Vital statistics

LOA (length overall)	30' 10"	feet	9.6	meters
LWL (waterline length)	30'	feet	9.1	meters
MAXB (maximum beam)	22' 5"	feet	6.8	meters
SA (sail area)	599	feet^2	55.6	meters^2
Draft:	1' 4" bdup	5' 6" bddwn	.45/1.7	
Designed displacement:	4,600	pounds	2086.5	kg.
Mast height above deck:	40'	bridge clearance: 46' (est) 14 m		
Auxiliary power:	outboard			

The F 31 is definitely a Class 2 boat the way it is, could be Class three with just a few minor modifications but shouldn't be referred to as Class 4 or 5. For those of you with a craving for performance without compromise, this is it!

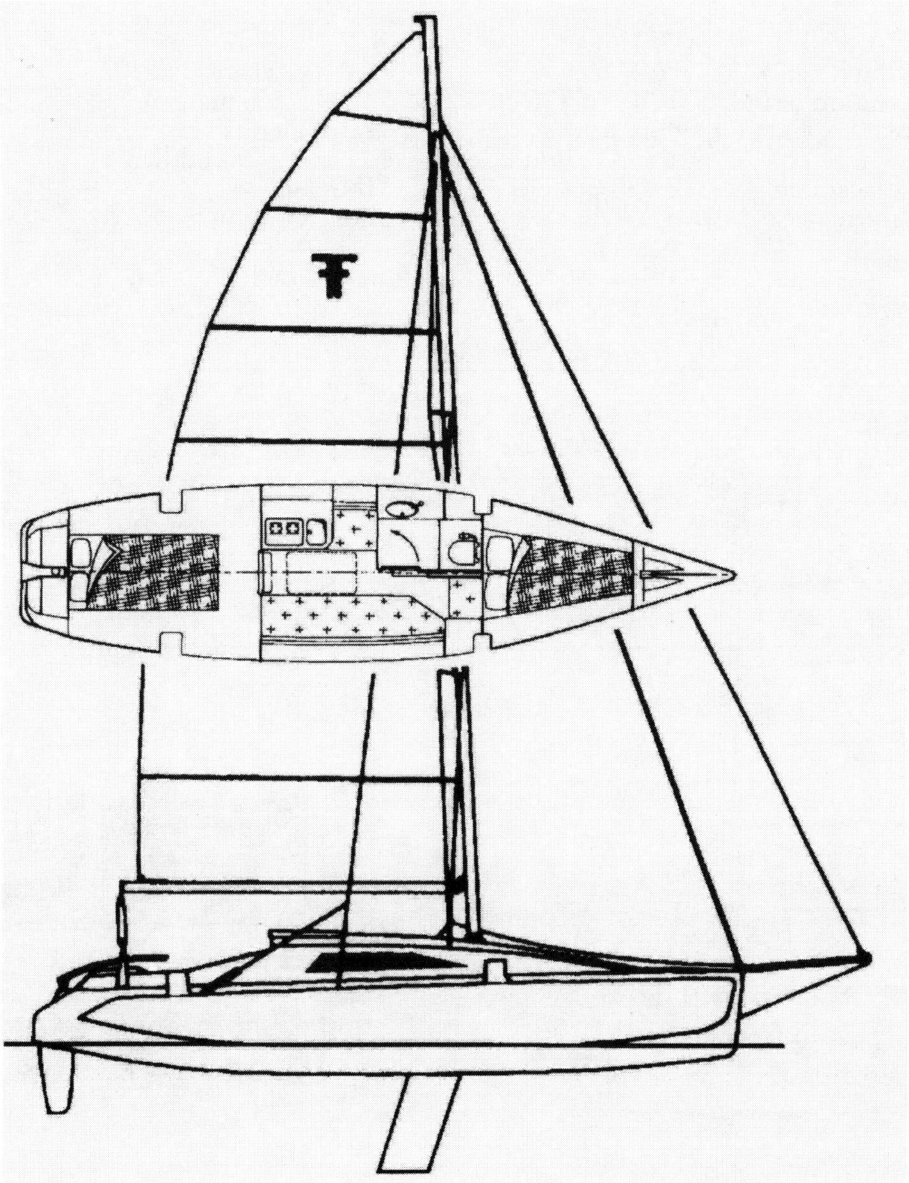

F 27

F27, Class 3, Racing/cruising

Designer: Ian Farrier

F27 Vital Statistics Sport Cruiser **TRAILERABLE !**

LOA (length overall)	27'1" feet	8.25 meters
LWL (waterline length)	26'3" feet	8.00 meters
MAXB (maximum beam)	19'1" /8'5" feet	5.82 / 2.57 meters
SA (sail area)	446 $feet^2$	41.30 $meters^2$
Draft:	1'2" / 0.36m bdup	4'11" / 1.50m bddwn
Designed displacement:	2600 pounds	1180 kg.
Mast height above deck:	37'0" / 11.28m	bridge clearance: 42' / 12.80m
Tankage:		
Water:	18 gal. 68.13	liters fuel: 3 gal. 11.36 liters
Auxiliary power:	Outboard	

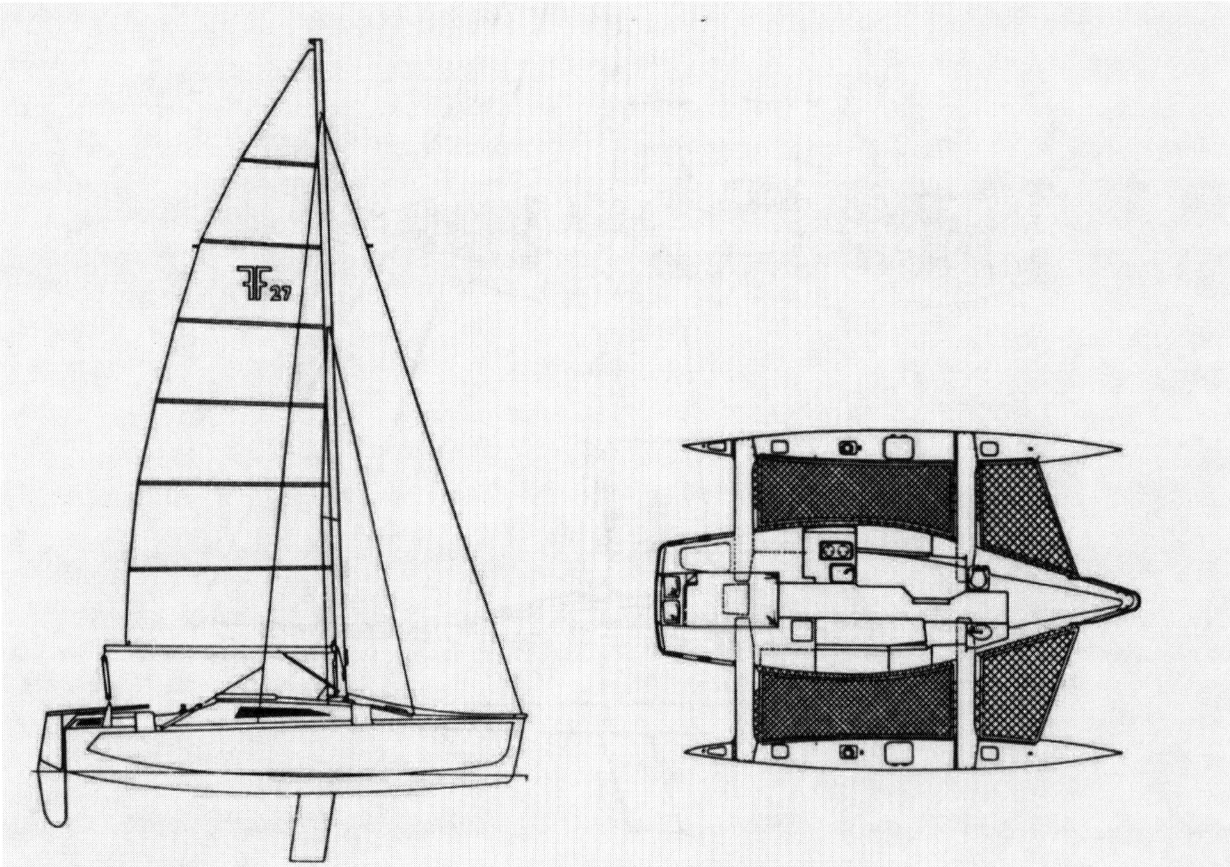

Telstar 26 Class 3, Racing/cruising

Designer: Tony Smith

The forerunner of folding trimarans, this slick vessel has solid wing decks and is road-legal everywhere having a folded width of eight feet.

A clever folding mechanism allows the amas to fold downward and tuck under the hull making a nice compact road package. Folding is done on the trailer rather than in the water. The hinges are at the joint of the deck and the main hull. Lifelines and stanchions are removed for trailering.

Other unique features are a pivoting centerboard that is slightly off-center of the hull. This allows a better passageway forward and prevents damage to the trunk and forcing debris into the trunk upon grounding. It does this with no noticeable difference in windward ability between tacks.

This was the first trimaran built without the large transverse bulkhead separating the forward and aft sections of the boat. It has a balanced dagger style rudder and a clever motor mount at the side of the cockpit closer to the vessel center of rotation. It has good panoramic view from the interior and a comfortable cockpit with good visibility.

This is a solid deck as opposed to an open deck trimaran. This gives it a catamaran-like ambience having a full deck layout including lifelines as in a cat. In addition, it has high buoyancy amas firmly planted in the water at rest, thus just about eliminating see-saw syndrome.

About 1500 of them were built between the late 1960's and 1980 when the molds were destroyed in a fire. Most having been built in England prior to Tony Smith moving his factory to America. Its age notwithstanding, these are prized vessels to this very day.

I surveyed two Telstar that sailed across the Atlantic.

Reviewing the **Telstar 26**

Sailing summary

Telstar 26

1=Poor 2=Fair 3=Average 4=Good 5=Exceptional

	1	**2**	**3**	**4**	**5**
Motoring			3		
Docking			3		
Backing				4	
Sailing				4	
Weather helm/lee helm				4	
Windward ability				4	
Ease of tacking				4	
Visibility from the helm				4	
Tracking				4	
Seakindly/ride				4	
Underdeck Slamming				4	
Convenient Deck Layout			3		
Interior layout			3		
Adequate storage areas			3		

Vital Statistics

Telstar 26

LOA (length overall) 26' 3" 7.99 m
LWL (waterline length) 24' 6" 7.47 m
BMAX (maximum beam) 16' 0" 04.8 m
SA (sail area) 350 ft^2 32.5 m^2
Draft: 1' 6"/.46m bd up 4' 7"/1.4m bd down
Designed displacement: 3,500 lbs 1588 kg.
Mast hgt above WL: 36' (est) 11 m
Tankage:
Water: gal. liters
Fuel: 6 gal. 22 liters
Auxiliary power: one 9.9 outboard

Type . . . . Open bridgedeck
Foredeck . . . . . Trampoline
Cabins . . . . . . Two
Galley up/down Down
Rudders . . . . . Transom hung daggerboard
Rig . . . . Masthead sloop

Telstar 26

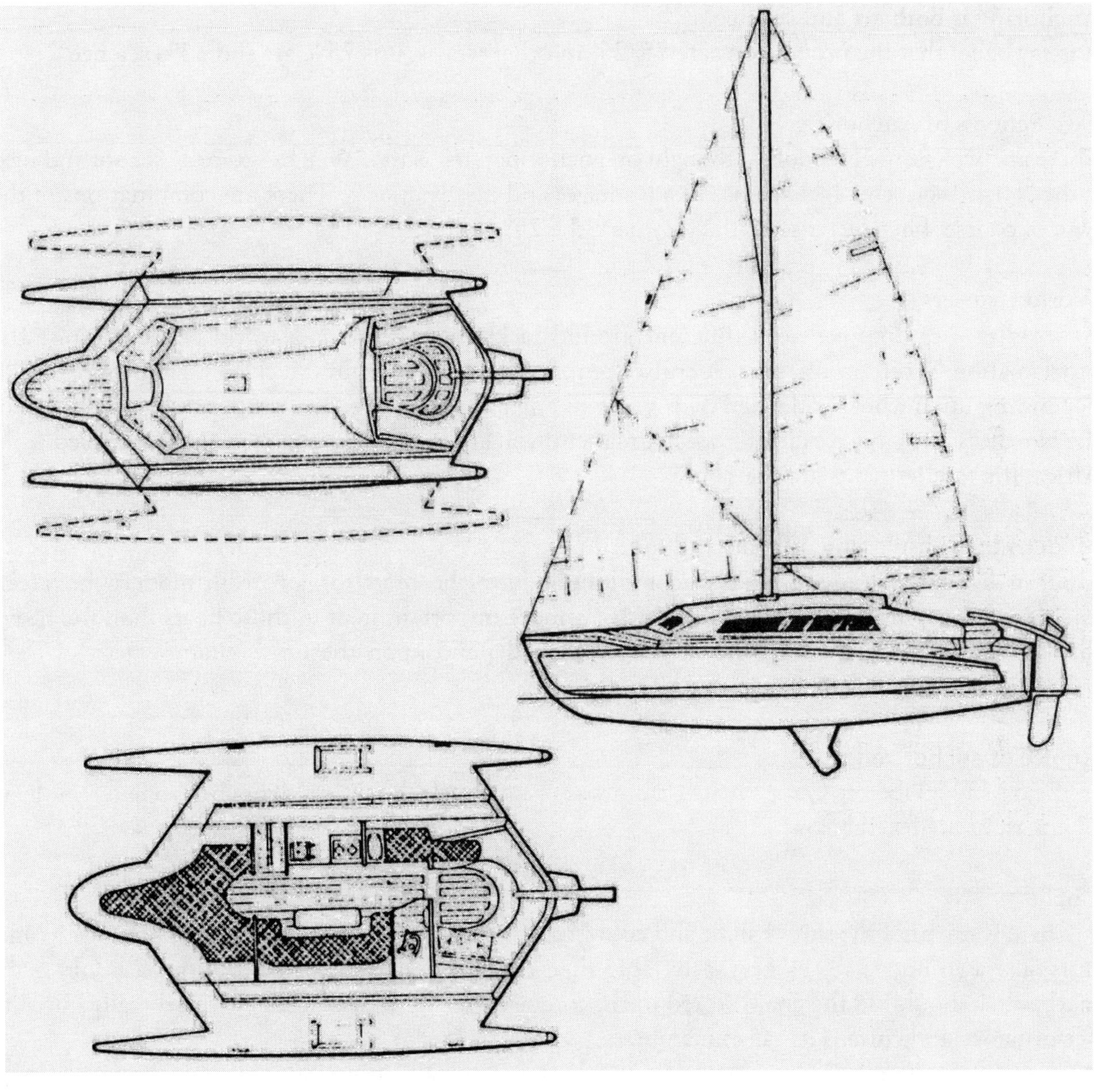

The Straight Skivvy on Anchoring

Anchoring is both art and science.
It is my belief that the proportions are 75/25. In other words, it is 75% art and 25% science.

Two Schools of Anchoring.
There are two essential schools of thought on anchoring, first is the "Weight is Great" school and next is the "High-Tech school." Each has its advantages and disadvantages. There are combinations of the two, of course. First, lets discuss the easy part, the 25% science.

World Cruisers
A world cruiser does *not* need different ground tackle because he is a world cruiser despite the proclamation of certain old salts. A cruiser requires the proper ground tackle and appropriate skill depending upon where and when he is going to anchor. If you need the proper tackle to anchor in the Northeastern United States, it doesn't matter if you are five miles from home or just arrived from Africa, the requirements are the same.

Modern light displacement boats and multihulls
Modern light displacement boats and multihulls have the most to gain from modern high-tech anchoring. Keeping weight out of the bows is a more important item in those boats than the heavy displacement, low freeboard types in which many depend upon the heavy chain as part of their ballast.

Choice of anchor rode
In the days BCG (before Coast Guard) chain was pretty much the universal anchor rode technology. The reasons are listed below.

Chain
- Chain was part of the ships ballast and could be off loaded into the long boats to lighten ship and thus raise a grounded vessel. Few yachts ship their chain into the bilge on a routine basis (although maybe they should) so the chain stored in the forepeak of a light displacement vessel really hurts its performance and curtails its safety margin.
- Chain was a valuable trade item and used as barter, especially in nations without iron technology.
- Chain was shot out of cannons to tear the masts sails and rigging off pirate and other ships.
- Chain was often used to blockade harbor entrances.
- A blacksmith and forge was an essential item on every seagoing vessel. Not only for chain, but the other black iron parts, guns, gun mounts *ad infinitum*.
- Chain is amenable to hauling by windlass. Deck level capstans with multiple spokes so several men could turn it at the same time were common items.
- Chain was easier to use than the twisted hemp rope available in that era.
- The hemp rope of that era was not as reliable or as strong as chain and was susceptible to sea critters

attack and fresh-water rot.

- Chain absolutely devastates the bottom. It destroys any and every thing, living or dead.

Examining modern three strand Nylon rope we find the following characteristics:

Nylon (three-strand recommended)

- High quality three strand Nylon anchor rode is two to three times as strong as equivalent sized chain.
- High quality three strand Nylon anchor rode is one quarter or less, the price of chain.
- High quality three strand Nylon anchor rode is a small fraction of the weight of chain.
- High quality three strand Nylon anchor rode is amazingly chafe resistant and practically indestructible underwater from chafe.
- High quality three strand Nylon anchor rode does not rust and stain decks and equipment.
- High quality three strand Nylon anchor rode is an exceptionally good shock absorber
- High quality three strand Nylon anchor rode is easier to handle when the windlass fails or you need extraordinary lengths for occasional deep anchorages.
- High quality three strand Nylon anchor rode is more likely to come up clean whereas chain can get extraordinarily filthy in the same anchorage.

To chafe or not to chafe, that is the question?

For those worried about chafe, let me put forth a few thoughts, observations and ideas.

First, there is little or no chafe happening underwater because of the lubricating and cooling effect of the water. Abrasion yes, chafe no. For those who always sail in the coral rock zones of the world, you must have noticed by now that these limestone rocks are soft rocks! Unlike the igneous granite and basalt found in other areas, these rocks, when pitted against high quality three strand Nylon anchor rode lose the battle. That is correct, try it some time. Hold a piece of rock underwater with your feet and saw a piece of nylon across it. You will be amazed to see that the rock will gradually get a groove in it and the nylon suffer little or no damage.

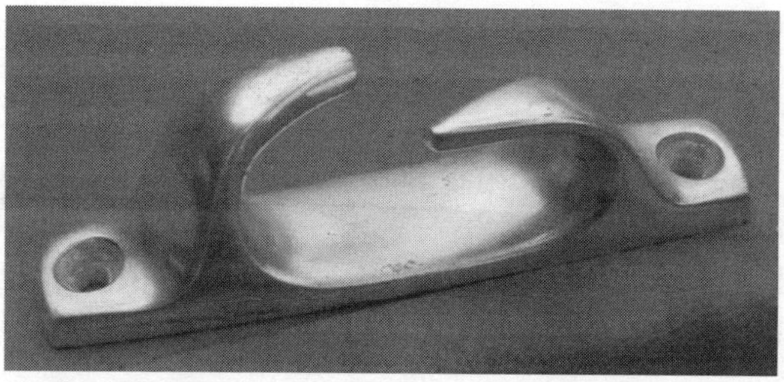

Typical skene chock. These will cut through a line like a knife in bad conditions

Chafe occurs at deck chocks and others fittings. The best way to avoid chafe is to replace solid chocks with roller chocks and/or relocate cleats so you get a fair lead for lines without the use of chocks. Trying to jury rig chafing gear is really poor business as stretchy lines manage to move the material no matter how well you try to secure it. It is almost comical to look at the various attempts at chafe protection hanging from lines after a blow.

You can eliminate almost all tendency to chafe by eliminating skene chocks, square chocks, hawse leads or any other rope lead item any distance from the cleat. These items can be replaced with either better placed cleats or roller chocks or roller hawse holes. Some hawse holes have permanent cleats built in.

The hardware store vinyl tubing may be virtually indestructible, but it is really not all that good as a chafe guard because the expanding and contracting nylon rode will often generate enough heat against the vinyl to damage itself. Melting rode loses strength fast. Better to get rid of the chock leads and not have to worry about chafe guard. The better chafe guards are leather and old fire hose.

Anchors

Ah yes, now we are down to the religion of anchors. Being an iconoclast on the subject, I might say that a good attitude toward anchor patterns is the old standby of prize fighters and stock brokers: "Bet on the champ until he loses." That way, you can get the benefit of engineering and field testing.

Then there are differences in anchor types and intent. The standard tests you see done over and over almost to boredom pretty much define the parameters under which the available anchors operate. What none of the anchor tests show is the performance of the anchors under field conditions and the handiness of the anchor to the cruiser user. I might respectfully suggest that the constant infusion of new anchors into an already crowded field, which are then hailed as the latest and greatest only to fall to obscurity is the result of a syndrome called "searching for the Holy Grail of anchors."

Miracles and other hopes

The miracle anchor being sought will answer all the ills of the present system. It will never drag, no matter how poorly it is handled by the inept, it will hold fervently no matter what the scope set, it will always reset without fail no matter what the tidal or wind conditions and lastly, it will easily break out when it is called back on deck.

Perhaps someday that miracle will occur. Maybe someone will figure out how to inject a swivel top Helix type mooring into the ground from our boats. Until that day comes, for us mere mortals in the hear and now, we better stick to what we have learned to love or hate.

Test results

It is interesting, in all these tests, some anchors win, some lose under similar conditions depending on who is doing the testing. It is also interesting to note that most of the anchor tests take place in laboratory conditions rather than life conditions. Because of these "trying to please all" conditions, both Fortress and Simpson Lawrence engaged me to do a real-life test procedure on their anchors.

Mom and Pop in the real world

The results of my three-year odyssey of testing anchors by Mom & Pop as they sojourn through various adventures is well documented in several articles in popular magazines. Suffice it to say that reliable old world Simpson Lawrence, gave birth to an updated CQR that fixed all the niggling little problems and produced what I believe to be the best all around plow type anchor, the DELTA. Of course, in religious fervor, zealots will hotly denounce my choice. Nevertheless, if you check the bows of the overwhelming majority of newer boats you will find a DELTA ensconced on the bow. There is a reason for that choice.

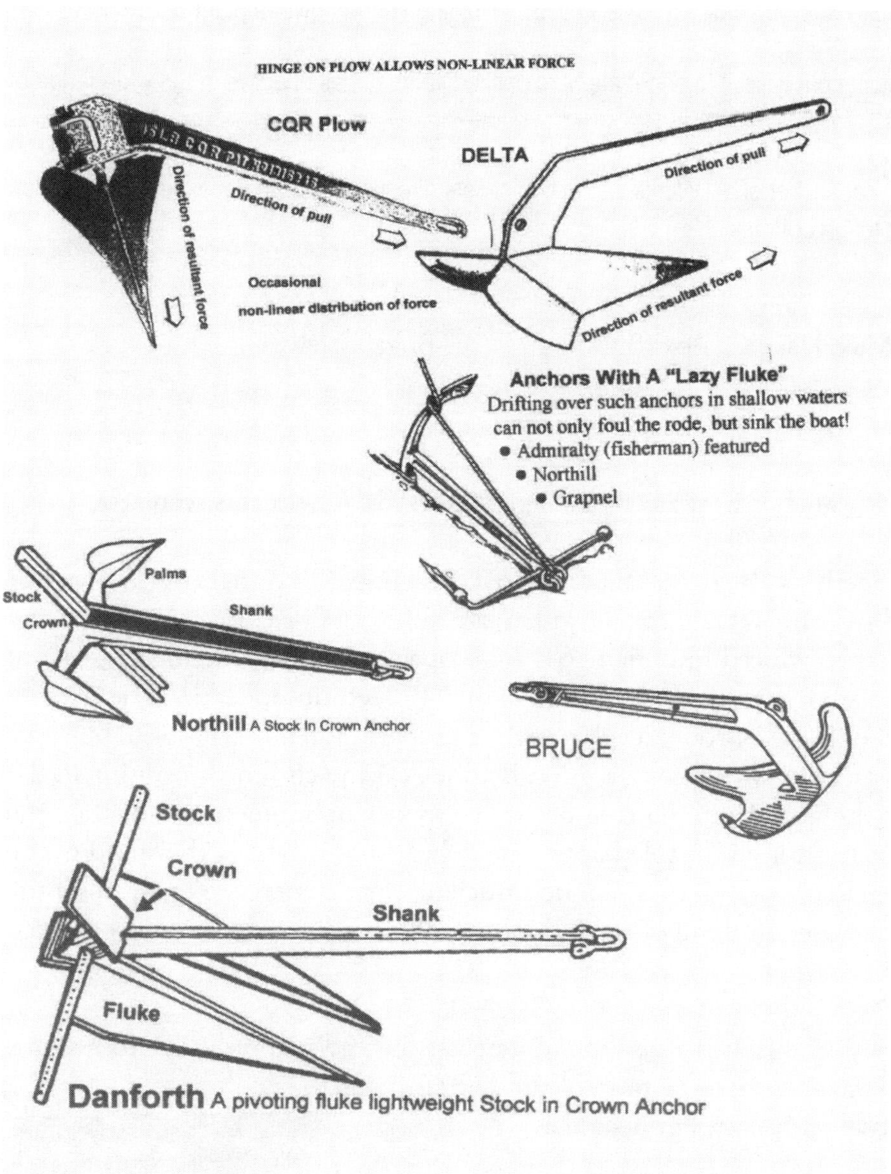

Fortress, on the other hand, took Robert Ogg's original Danforth concept and upgraded it to modern materials and clever assembly delivering a truly fine anchor that does what it is supposed to do and, in my opinion, is the best secondary anchor you can have on your boat for a host of reasons. I emphasize secondary. I shudder when I see a Fortress in the bow position as a primary anchor because a primary anchor should cover the widest range of bottom conditions and should be able to unequivocally reach the bottom where it was dropped not matter what. Fortress may just plane or skid if dropped from a moving boat in an emergency. A DELTA or it's ancestor CQR will reach bottom no matter what.

With this in mind, let us reason together about some of the popular anchors and the brash new breed that would like to usurp their place in the pantheon of successful marine yacht anchors and which ones best suit our needs in this new high tech era.

1. CQR versus DELTA
Note the problems caused by the hinge. In addition, the CQR has the most of its weight is in the shank , thus the rest of anchor just drags along flickering at the hinge until it hooks into the bottom by chance. The DELTA has its weight in the tip thus concentrating the digging force. It has more fluke area, better fluke angle and a thin shank that will penetrate to bottom. DELTA is a definite high tech winner. A 22 lb DELTA has the same rated holding power as the CQR 45.

2. The old Admiralty needs a minimum weight of fifty (50) pounds to have any effect at all. With its stock-in-head, it is the klutziest thing ever developed by mankind. Low tech, old hat, it is good for special purposes.

3. Northill. Great, but has a lazy fluke like the admiralty. Mostly used in folding version by sea planes.

4. Bruce. Very expensive, heavy, blunt front difficult to penetrate hard packed sand; not as effective in the wide range of the DELTA so why bother?

5. Danforth and imitators. Most popular anchor of all. Works well in original set direction only. Do not expect it to reset on direction change. Poor in grass or tough bottoms.

The list of also ran:
Barnacle, Wishbone, Fluke, Max, Hans C, Box, Bulwagga, FOB, Bugel, Claw, Digger, others. . .

Scope and dragging

Why do people some times drag anchor in perfectly benign conditions? Check this scenario.

Vessel "A" is anchored properly with an acceptably sized pivoting fluke, lightweight anchor (commonly called "Danforth" type) in 10 feet of water with a five foot tidal range. Because he is using rope, conventional wisdom has convinced him he needs 7:1 scope. Going by the figures, 10 feet of water plus five feet of tide plus four feet for freeboard gives him 19 feet multiplied by 7 he needs a total rode of 133 feet. He pays out 140' of rode, sets his anchor properly and all seems well, but is it? At midnight, the tide is low slack. Now his scope is 12:1. By one AM the current is running in the opposite direction and his ten-thousand pound boat is beginning to pick up momentum. By the time his boat reaches the end of its leash it has traveled approximately 280' at whatever the speed of the water is flowing, which is more than enough to generate sufficient kinetic energy to easily upset the anchor. If the anchor is upset encased in a ball of mud, or all bottom conditions are not just perfect, we have a "mysterious" case of dragging.

Obviously, this case is not so mysterious. It is a case of too much scope and inappropriate technique. While different anchor tackle may change the ratio of times dragging to times not dragging, it still misses the point. For instance:

Suppose vessel "A" was using a big heavy plow and an all-chain rode instead of his pivoting fluke lightweight and rope rode. Chances are that if conditions were light, the plow and chain may not upset and drag simply because of the weight involved. However, my experience shows that when the going gets rough such as a sudden thunder storm, the "weight is great" contingent suffers as badly as the "modern high-tech" group mainly because the sudden loads generated when they come up short with no shock absorbing rode, jerks the anchor out of the ground. Experience also shows that in high wind conditions, the Danforth will hold better longer than the CQR simply because it digs deeper longer, whereas the CQR simply lays on the bottom with one ear sticking up and never really digs in.

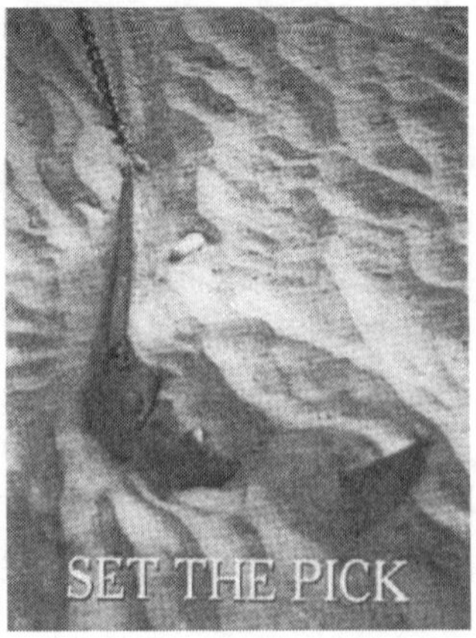

Photo from major publication article correlates with my findings completely

In my study of underwater anchor behavior, neither group actually resets. Both just stay set, and facing their original direction. Unless a person has too much scope or plain old bad luck, they usually never realize how close to dragging they came. The concept that anchors migrate around or "reset" on the tide in many cases is not really accurate. Recent studies by Practical Sailor prove that point yet again.

There is a tried and proven way to avoid this problem. It is useable with old world, *weight is great* technology or you can treat yourself to the low-cost light-weight world of *high-tech.* Since this book is about modern light displacement boats, let us explore the world of high tech anchoring. We agree with the old salts, "if it is heavy enough it will work," we simply want a better more practical way to achieve the same goal. That goal being getting a good night sleep presuming we will awake in the same location we anchored.

Now that we have reviewed the scientific 25% of anchoring, let us begin the far more relevant 75% that is art, or skill as the case may be.

What?

The essential difference between the way our modern light displacement vessels and the old heavy displacement, low freeboard, full keel boats lay to an anchor is lee helm. As soon as sails are dropped most modern boats have a tendency to fall off away from the wind. The reason is the high bows and high freeboard move the center of effort (CE) forward of the center of lateral resistance (CLR). High aspect fin-keel spade rudder boats, which represent the vast majority of monohulls circa 2002 are the most likely to have this problem as do most cruising catamarans.

For those of you with cruising experience I am sure you can recall seeing annoying boats in an anchorage that just keep anchor-sailing all the time. Boats that lurch back and forth from one tack to another are worse than annoying, they represent a potential danger to themselves, to you and to all the others in the anchorage.

Tether is the magic word

The single easiest way to avoid this entire problem is to anchor with two anchors in what is commonly called a "Bahamian Moor." With a Bahamian moor, anchors are never asked to reset on the tide. Nor can your vessel develop enough momentum to upset an anchor because your boat is "tethered" in its original location. Next time you watch a program where they handle animals, note the way they tether them between two people on opposite sides. That way the animal cannot move in any direction. The Bahamian moor does the same thing for your boat.

Why?

Why do we use multiple anchors? Reduce swing, eliminate anchor sailing, eliminate the need for the anchor to "migrate" around and reset at a current or wind shift and expect it to reset, safety, increase odds of secure anchoring, increase scope for poor holding without the negative side effects of anchor-sailing caused by too much scope.

It is difficult to estimate how many cruising boats we have helped rescue from anchoring related problems over the years and further, how many needed rescue of which others helped and we were unaware. Based on our experience, these rescues fall into two major categories. First, are rodes wrapped around appendages like rudders, keels and propellers. Second, grounding caused by not understanding the distance they would cover across the bottom on tide change when swinging to a single anchor.

How?

How multiple anchors relate to rode and scope:

Too much scope is worse than too little, especially in a modern, multihull or other light displacement boat. As scope increases beyond the essential 5:1, anchor-sailing increases exponentially. In many areas of extreme tidal variation, your scope changes radically depending upon state of tide.

For instance, if you anchor in the Ogeechee River in Georgia, the tide range is thirteen feet. If you pick a spot with six feet at MLW, it will be 19 feet at MHW. Anchoring for high tide, you would need 95 feet of rode for 5:1 scope. At low tide, your scope would be 16:1 which is clearly unacceptable for a host of reasons including swinging radius with a good possibility of going aground, anchor-sailing dragging your rode through the mud and the possibility of snagging your anchor with your own rode or wrapping your rode around underwater appendages. A Bahamian moor moderates those possibilities.

In the accompanying diagrams, you see the result of being tethered in one spot. The bow of the boat simply pivots around a point

The Classic Bahamian Moor

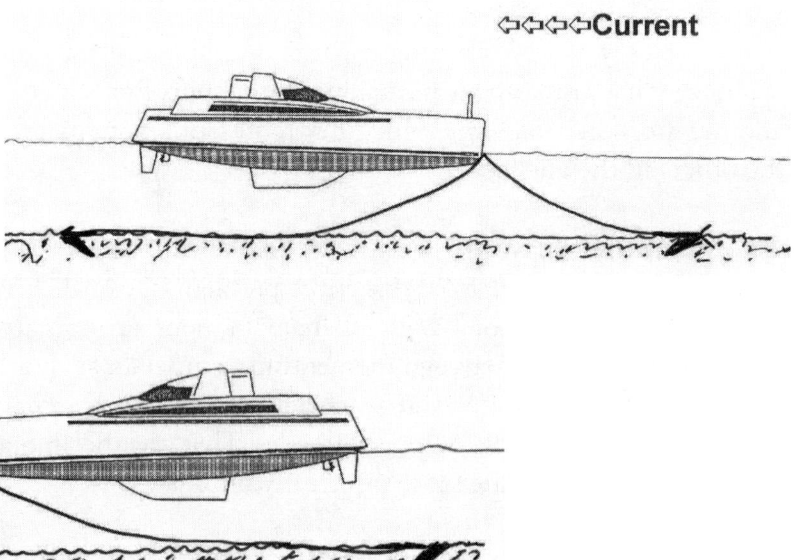

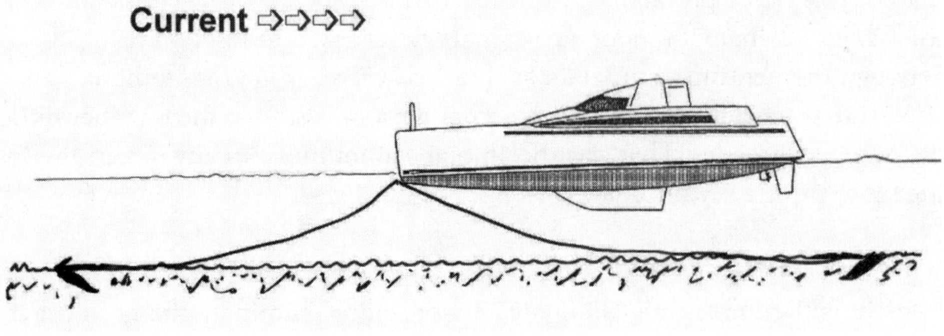

The Bahamian Moor

The mainstay of cruising anchoring
(two anchors opposed 180^0)

- Use any time you need to anchor without excessive swing.
- Use anytime you anchor in swift, changing currents.
- Use every time you anchor in questionable holding ground.
- Use in conditions where you need excessive scope due to poor holding, large tide range, erratic currents or wind against the tide situations.
- A boat single anchored to a single anchor swings in a circle of which the radius is the length of the rode plus the length of the boat.
- A Bahamian moored boat swings in a circle approximately the length of the boat.
- Excessive scope on a single anchor allows anchor-sailing, dragging without re-setting and major annoyance to your anchored neighbors and danger for all.
- The most often used system in tight places like the United States southeast coast, the Bahama Islands, and other waters noted for swift changing tidal currents.
- The possibility of chafe, abrasion and damage to the benthic resources are dramatically reduced because rodes remain more or less in a straight configuration and are not constantly being dragged across the bottom in big bights the way a farm tractor breaks up the clods with a chain.

Tricks of the trade

Procedure for launching a second anchor

1. Launching from a dinghy.

Many users find the most effective way to create a Bahamian moor is to carry the downstream anchor out in a dinghy. This is where ultra light weight high-tech is at its best. An aluminum Fortress with a nylon rode is infinitely easier to carry out either by dinghy, by wading or by floating it out on a fender than some heavy weight anchor with a full compliment of chain. This is especially useful in areas of large tidal range where you will need above average scope. It gives you the option of easily laying out the anchor exactly where you want it.

2. The drop-back method.

Set your first anchor up-current. Then allow your boat to drift back the appropriate distance, usually the same 5:1 as your upstream anchor. Carefully lower the anchor and being exceptionally careful not to snag the rode move yourself back to your original position equally between the two anchors. Set the downstream anchor either with your engine or an appropriate winch.

3. Walking the anchor to the stern of the boat, method.

As with many catamarans, you will be anchoring in very shallow water, usually this is the preferred location. Assuming a 40 foot boat anchored in five feet of water, forty feet would be sufficient scope. Thus you can easily carry your lightweight anchor back to the stern being careful to be outside the shrouds and pulpits, lower it carefully and set it with a winch up forward.

The downside to any multiple anchoring system.

1. The obvious first objection is handling two anchors or more anchors. If you are using lightweights with nylon rodes the difficulties are somewhat ameliorated.

2. The most pressing nuisance is twisting of the rodes when anchored in the same place through several tide changes. Sometimes this can be eliminated by locking the helm over enough to determine the direction of the swing. Next step is to turn the helm in the opposite direction for the following tide change. Occasionally I observe people actually motoring their boats in circles to untwist the rodes. Usually it is possible to just work the twisted rode off the one still holding because the downstream rode will be or can be made slack and the one still holding will be taught. It is not unusual to retrieve the downstream anchor by dinghy.

Augmenting your Bahamian Moor

The Kellet. (Also called Angel, Sentinel or Bouncer)

The objective:

- Increase effective scope.
- Decrease anchor sailing.
- Help prevent rode wrap around underwater parts.
- Dramatically increase overall holding power.

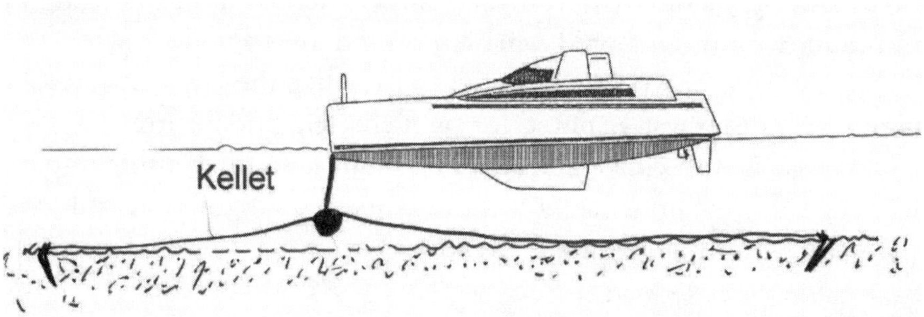

Creating a kellet

A kellet is simply a weight that is placed along the anchor rode. It effect is to pull the rode down towards the bottom thus creating additional catenary in the rode which effectively deceases the scope angle thus increasing the power of anchor. Two side benefits are keeping the rode down away from powerboat props and helping moderate anchor sailing.

- A kellet can be as lavish as the professionally built *Rode Rider*, a remarkable bronze and lead affair with a built in roller or it can be as simple as a cement block tied to your rode with scrap line. In either case, the effect is the same. Keep in mind limestone rock and cement lose about a third of their weight in water. (Some people believe they can hook two anchors to the same rode. In effect, they have put a permanently spaced kellet on the rode because it is impossible to set both anchors and what most often happens is the inner anchor cannot set and simply acts as a kellet)

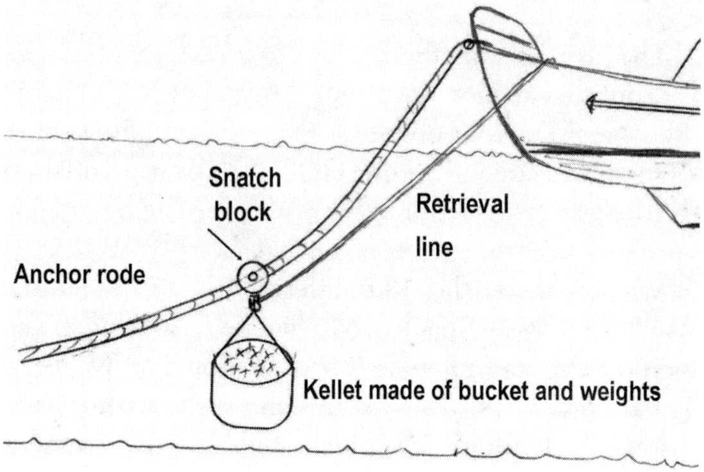

• Most people make their own kellet. In my case, I use a canvas bucket filled with dive weights or old CQR anchor parts or whatever is handy that is heavy. The bucket is clipped to the rode with an ordinary snatch block and lowered to the appropriate depth with an ordinary piece of line, or you can use a dockline. To make a permanent kellet, many people use a plastic bucket filled with old heavy stuff, lead wheel weights being preferred, cemented in place with a piece of chain through the center to attach the snatch block. A short length of 2" PVC pipe with appropriate end caps and a chain running through center, filled with old tire weights makes a snazzy kellet. Kellet weights usually begin at approximately the same weight as the main anchor, or in the case of a Fortress, about double. The kellet must be heavy enough to keep the rode down.

• Using a kellet is simplicity itself. Set your anchor. On the rode, fasten your snatch-block with attached kellet and retrieval line. Lower kellet to preferred position. If you are using the kellet around multiple rodes, use a short piece of chain or better, one of those nifty stainless steel rings sold for the purpose of retrieving stuck anchors in place of the snatch block. Anything that will encircle all the rodes and slide down without much fuss will work.

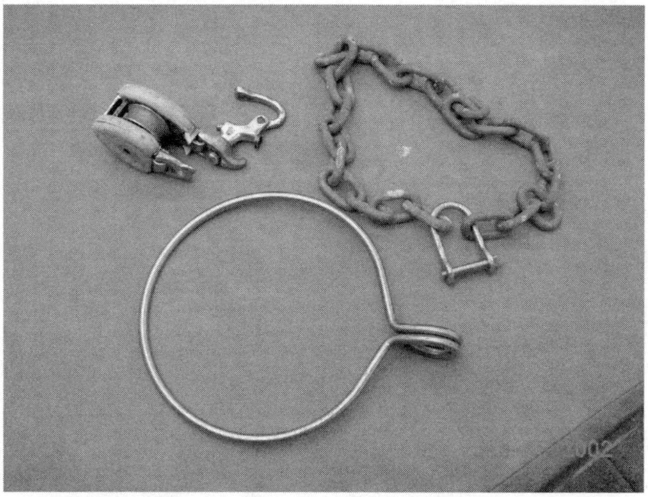

Simple kellet attachments. The chain or the ring can be used on multiple rodes either rope, chain or a combination and the snatch block on single rope rodes

• Setting the kellet, knowing how deep to set it. Back in the scientific range, having the kellet in the center of the rode's catenary will provide the maximum force. However, from the user standpoint that may not give you the best performance for the situation. I have usually found that the depth of the water in which you are anchored is a better starting point. Experiment for the best location for your own boat in the conditions in which you are anchoring.

• For those looking for a way of moderating anchor sailing but using a single anchor, this may be your answer. You sure can try it with simple jury-rig stuff before committing to anything more elaborate.

Of moorings temporary or permanent

A Bahamian moor is simply put, a two anchor mooring. For those traveling in hurricane country, often greater protection is required. In that case, the *star mooring* is often the answer. A star mooring is constructed just like a two-anchor mooring except it has three or more anchors. In a three anchor star, the anchors are set 120 degrees apart and joined at the center. Rodes can be shackled or tied. A kellet in the center really creates an almost unbeatable mooring. Many jurisdictions use moorings constructed this way as permanent moorings. The best anchors for that type of mooring are Danforth.

Hopetown, Abaco, Bahamas

Corinne and I survived this hurricane with no damage on a mooring I constructed out of my three basic anchors, a DELTA 22 and two Fortress FX 11. Rodes are all nylon with six ft. of 3/8 chain on the DELTA, per Simpson Lawrence instructions and fifteen feet of 1/4" chain on the Fortress per their instructions.

The kellet was about fifty pounds of broken rocks and blocks we scrounged from demolition sites on shore, held in an old canvas ice bag. (Concrete loses half its weight in water)

We experienced winds up to 85 knots from the NE, calm as the eye went over us and 45 knots from SW after that. Tidal surge went several feet over the noted docks.

Hurricane Erin
La Forza at Hurricane Mooring, 01 August 1995

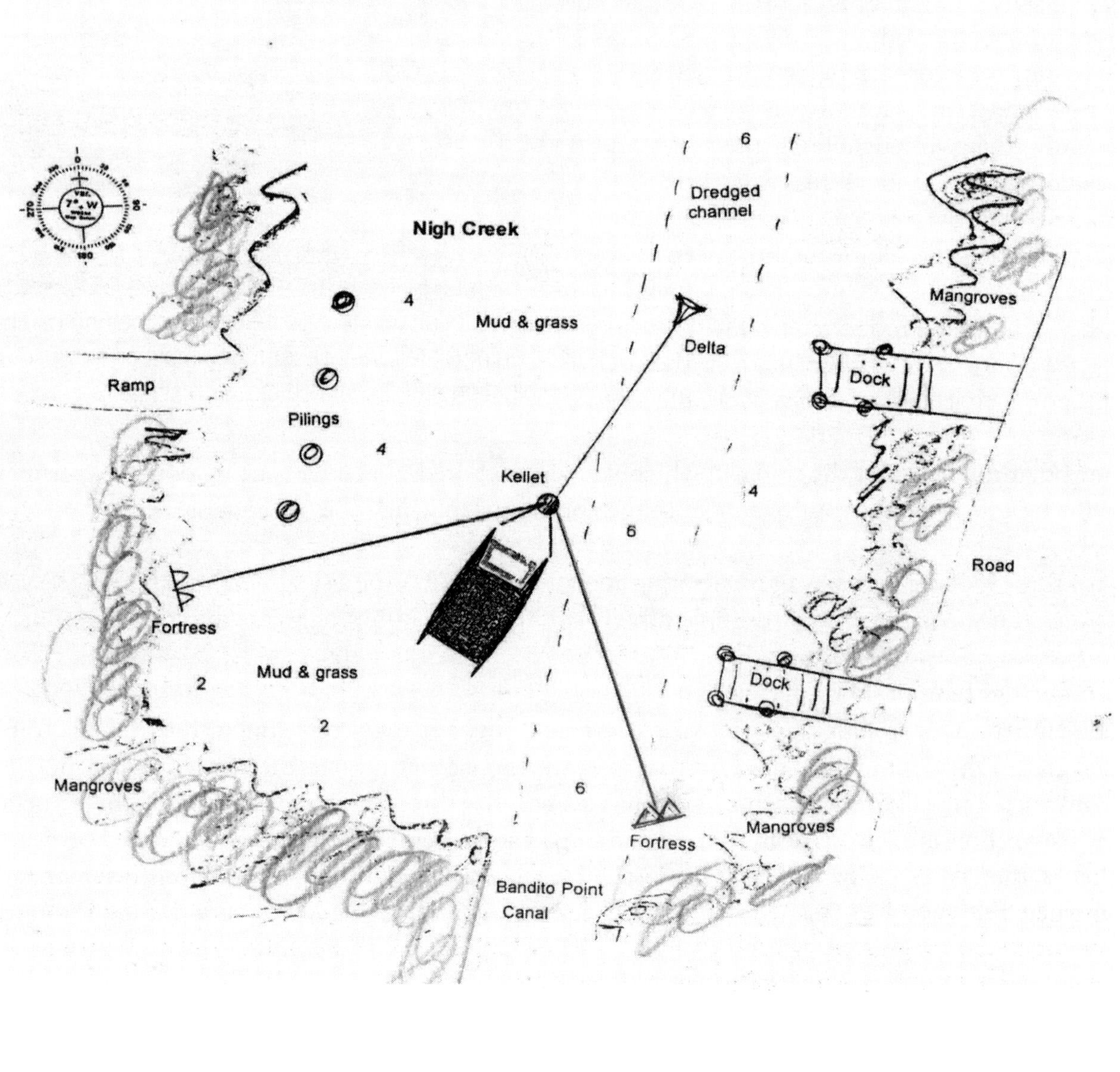

With a star mooring, you are always tethered to two or more anchors. The Society of Naval Architects and Marine Engineers (SNAME) shows a star mooring with five anchors but even with three anchors as could most reasonably be expected to be carried aboard a yacht you would always be tethered to two.

In the diagram above, you can see how vital it is to keep the boat in the same location. There simply is no possible way to anchor on a single anchor without being blown ashore in a location of that nature. In so many places this is the critical component of a good nights sleep.

Emergency anchoring no-no-no's

1. Myth one, two anchors on one rode. This just does not work. A simple vector diagram from a

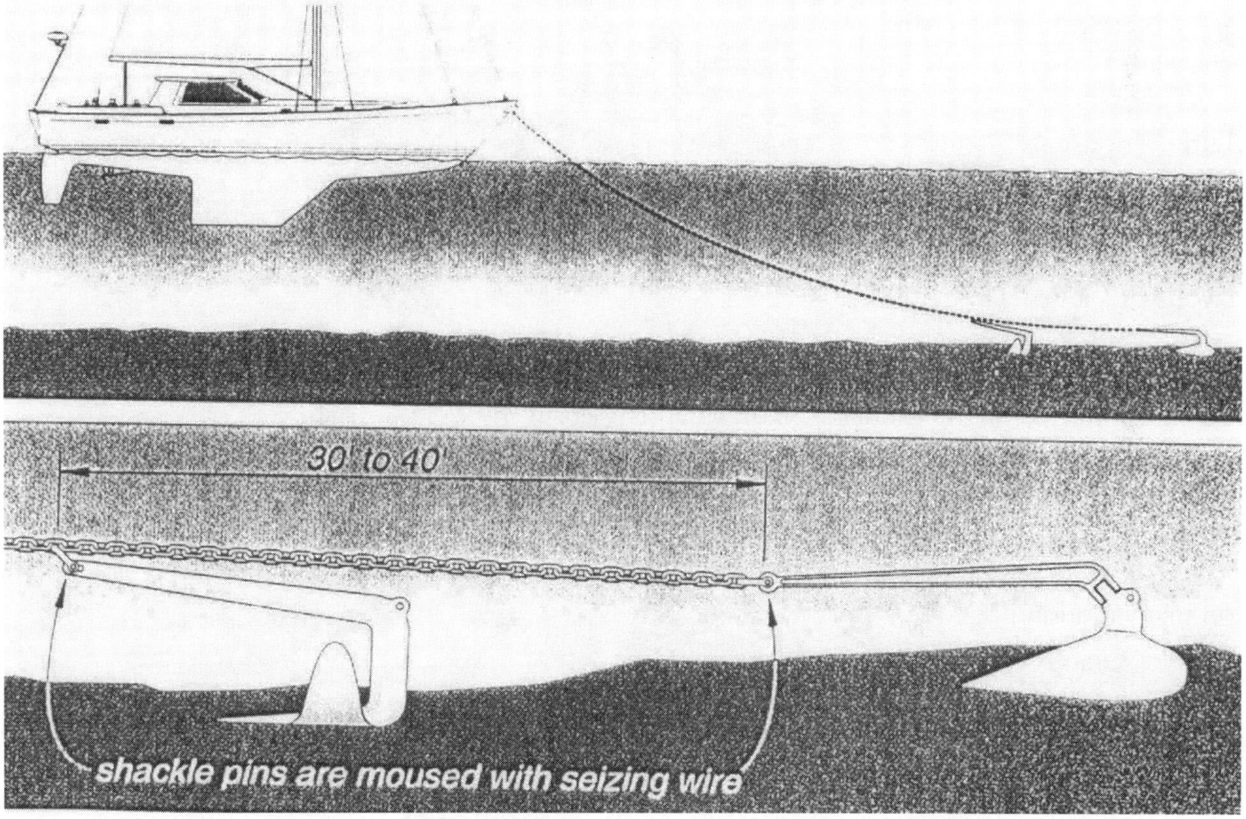

Copied from a popular magazine. Obvious problems with entire concept including trying to get this mess back on board.

school child can put that concept to rest.

2. Common knowledge says the Bruce will set before the plow. In any case, with any two different patterns, how are you supposed to get them both set? Obviously, one would set before the other even if such a combination is possible. If this system actually worked, would it not be recommended by manufacturers, organizations such as ABYC, SNAME, etc.? How come this theory is strictly the province of bar-room speculators and pop magazines?

Hurricane prediction is not a precision science. Nobody in the business would have the temerity to predict exactly where the hurricane will strike thus it is impossible to predict exactly from which direction the winds will blow.

A couple of miles off means a couple of degrees of angle off. Thus any pundit who describes exactly to the letter just how multiple rodes and multiple anchors should be laid out ought to go back to school. (That is precisely why you use a star mooring so you are protected from any wind direction.)

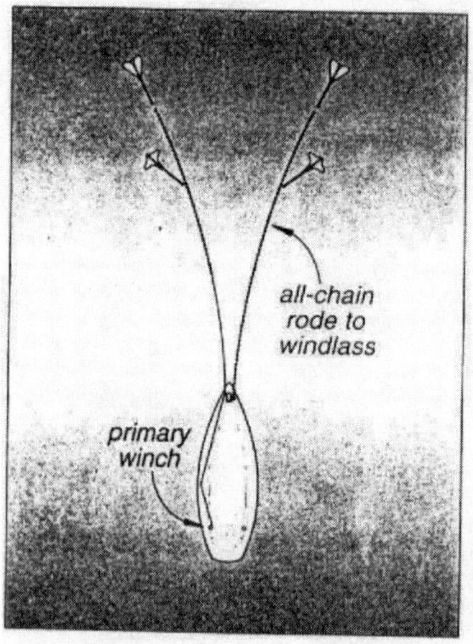

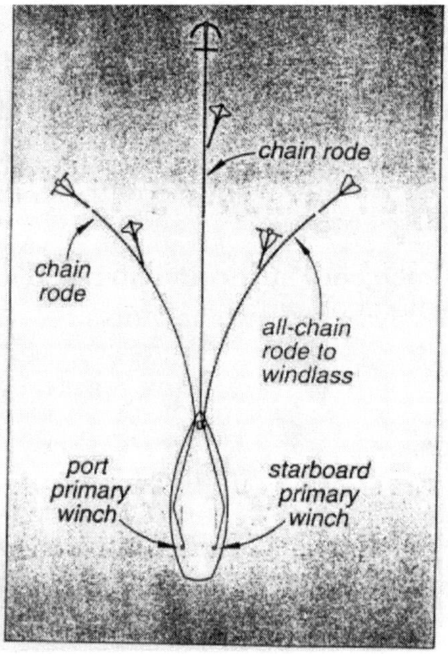

Utter nonsense from a popular sailing magazine. First, who has that much tackle on board? Second, the layout precludes wind from any other direction save straight ahead. Third, how on earth are you going to get those anchors located like that? Forth, what about other boats or obstructions in the area? Fifth, the *last* thing you want is a solid chain rode to the windlass. Sixth, what happens if the wind does shift?

Bridle your boat

A bridle is an essential piece of anchoring equipment for most modern catamarans. It can be a very useful item on many monohulls.

What is a bridle?

A bridle is a rope construction that secures your anchor rode on the centerline of the boat a measured distance in front of it.

Purpose of bridles

1. Maintains an actual center point of pull sufficient far in front of the boat to offset the natural tendency to anchor sail. Wider boats have a greater tendency to anchor sail thus a bridle becomes a necessity.

2. Shock absorber for chain rodes. Anyone using chain rode needs good snubbers. A nylon bridle makes an ideal snubber.

Length of bridle legs
Most people simply use their boat's beam measurement as the correct one for the bridle legs, thus constructing and equilateral triangle. This works about 90 percent of the time. The other ten percent requires a little experimentation.

Other workable ideas
The French chain pile moor. An interesting development from catamaran sailors with large amounts of chain rode. It consists of setting your anchor in the usual way, which consists of about 5:1 scope and setting the bridle. Then, after your position in the current is confirmed, you dump the balance of your chain in a big pile on the bottom. This puts all that weight on the bottom where it can do some good rather than leave it in the boat where it may do some harm.

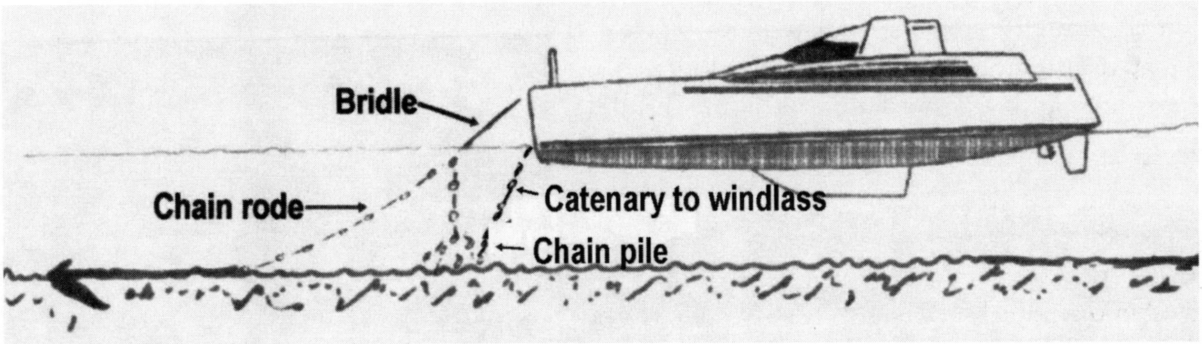

Bridle using a fixed center bridle
All the production catamarans that come with standard equipment all-chain rodes have a fixed center bridle as part of the system. These bridles invariably have a snap shackle to attach to the chain This is a potential problem hookup that needs close monitoring. Some use a standard chain hook which is essentially a piece of steel with a deep open slot. This is actually stronger, better and easier to use but, it is not permanently secured and requires continued tension to remain attached to the chain. Thus, there is always a chance of accidently dropping the bridle.

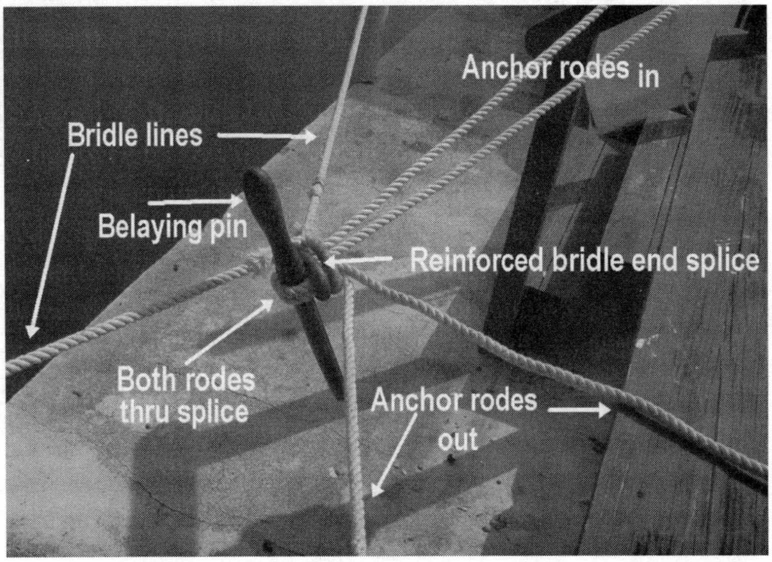

With rope rodes
For those who are using rope for their anchor rodes, you can do even better than the chain gang with a bridle of this type. Make up the bridle by splicing or even tying a central eye in the middle of the appropriate length bridle. Set your anchors, either single or

Bahamian moor, Then, from the lazy side of the rode, push a bight of rode through the loop and use a belaying pin to hold it in place. Once in place, let out a little more scope to tension the bridle. Always work from the portion of the rode between the cleat and the bitter end. (The lazy side) Never try to work with a rode in tension for this application. Do not forget to secure your belaying pin to the bridle with a lanyard, just in case!

Bridle using the anchor rodes as half the bridle

Many cruisers, especially in smaller boats, will opt to use the anchor rode itself as one side of the bridle. There are several successful ways to do that.

Prussic hitch

First, is the use of a line, often a dockline, with a sliced eye on one end. The sliced eye is turned into a prussic knot. After the anchor is set, the knot is constructed around the lazy side of the anchor rode and then secured to the appropriate mooring cleat or designated fastening on the opposite hull. Then the anchor line itself is payed out until the bridle is formed, then cleated on the opposite hull.

Genesis of a prussic hitch. Wrap the eye around the anchor rode pulling it through as shown.

Second step is to feed the lazy end up through the loop and tighten it forming what looks like a double-sided clove hitch

The finished prussic hitch. It will slide easily in either direction when tension in released but hold securely when tension is applied. Mast climbers can be made from two short line with eyes on both ends. One goes on the halyard, the other on your foot. With two lines you can walk right up a halyard.

Standard knots

Personally, I use a slippery timber hitch at the end of a precut line with a spliced eye. That way, I have the option of tying to the set rode then releasing more scope. Otherwise, a regular timber hitch works well. After a few years of using it, you gain enough confidence in your ability to make the hitch slippery. That is a great aid a retrieval time when a slight tug on the free end of the hitch drops your bridle. In the fifteen years we lived aboard and cruised, I estimate I have anchored with a bridle on one or both anchors no less than 1000 times. This does NOT include anchoring other boats on races and deliveries or family weekends in non-liveaboard times.

Slippery timber hitch. A quick pull on the loose end releases the knot. Takes a little extra care when making the knot to keep it tight.

Common timber hitch. Think of it as a clove hitch with one extra turn around the rode. Very useful knot for pulling most anything. Often called a timber hitch because it used extensively by the logging industry.

Surveying and Surveyors, What, Why, How, Who and When?

What is a yacht survey?
A yacht survey is an inspection by a professional. Depending upon the type of survey contracted, the survey may determine the condition of the vessel, its compliance with regulations and its value. The most common types of yacht survey are: Condition & Value (C&V) which may include or specify damage, insurance/financing, pre-purchase.

What should you expect from a prepurchase C&V Survey?
For your survey dollar, you will receive a clearly written, concise report of findings about the condition and state of repair of the vessel, its systems, subsystems, its equipment and all of its parts. Most reports will include an appraisal and a statement of financing and insurability risk.

A survey is an independent professional evaluation
It is not advisable to purchase a boat, new or used, without an independent evaluation from a competent, experienced marine surveyor. Are you surprised by the recommendation to survey new boats? If you follow the trade papers and note the overwhelming amount of warranty work involved with boats, you will understand my recommendation.

The surveyor is an independent, experienced, knowledgeable professional who has an unbiased view of your prospective boat and will dutifully examine every square inch of it and issue a report of survey. A survey compares to a premarital physical, a transcript of school records and an FBI clearance investigation all combined.

Private detective
The surveyor compares to the private detective, or sleuth. He will find evidence of previous failure and repair, inconsistencies in equipment and installation, jury-rig and/or haphazard installations, weak, inadequate or wrongly applied materials or installations.

The survey report itself will read quite a bit like a physical examination. It will list in detail the condition of every aspect of the boat, it's weaknesses, its lack of proper and required equipment. Within certain constraints and guidelines, it describes recommendations for repair, replacement, or redesign. However, do not expect a repair estimate. Estimating the cost of repairs is clearly beyond the scope of prepurchase survey. Design features are the province of a naval architect and navigation limits are the responsibility of the designer or builder.

Pass or fail?
A vessel does not "pass" or "fail" a survey, except in a predetermined sense of suitability, or to confirm the sellers claims. For instance, suppose you want to buy a boat for circumnavigation. The surveyor can evaluate the scantlings, construction and equipment based upon the claims of the designer/builder and note them accordingly in his report. This report will then give his opinion as

to the vessel's suitability for its intended use, and may include the capabilities of the intended purchaser. Or, perhaps the seller has made a claim of being "free of rot." Your survey can, within certain limits, confirm or deny this claim. Therefore, pass or fail.

Beyond the scope

Recommendations for additional equipment, modifications, refit and redesign are normally beyond the scope of a prepurchase survey. Negotiate those things into the survey format prior to engaging the surveyor. However, before doing this, make sure your chosen surveyor has the skill to properly undertake the responsibility. Such items usually require the services of a Naval Architect or a Marine Engineer. However, if your surveyor is truly knowledge and experienced in the field he or she can be a valuable resource for such items.

Confirm or deny

A survey report will also confirm or deny claims made by owners, brokers and other parties with vested interests. Since the surveyor works for you, and has no interest in any other party, he or she can, should and usually does represent the boat. This representation is the most important consideration of the contract. You, the purchaser, are disqualified from objective evaluation because you are in love. The other side of the bargain, owners, brokers, etc. are disqualified from objective evaluation because they are in need: of cash, that is. Therefore, your most trustworthy, unbiased, objective evaluation will be from your independent surveyor.

Basis for financing and insurance

A Report of Survey is usually the basis for the acquisition of both financing and insurance. Both institutions rely heavily upon such documentation and usually accept the surveyor's opinion as to the value and insurance and financial risk involved in the vessel.

Appraisal

Most survey reports will contain an appraisal. Therefore, it is important that you obtain a surveyor familiar and experienced with the type of vessel you are purchasing. The surveyor must know how to obtain comparable sales pricing information beyond the BUC, ABOS and NADA listings. This is especially important when purchasing multihull sailboats.

There is precious little information given to traditional sources, such as BUC or NADA, about multihulls. BUC is apt to have recorded only those sales through nonstandard channels such as foreclosures, estate sales, divorces, other distress situations and brokers who do not normally handle that class of vessel. Many multihull brokers I routinely deal with are notoriously lax in reporting their sales to BUC, therefore, it is understandable that BUC has little valuable pricing information. Therefore, it is more important than ever that the surveyor you engage knows how and where to research and document multihull pricing and appraisal information.

Bargaining chip

A survey report is usually used as a bargaining chip between buyer and seller. "Subject to Survey" is the clause most often found in contracts for sale of a vessel. Many contracts contain a clause allowing for a percentage differential of price based on the survey. For instance, many consider that the purchaser absorb repairs and replacements determined by survey, up to 5 percent of the purchase price. Beyond 5 percent, the difference is either split between the parties or absorbed by the seller.

You can see how this "Report of Survey" is a most important document for anyone purchasing a boat.

Why a survey?

Beyond the previous mention of needing the survey for insurance or financing reasons, the compelling reason for using an independent specialist is unique ability. Ability not only to find an item in need of repair but to evaluate the vessel based on its lineage, its intended purpose and the capabilities of the aspiring owner(s).

A surveyor can save you many times the amount of the cost of the survey. Just because a person is a well known racing sailor does not mean he can understand what is happening to a vessel. Massive rot in the aft crossbeam transom; serious hull blistering penetrating into the laminate; jury rigged engines and fuel tanks; illegal equipment installations and a host of other problems that slid right past the purchasers. A survey, any survey, even a survey by a marginal surveyor most likely will uncover at least some of these problems. There is no doubt in my mind that a good, prepurchase survey will save many times its cost.

It is interesting how much information about a boat a surveyor discovers that was unknown by the owner, though the owner is a competent, thorough, caring person. It is a rare person that knows everything there is to know about his vessel.

When surveying new boats, the surveyor provides yet another function, that of a watchdog. You may avoid potential warranty problems by survey. Most surveyors can explain the vessel's paper trail and advise about registration, documentation, sales tax, liens, etc. and where to go to facilitate or expedite this important aspect of sale.

A survey is also a useful tool to add to the vessels permanent log. A survey and an inventory list will often satisfy both U.S. Customs and foreign governments about the origins and ownership of the ship's equipment. For those of you who travel to foreign lands, adding the paid bills and the designers plans to your official ships papers is a good idea.

U.S. Customs will take their drills and chain saws and cut into all concealed compartments on a vessel. A survey may find hidden compartments that you may not have even suspected. Your surveyor can recommend ways to make these visible and/or accessible to satisfy customs regulations. Having hidden compartments is cause for seizure of the vessel, according to United States Customs.

There are other reasons for getting a survey, such as insurance renewal, pre-race, pre-yacht delivery and acceptance of delivery from either a yacht delivery company or a trucking company.

Choosing a surveyor
Who is Surveying and how did he get his qualifications?

The first problem most new purchasers have is finding a suitable surveyor. It is necessary to have some working knowledge of how one gets to be a surveyor to find one, then evaluate capabilities and qualifications. Since there are no legal requirement nor professional license requirements except certain local occupational permits, the surveyor's professional association and affiliations become of paramount importance.

It is estimated that there are about fifteen hundred independent surveyors in the United States (US). That is a surprisingly small number. There are only three (3) professional organizations for surveyors in the US, *The Society of Accredited Marine Surveyors* (SAMS), *The National Association of Marine Surveyors*, (NAMS) and *The Marine Surveyors Guild* in Louisiana. Most Guild members are cargo and work boat specialists. The same relationship of cargo specialists to yacht surveyors exist with NAMS. SAMS has a higher percentage of yacht-oriented surveyors and is a world-wide organization. SAMS also has stringent membership requirements which include continuing education.

BOAT/US, one of the nation's largest insurers, has a non-exclusive list of surveyors from whom its insurance underwriters will accept surveys. This is a practical reference and cross-check. Without using such outside help, you are subject only to referral and personal recommendation. You must be extremely careful from whom you accept personal recommendation. There are simply too many people willing to bend a bit for a recommending seller. It is a very difficult position for a surveyor to find himself in! Be wary of surveyors recommended by the selling broker. A broker should really give you a list of surveyors known to him to be experienced in your particular type of vessel. You make a choice between them and then cross check with other recommendations. Check his or her affiliation with a professional organization. Always interview your prospective surveyor.

Before you contract with a marine surveyor to survey a multihull sailboat, he should have the following qualifications

1. Actual, first hand knowledge about the type of boat to be surveyed.
2. A working knowledge about the materials used and the principles of construction used in the vessel.
3. Some background experience with multihulls.
4. Knowledge and sources for an appraisal.

Perhaps we might compare career experience levels to the Railroad Car inspector and the Aircraft Inspector. The railroad car inspector upon examining the aircraft might look with disdain upon the lightly constructed flimsy aircraft, whose wings wobble with a touch of the hand. The aircraft inspector looking at the railroad car might surmise the thing to be crude, clumsy, heavy and unnecessarily weighty. Simply a matter of perspective on both parts and the same applies between ballasted monohulls and multihulls. Perspective, experience, attitude and knowledge.

Many surveyors, like other professionals, will join and participate in professional organizations. Before you engage them, ask them to which organization they belong? These professional organizations, have strict entry requirements and screen their members.

Other organizations, like The American Boat and Yacht Council (ABYC) and The National Fire Protection Association (NFPA) are informational and standards setting organizations. Surveyors are encouraged to belong and contribute. The Society of Naval Architects and Marine Engineers (SNAME) has been around well over a century also screens its members and provided valuable technical information to surveyors.

I have found the single most useful organization to be the BOAT/US Exchange for Marine Professionals. The Exchange is the most prolific source of information about what is happening in the field of yacht survey, casualty, litigation, insurance, investigations and USCG recall. I have found the most vigorous, enthusiastic and active organization for promoting professionalism among yacht surveyors to be SAMS, The Society of Accredited Marine Surveyors.

How a survey is contracted

I always suggest a written contract for survey that explains the exact terms conditions and expectations of the survey, to the purchaser of the service. However, in the real world, this seldom happens. Most contracts are last minute, rush items undertaken over the phone. Therefore, it is more important than ever that you have an idea about the questions a surveyor will ask before agreeing to do your survey and you should be prepared to answer them.

The following is a sample list of questions. It does not pretend to be complete or all inclusive:

1. What kind of boat is it? Who is the manufacturer. What is the length, beam, draft, year of manufacture, construction, equipment, rig, etc.
2. The boat's name and description, name and phone number of the present owner or managing broker.
3. Details of sale, including price agreed upon.
4. Legal details: import duties paid, documentation numbers or state titles if any, mortgagees, liens, etc.
3. Location of the boat.
4. What arrangements, if any, have been made for haulout? Who is responsible for haulout, where and when? (most surveys, the owner arranges and manages the haulout, the purchaser pays, but this is not carved in stone)
5. Who will be present when the survey is performed? Who will take care of the yard bills? If sea-trials are involved, who will be captain?
6. What is going to be covered under the survey, and what is going to be excluded. (in most cases, engines, electronics, extraordinary equipment such as air conditioning and SCUBA compressors are not covered, but are handled by other specialists)

Haul that boat

Even if the owner can show you paid bills for having the bottom done last week, you should still haul and inspect. You just never know what the yard missed or what should have been repaired but was not. Nor do you want to accept a third party assessment of osmotic blistering .You also need to know about hull-fairness, zincs, through-hull fittings, shafts, props, struts, cutless bearings, speed scoops, pintles, gudgeons, rudders, distortion, condition of paint, hogging and alike.

The underwater parts represent the most difficult, costly and inherently dangerous areas of the boat. They should never be glossed over or taken for granted. The assessment of a boatyard is different than the assessment of a surveyor and the needs of the purchaser.

A prepurchase survey should always be done out of the water. If an owner refuses to allow your surveyor to survey out of the water, alarm bells should go off in your head. Unless it is the most incredible deal imaginable, you are probably better off just walking away and finding another vessel.

The walk-around

A good surveyor likes to be present, if possible, when the boat is hauled. As the boat is lifted in the slings or on the railway, he specifically watches for excess mobility, excess flexing, bending, hogging and other signs of structural inadequacy, weakness or damage. Once the boat is out of the water, before he touches a mallet or a probe, he will first, very carefully study the vessel's lines from a distance, walking around the boat and observing it from all angles. What he is looking for are bumps, hollows, bulges, unfairness, hogging, hard spots which may be telltale signs of delamination, broken frames, poor construction, inadequate scantlings, repaired damage or over stressed rigging, or combinations of any of the above.

If he spots any suspicious areas, he will usually mark them with chalk for later comparison from inside the hull. Then he will pick up his mallet and probe and get close to his work.

Inspect the bottom

He will sound the entire hull with a mallet pausing to scrutinize every thru-hull fitting. He will carefully inspect the rudder(s) and all its connections, pintles, gudgeons, rudder post, bearings, often turning up both excessive wear and galvanic corrosion damage overlooked by both owner and boatyard. He will poke around centerboard/daggerboard trunks looking for stress cracks, damage, pivot pin wear, leakage and will note the shape and fairness of the boards. He will then sound the propeller, inspect the shaft, struts, cutless bearing and fittings taking careful note of their condition, composition and size. If he knows anything about multihulls at all, he will know exactly where to check for stress cracking of the structure at the major joins. However, even the best surveyor may not be able to distinguish between cosmetic cracking of faring compound and actual laminate damage. Destructive testing or special expertise may be called for. Then he will meticulously inspect the hull-deck joint, a classic recipient of many problems of which leakage is only one.

Mechanical equipment

Most surveyors will take responsibility for the outside of engines, generators and other major equipment installations. For instance, I will check the vital engine fluids looking for signs of failure. Check the drive belts looking for wear and adjustment. Start the engine(s) check the exhaust, looking for tell tale signs of problems. I always recommend a laboratory oil analysis to double check assumptions about the engine and to create a baseline for the new owner. I will check the motor mounts, engine beds, stuffing box and all the connective cables and hoses.

The same process will cover the balance of the installed equipment such as refrigeration, air conditioning and electronics. Your surveyor can check for operation and appropriate installation but not adjustment, accuracy or aspects of application and usage, that is the province of specialists.

Safety equipment

Conforming to United States Coast Guard regulations and compliance with ABYC standards is a vital part of any survey. Flares, fire extinguishers, PFDs, signaling devices, etc. will be inspected.

Sails

A full inspection and evaluation of sails is the job of a sail loft, especially on a racing vessel where the sail inventory is a major portion of the price. Many surveyors will unfurl a jib, raise a main or inspect bag sails in the bag if negotiated by the purchaser. Many times sails can be spread on a lawn or driveway to be inspected. Clear this with your surveyor before signing any agreement.

Sailboat mast and rigging

Most surveyors will inspect at deck level only. Most believe that rigging is the province of expert riggers. However, as with engines and other equipment, a great deal can be learned from the attachments of the rigging to the hull. The surveyor usually advises the client about the ages of rigging and the usual replacement and maintenance procedures. Many surveyors will include a disclaimer about stainless steel that is hidden from view.

Electronic testing

Testing with a Moisture Meter or an Electronic Sounding Device may or may not be part of a survey. Many people have become infatuated with these instruments and insist upon their use. Surveyors are usually more than willing to comply since not only does it transfer some of their judgment, with its concomitant responsibility, to a machine, but may allow another charge in the billing. For most applications, a good surveyor need not perform tests using this equipment. However, there are times when they are indispensable and the prospective purchaser should understand the need for additional payment. Like GPS or computers, this is yet another symptom of modern society transferring art to science in its endless quest for a risk free environment.

Non-destructive testing

Thus far, we have only discussed survey using nondestructive testing, which is testing by observation, sound or radio frequencies.

Destructive testing is quite another matter. Before a surveyor can probe a single blister, he must first get permission from the owner of the vessel. Often, permission for such superficial destructive testing as probing blisters is implied and the owner may benefit as much or more than the prospective purchaser. Often, there is outward evidence of inward problems that cannot be resolved without destructive testing.

For instance, the cracking of fairings in high stress areas or the condition of suspect fuel tanks that are inaccessible. Often, a surveyor will want to reserve judgment until he sees what it looks like inside. Destructive testing that involves specialists like fiberglass or fuel tankage are contracted for separately. There is a clear conflict of interest if the surveyor himself undertakes such tests and the needed repairs after testing. It is not at all unusual for the closing of a sale to hinge upon the results of such tests.

Checking the papers

After he has laid his hands on every square inch of the vessel, run his hands over the rigging, (although probably not the mast above deck level, which unless specifically agreed beforehand, is the job of a rigger) checked the halyards, sheaves, sails and all the myriad other parts, equipment and safety gear, he will ask about the ships papers. The registration, title, USCG document, import duty receipts, receipts for equipment, instruction books, warranties, owner's manuals etc. This is for your protection. You would not be the first person to buy a boat and find that it had a lien on it that was undisclosed by the seller.

The report

When the surveyor is finished with the hands on part of his inspection, an entirely different task emerges which may be even more demanding than the first part. That part is preparing and executing the final *Report of Survey*. Many surveyors specify delivery of the report with a certain period. Usually this is important because other parties, such as finance institutions, are awaiting the report to act.

Research

The surveyor must now do some research about the vessel, its design parameters and its financial worth. He must make some decisions based not only upon his on sight observation and critique, but upon his knowledge and experience in the field of survey.

SUMMARY

To summarize; a prepurchase survey is as necessary when buying a multihull as a physical examination is to your health. Understand that the surveyor is a diagnostician, not a practitioner. Contract the surveyor on his track record. Do not accept in the water prepurchase surveys. Use the established surveyor organizations as part of your background check.

Bibliography

Each and every author in this bibliography contributed to my present knowledge. Since this was a book about cruising, I decided to limit the bibliography to those books that were directly related to that subject.

You will note that most of the volumes about catamarans appear rather dated. It is an interesting phenomenon that there was an enormous outpouring of literature beginning in the late 1950s and 1960s and tapering off in the 1980s. By the 1990s, only a few volumes were published, two of note by Kevin Jeffery and myself, *SAILOR'S MULTIHULL GUIDE* in the late 90s and my classic volume, *Cruising On More Than One Hull* in 1992. Most post 1980's multihull development is detailed in periodicals, foremost among them is MULTIHULLS Magazine, Boson, MA.

Information about publications and ISBN numbers where known, are included to be helpful in your research.

Bibliography

A

Aebi, Tania, with Bernadette Brennan, *Maiden Voyage*, Simon & Schuster, NY, 1989, ISBN 0-6711-666653-3.

Amateur Yacht Research Society, *Cruising Catamarans*, John Morwood, Editor, F. J. Parsons, London, 1972

Andrews, Jim, *Catamarans for Cruising*, Hollis & Carter, London, 1974, ISBN 0~3700~102944~0.

Angel, Nicolas, *Capsize*, Norton, New York, 1980 translation, originally published by Pen Duick, Paris, 1979, ISBN 0-393-039264-7.

Ashcroft, Tami Oldham, *Red Sky in Mourning*, Bright Works Publishing, Friday Harbor, Washington, 1998, ISBN 0-9655837-7-5.

AYRS members, *Self Steering*, The Amateur Yacht Research Society, Woodacres, Hythe, Kent, England, 1967.

AYRS members, *Cruising Catamarans*, The Amateur Yacht Research Society, Hermitage, England, 1972.

B

Babson, Stanley M., *Where Sands Are Pink*, Milstan Enterprises, New Canaan, CT.

Bailey, Maurice & Maralyn, *Staying Alive,* David McKay Inc.,New York, 1974 originally published as *117 Days Adrift*, Nautical Publishing Co. Ltd., Lymington, England 1974, ISBN 0-679-50458-3.

Second Chance, David McKay Co., New York, 1977, ISBN0-679-50752-3.

Bamford, Don, *Anchoring,* Seven Seas Press, Newport,R.I..

Beiser, Arthur, *The Sailor's World,* Random House, New York, ISBN 0-394-46852-X.

Beyn, Edgar J., *The 12 Volt Doctor's Practical Handbook*, Spa Creek Instruments Co.,Annapolis, Md. 1983,ISBN0-911551-07-7.

Bitchin, Bob, *Letters from the Lost Soul,* Sheridan House, Dobbs Ferry, NY, 2000, ISBN 1-57409-112-3.

Blackistone, Mick, *Sunnup to Sundown,Watermen of the Chesapeake*, Acropolis Books, Ltd., Washington, DC, 1988, ISBN 0-87491-891-X.

Bjelke, Rolf & Shapiro,Deborah, *Northern Light*, Clarkson N. Potter, Inc., NY, 1986, ISBN 0-517-56406-8.

Boehmer, Richard, *Multihull Ocean Racing,* Boehmer Publishing, 1977.

Bombard, Alain, *The Bombard Story*, Grafton Books, London, 1986,, translation, originally published 1953, Andre Deutsch Ltd., Editions de Paris, Paris.

Brown, Jim, *The Case For The Cruising Trimaran,* International Marine, Camden, ME, 1979, ISBN 0-87742-100-5.

Buckley, Jr.,William F., *Airborne*, Macmillan Publishing, NY , 1976, ISBN 0-02-518040-1.

Butler, William A. and Simonne S. Butler, Exmart Press, Miami FL 1991 ISBN 0-9632519-0-2.

Burke, Katy, *The Complete Live-Aboard Book*, Seven Seas Press, Newport, RI 1982:

Managing Your Escape , Newport, R.I. Seven Seas Press, 1984, ISBN 0-915160-67-9.

C

Calder, *Nigel, Boat Owner's Mechanical and Electrical Manual.* McGraw-Hill, 1990 ISBN 0-87742-982-0:

Cuba: A Cruising Guide Imray Laurie Morie & Wilson Ltd. 1997 ISBN 0-85288-370-6.

Callahan, Steven, *Adrift*, Ballantine Books, New York, 1986, ISBN 0-345-34083-3.

Cantrell, Debra Ann, *Changing Course*, McGraw Hill, 2001, ISBN 0-07-1360875.

Carpenter, Wayne, *Voyage Of The Kristina,* out of print.

Carter, Ken, *Chesapeake Reflections*, Amantha Publishing Co., Marathon, FL 1991, ISBN 0-9628793-4-7.

Bibliography

Chandler, Roy F. *A 30', $6,000.00 Cruising Catamaran*, Bacon & Freeman Publishers, Orwigsburg, PA.

Charles, Simon 1995-1996 *Cruising Guide to Cuba,* Cruising Guide Publications, Dunedin, FL 1994 ISBN 0-944428-26-6: second edition 1997, ISBN 0-944428-36-3.

Chichester, Sir Francis, *Gipsy Moth Circles The World*, Coward-McCann, Inc., NY, 1967.

Chiodi, Charles, *The Capsize Bugaboo*, Chiodi Advertising & Publishing, Inc., Boston MA 1980
The Symposium Book II, 1985.

Choy, Rudy, *Catamaran's Offshore*, Macmillan, NY, 1970.

Coles, K. Adlard, *Sailing Years*, John De Graff, Inc. Clinton Corners, New York, 1981, ISBN 0-8286-0089-9.

Cornell, Gwenda, *Pacific Odyssey*, Adlard Coles Limited, London, England, 1985,ISBN 0-229-11758-9:
Cruising With Children, Sheridan House,1986, ISBN 0-229-11790-2.

Cornell, Jimmy, *World Cruising Handbook*, International Marine, 1990, ISBN 0-87742-297-4:
World Cruising Routes, International Marine,1990:
Ocean Cruising Survey, Sheridan House, 1986, ISBN 0-87742-250-8.

Cotter, Edward F. *Multihull Sailboats,* Crown, NY 1963-1971:
Sailing and Racing Catamarans, Chilton Books, NY 1963

D

Dear, Ian, *The America's Cup*, Dodd, Mead & Co.,New York, 1980 , ISBN 0-396-07848-6.

deRoos, Willy, *Northwest Passage*, International Marine, Camden, ME, 1980, ISBN 0-87742-087-4.

Donaldson, Sven, *Understanding The New Sailing Technology*, G. Putnam & Sons, N.Y. 1990.

Dye, Frank & Margaret, *Ocean-Crossing Wayfarer*, David & Charles, North Vancouver, BC, 1977 out of print.

E

Edwards, Peter, MD, *Advanced First Aid Afloat*, Cornell Maritime Press, Centerville, MD, 1987, ISBN 0-87033-376-3.

Ellam, Patrick and June, *Wind Song*, International Marine, Camden, ME, date unknown, ISBN 0-87742-061-0.

F

Francis, Clare, *Woman Alone*, David McKay Co., New York, 1977, ISBN 0-679-50758-2.

G

Gerbault, Alain, *Firecrest Round The World*, David McKay Co., New York, 1981, ISBN 0-679-50978-X.
Gibbons, Rod, *The Cruising Catamaran Advantage*, Island Educational Publishing, Manchester, WA, 1988, ISBN 0-929458-11-7.
No Cost Boating for Power and Sail, 1991, ISBN 0-929458-12-7.
Gibbs, Tony, *Practical Sailing*, Hearst Corp., New York, 1971, ISBN 0-910990-00-X.
Goerch, Carl, *Ocracoke*, John F. Blair, Winston-Salem, NC, 1956, ISBN 0-89587-031-2.
Gougeon, Meade, *Gougeon Brothers On Boat Construction*, Gougeon Brothers, Inc., Box X908, Bay City, MI. 48707,1990.
Graham, Robin Lee, *Dove*, Harbor & Row, NY, 1972, ISBN 06-011603-X.
Greenwald, Michael, *Survivor*, Blue Horizons Press, San Diego, CA, 1989, ISBN 0-931297-02-6.
Groser, John, *Atlantic Venture*, Ward Lock & Co. Ltd.,London, 1968.

H

Hackler, Lew R., & Casey, Don, *Sensible Cruising*, Seascape Enterprises Colonial Heights, VA. 1987, ISBN 0-931595-01-0:
Let's Name It 1987, Hackler,L.R. & Corcoran, J. ISBN 0-931595-02-9:
* *Boating With Cap'n Bob and Matey*, 1989, ISBN 00-931595-03-7:
Discovery, With Cap'n Boat & Matey, 1990:
Fishing, With Cap'n Bob & Matey, 1990:
The Cat Who Learned To Sail, 1991.
Hanks, Jr., Douglas, *Muskrat*, Mustkrat Publishers, Easton, MD. 1987, LIC TXU-2270-198.
Harris, Robert B. NA, *Modern Sailing Catamarans*, Charles Scribner's Sons, NY 1960, LIC 60-14017.
Harvey, Derek, *Multihulls For Cruising And Racing*, (First published in Great Britain, Adlard Coles, 1990) International Marine, Camden, ME, 1991, ISBN 0-87742-293-1.
Hassel, Mark, *Love For Sail*, Out of print.
Herd, Shirley, *Seawoman's Handbook*, S.Deal & Associates, San Diego, CA, 1989, ISBN 0-9300006-03-8.
Hinz, Earl R.,*The Complete Book of Anchoring and Mooring*, Cornell Maritime Press, Centerville, MD, 1986, ISBN 0-87033-48-8:

The Offshore Log, 1968:
Sail Before Sunset, 1979:
Landfalls of Paradise:
The Guide to Pacific Islands, 1980.

Hiscock, Eric C., *Come Aboard*, Oxford University Press, Oxford, England, 1978:
Wandering Under Sail, 1977:
Cruising Under Sail, 1965:
Around the World in Wanderer III, 1956:
Voyaging Under Sail, 1970:
Beyond the West Horizon, 1963:
Atlantic Cruise in Wanderer III, 1968:
Sou'West in Wanderer IV. 1973:
Two Yachts, Two Voyages, 1975.

Hitz-Holman, Betsy, *Sitting Ducks*, Seven Seas Press, Newport, R.I., 1983, ISBN 0-915160-60-9.

Hoyt, Garry, *Ready About!*, International Marine, Camden, ME, 1986, ISBN 0-87742-229-X.

J

Jeffrey,Kevin,*Free Energy Afloat*, Seven Seas Press, Newport, R.I., 1985, ISBN 0-915160-78-1.

Jeffrey, Kevin, with Nan Jeffrey, *Boatowner's Energy Planner*, International Marine, Camden, ME, 1991, ISBN 0-915160-63-3.

Jeffrey, Nan, with Kevin Jeffrey, *Adventuring With Children*, Avalon House Publishing, 1990, ISBN 0-9627562-0-2.

Jobson, Gary, *How to Sail*, Ziff-Davis, NY, 1980, ISBN 0-087165-061-4:
Sailing Fundamentals, Simon & Schuster, Inc. New York, 1987, ISBN 0-671--605098-4.

Jones, Thomas Firth, *MULTIHULL VOYAGING*, Sheridan House, NY, 1995, ISBN 0-924486-56-2

Jones, Tristan, *The Incredible Voyage*, Sheed Andrews and McMeel, Inc., Mission, KS, 1977, ISBN 0-8362-0703-3:
ICE, Universal Press Syndicate, Mission KS, 1978, ISBN 0-8362-6302-2:
Dutch Treat, Universal Press Syndicate Co.,New York, 1979, ISBN 0-8362-6107-0:
Saga of a Wayward Sailor, Sailors Bookshelf Publishing, Hillside, NJ,1979, ISBN 0-943869-00-5:
Adrift, Macmillan Publishing Co.,New York, 1980, ISBN 0-02-559860-0:
Aka, Macmillan Publishing Co., New York, 1981, ISBN 0-02-559870-8:
A Steady Trade, Martin's Press, New York, 1982, ISBN 0-312-76138-4:

Yarns, Sail Books, Boston, MA, 1983, ISBN 0-914814-41-9:
Heart Of Oak, St. Martin's Pres, N.Y. 1984 , ISBN 0-312-36598-5:
Outward Leg, William Morrow, 1985, ISBN 0-688-08255-6
The *Improbable Voyage*, The Bodley Head Ltd, London England, 1986, 0-370-31014-4:
Somewheres East of Suez, William Morrow Co., New York, 1988, ISBN 0-688-07750-1:
One Hand for Yourself ,One for the Ship, Macmillan, N.Y. 1982, ISBN *0-924486-03-1*
To Venture Further, Grafton Books, *1991*, ISBN 0-688-08022-7:
Seagulls In My Soup, Sheridan House, 1991, ISBN 0-924486-17-1.

K

Kanter, Charles E. *Cruising on more than One Hull*, 1992, *SAIL*co Press, ISBN 0-9618406-2-5.
Kanter, Charles E. and Corinne C Kanter, *Cruising is Contagious, SAIL*co Press, Key Largo, FL, 1999 ISBN 0-9618406-5-X.
Kanter, Corinne C, *The Galley K.I.S.S. Cookbook, SAIL*co Press, Key Largo, FL, 1987, ISBN 0-9618406-0-9:
The CRUISING K.I.S.S. Cookbook, *SAILco* Press, Key Largo, FL ISBN 0-9618406-3-3.
Knight, Kathryn Lasky, *Atlantic Circle,* Norton, NY, 1984.
Kyselka, Will, *An Ocean In Mind*, University of Hawaii Press, Honolulu, HI, 1987, ISBN 0-8248-1112-7.
King, William Commander, *Adventure in Depth,* G.P. Putnam's Sons, New York, 1975, ISBN 399-11493-9.
Kelsall, Derek, *Catamaran Sailing*, The Crossword Press, London, 1992, ISBN 1-85223-708-2

L

Leslie, Anita, *Francis Chichester,* Walker, & Co., New York, 1975, ISBN 0-8027-0510-3.
Lewis, David, *Ice Bird*, W.W. Norton & Co., New York, 1975, ISBN 0-393-03185-3:
We, the Navigators, University of Hawaii Press, Honolulu, HI, 1972, ISBN 0-8248-0394-9.

M

McCall, Fiona & Howard, Paul, *All in the Same Boat*, McClelland & Stewart, Toronto, Ont., Canada, 1988, ISBN 0-7710-5437-8:
Still in the Same Boat, 1990, ISBN0-7710-5440-8.
Marchaj, C.A., *Sailing Theory and Practice*, Dodd, Mead & Co., NY, 1964, LI C 64-13694:
Seaworthiness, The Forgotten Factor, Adlard Coles, Great Britain, 1986.

Mele, Andre, *Polluting For Pleasure*, WW Norton, N.Y. 1995, ISBN 0-393-03510-7.
Moeller, Jan and Bill, *The Intracoastal Waterway, Norfolk to Miami*, Seven Seas Press, NY, 1979 ISBN 0-915160-23-4.
Moitessier, Bernard, *Sailing to the Reefs*, Hollis & Carter, London, 1971, ISBN 0-370-01307-7:
The Long Way, Doubleday & Co.,Inc., Garden City, New York, 1975, ISBN 0-385-03867-4.
Munroe, Ralph Middleton, Gilpin, Vincent, *The Commodore's Story*, Historical Association of Southern Florida , Miami, Fl. 1930, ISBN 0-935761-00-4.

N

Neumeyer, Ken, *Sailing The Farm,* Ten Speed Press, Berkeley, Ca., 1981, ISBN 0-89815-051-5.
Nicolson, Ian, *Surveying Small Craft, Sheridan* House,Dobbs Ferry, New York, 1984, ISBN 0-911378-47-2.

P

Palley, Reese, *Unlikely Passages,* Seven Seas Press, Newport, RI, 1984, ISBN 0-915160-83-8.
Papy, Frank, *Cruising Guide to the Florida Keys*, 1977 thru 2002 ISBN 0-9619838-5-X.
Sailing, Impressions, Ideas, Deeds, 2002 ISBN 0-9619838-6-8.
Pardey, Lin & Larry, *Cruising in Seraffyn*, Seven Seas Press, NY, 1976, ISBN 0-915160-19-6:
The Care And Feeding Of The Offshore Crew, Pardy Books, CA. 1980, ISBN 0-9646036-0-8:
Seraffyn's Mediterranean Adventure, Sheridan House, NY, 1991 ISBN 0-924486-15-5:
Cost Conscious Cruiser, Pardey Books, Arcata, CA. 1999, ISBN 0-9646036-5-9.
Parsons, Kathy, Spanish for Cruisers, Aventuras Publishing, Hallettsville, TX. 2000 ISBN 0-9675905-0-7.
Payson, Herb, *Blown Away*, Sheridan House, NY,1995 ISBN 0-924486-95-3:
You Can't Blow Home Again, Hearst, N.Y.1984, ISBN0-688-04069-1:
Advice To The Sealorn, Sheridan Hse, N.Y. 1997 ISBN 1-57409-002-X.
Philbrick, Nat, *The Passionate Sailor*, Contemporary Books, Inc., Chicago, IL, 1987, ISBN 0-8092-5018-7.
Pretzer, Roger, *Marine Metals Manual*, International Marine, Camden ME, 1976.
Puleston, Dennis, *Blue Water Vagabond,* Rupert Hart-Davis Limited, London, England, 1955.

R

RCC Pilotage Foundation, *The Atlantic Crossing Guide*, W.W. Norton & Co., New York, 1983, ISBN 0-393-03283-3.

Riggs, Doug, *Keelhauled,* Seven Seas Press, Newport, R.I., 1986, ISBN 0-915160-85-4.
Robertson, Dougal, *Survive the Savage Sea*, Granada Publishers Ltd., London, 1985, ISBN 0-246-12509-8.
Roth, Hal, *Two on a Big Ocean*, Macmillan Co., New York, 1972, Library of Congress 70-171991:
Two Against Cape Horn, McLeod Limited, Toronto, 1978, ISBN 0-393-03223-X.
After 50,000 Miles, Norton, N.Y. 1977:
Chasing The Long Rainbow, Seaworthy, Port Washington, WI, 1998, ISBN 0-9639566-6-3.
Rousmaniere, John, *Fastnet Force 10*, Norton, N Y, 1980, ISBN 0-393-03256-6.

S

Sandstrom, Joanne, *There and Back Again,* Earendil Press, Oakland, CA, 1983, ISBN 0-914577-07-7.
Seyfarth, Fritz, *Tales of the Caribbean,* Spanish Main Press, USVI., 1978, ISBN 8286-0081-3.
Shane, Victor, *Drag Device Data Base*, Para-Anchors Intertnational, Summerland, CA 1991.
Shapiro, Deborah, Bjelke, Rolf, *Time On Ice*, International Marine, Camden, Maine, 1998, ISBN0-07-006399-0.
Simpson, Bob, *When the Water Smokes*, Algonquin Books, Chapel Hill, NC, 1983, ISBN 0-912697-03-3.
Slocum, Joshua Captain, *Sailing Alone Around The World*, Norton, New York, 1984, ISBN 0-393-03277-9.
Smeeton, Miles, *The Sea was our Village,* Gray's Publ. Ltd., Sidney, British Columbia, Canada, 1973, ISBN 0-88826-044-X.
Smith, Hervey Garrett, *Boat Carpentry*, Van Nostrand, NY, 1965.
Snaith, William, *On the Wind's Way*, G.P. Putnam's Sons, NY, 1973, ISBN 399-11227-8.
Society of Naval Architects and Marine Engineers, *Sailing Yacht Capsizing,* 1983.
Spurr, Daniel, *Steered by the Falling Stars* McGraw-Hill, 1992 ISBN 0-87742-332-6.
Stadler, Michael, *Psychology of Sailing,* International Marine, Camden, ME, 1987, ISBN 0-87742-963-4.
Street, Jr., Donald M., *The Ocean Sailing Yacht,* Norton , N. Y. 1973, ISBN 0-393-03168-3.
Stuermer, Gordon and Nina, *Starbound,* David McKay Co., New York, 1977, ISBN 0-679-50778-7.
Sullivan, Amy & Donnelly, Kevin, *Cruising 101*, Free Fall Press, San Diego, 1997 ISBN 0-9658247-8-0.
Swale, Rosie, *Children of Cape Horn,* Walker & Co., New York, 1974, ISBN 0-8027-0465-4.

T

Tabarly, Eric *Lonely Victory*, Sauvenir Press, 1965.

Bibliography

Tangvald, Peter, *Sea Gypsy*, William Kimber, 1966, London,

At Any Cost, Cruising Guide Publications, Dunedin, FL 1991, ISBN 0-944428-09–6.

Teller, Walter Magnes, *The Voyages of Joshua Slocum*, Rutgers Univ. Press, New Brunswick, N.J. 1958, ISBN 8135-0297-7.

Tetley, Nigel, *Victress Round the World* (Trimaran Solo), Nautical Publishing Co., Lymington, Hampshire, England, 1970, ISBN 245-59950-9.

Tomalin, Nicholas, & Hall, Ron, *The Strange Last Voyage of Donald Crowhurst*, Stein & Day, N.Y. 1970,ISBN8128-1301-4.

V

Vaitses, Allan H. *What Shape Is She In?*, International Marine, 1985, ISBN 0-87742-192-7.

Van Sant, Bruce, *The Gentleman's Guide to Passages South*, Salinas, Sept. 1988.

Vignes, Jacques, *The Rage to Survive*, William Morrow & Co. Inc., New York, 1975, ISBN 0-688-02992-2.

W

Warren, Nigel, Metal Corrosion In Boats, International Marine, 1980, ISBN 0-87742-234-6.

Ward, Buddy Captain, *Tales of the Anna Karrue*, Tradd Street Press, Charleston, SC, 1988, ISBN 0-9337684-25-2.

Wharram, James, *Two Girls Two Catamarans*, Abelard-Schuman, 1969, London, ISBN none.

White, Rick, Wells, Mary, *Catamaran Racing For The 90's*, RAM press, Key Largo, FL 1992 ISBN 1-880871-00-9.

Weld, Philip S., *Moxie*, The Bodley Head Ltd., London, England, 1982, ISBN 0-370-30492-6.

White, Charlie, *Living off the Sea*, Maclean Hunter, Vancouver, BC, 1985, ISBN 0–88896-152-9.

White, Chris, *The Cruising Multihull*, International Marine, Camden, ME, 1990, ISBN 0-87742-264-8.

Wilson, Sloan, Away from It All, G.P. Putnam's Sons, NY, 1969.

Y

Young, Claiborne C., *Cruising Guide to Eastern Florida*, Pelican Publishing Co., Inc. Gretna, LA 1989:

Power Cruising, 1990:

Cruising Guide to Coastal North Carolina, Cruising Guide to Coastal South Carolina and Georgia, Cruising Guide to Northern Gulf Coast. Cruising The Florida Keys.

GLOSSARY OF TERMS

One problem in writing a glossary is that language is dynamic and constantly changing. This process keeps lexicographers employed. What this means to you and me is that we can't keep abreast of when common usage actually changes the authoritative use of the word. Such a word, in general usage, is "gay!" If you check in Webster's, you will find "gay" means: "happy and carefree." Not anything at all its present implication.

In our parlance, we have the noun "cutter." According to an overwhelming amount of common usage, a "cutter" is a sailboat with two (2) headsails. It is not uncommon at all to see, written by people who really ought to know better, that a boat is "a cutter rigged ketch," or it is cutter rigged simply because it has an inner staysail added to its rig. This is not technically correct. The correct definition of "cutter," is a single masted sailing vessel with its mast stepped at the 50% mark in its length on the waterline. Based on this theses, I will try to give the most accurate meaning possible to the definitions, especially those that pertain directly to multihull sailboats.

Aft: Stern, back, rear, *after*, the rear section.
Ama: The outer floats (hulls) of a trimaran.
Anchor: "To hold something in position." A device to hold a vessel in position in an anchorage. At anchor, anchored.
Apparent wind: The direction from which the wind appears to blow. On a moving boat, it is not necessarily the same as the true wind. The apparent wind is either more or less than the true wind, depending upon which direction you are going.
Aka: The cross beams connecting the ams to the main hull of a trimaran. Note: These terms, Aka & Ama are derived from the Polynesian. Many designers and writers prefer them to the more mundane crossarm or crossbeam or outer float. Time will tell if they became the standard generic notation.
Aspect Ratio: The ratio of the length of the luff divided by the length of the foot of a sail, or the length of a keel and divided by the depth of the keel.

Back: To force a sail against the wind when maneuvering. A jib is "backed" when you want to force the bow to fall off the wind. (see backwind), *sails are aback, archaic.* to allow or cause the boat to move backward with the bows pointed into the wind.
Backwind: To impair the effectiveness of a mainsail by sheeting in the jib so far as to deflect a stream of air against the lee side of the mainsail, thus destroying its partial vacuum. Also, see *back*.
Backstay: A supporting wire cable for a mast, leading to deck aft of the mast, as opposed to the headstay which supports the mast from forward. Many catamarans have twin backstays leading from the masthead to the outer aftermost corners of the boat. Many catamarans have no backstays at all, but rely upon shrouds and the mainsheet for mast support.
Bahamian moor: Two anchors laid out in 180 degree opposition.
Ballast: Weight on a boat to give it stability, counterweight.

Barberhauler: A device of lines and blocks for changing the jib lead point, to haul it in or out.
Batten: (noun) A thin strip of wood or plastic inserted in a sail pocket to stiffen and hold the sail shape. A fully battened sail has battens extending across the full width of the sail.
Batten: (verb) To "batten down" the hatches... to securely close.
Beam: The width of a boat at its widest point. expressed as BMAX. Also Hull Beam, which is the width of an individual catamaran hull, expressed as HB.
Beam reach: The point of sailing with the wind directly abeam, or 90 degrees from the centerline.
Beat: (beating) Close hauled, going to windward, generally denotes as close to the wind as that vessel is capable of achieving. (see chapter on *Windward ability* for more detail)
Bilge: The bottommost portion of the interior of the hull. That portion below the cabin sole.
Boom: A spar to which the foot of a fore-and-aft sail is attached.
Boomkin: A spar, a bowsprit like extension from the stern of a vessel usually for attaching a mainsheet or mizzen sheet.
Boom Vang: Most often a block-and-line device, but occasionally a hydraulic or rigid spar, attached between the underside of the boom and the hull or foot of the mast, that prevents the boom from lifting and helps create proper sail shape. It also may be used as a "preventer," which prevents the boom from gibing accidentally.
Bow: The forward or front end of the boat.
Bow wave: A continuing wave that forms on each side of the bow when a boat is in motion.
Bowsprit: A spar extending forward of the bow of a boat to take the tack of a sail, thus extending the base fore triangle of a sailboat.
Bridgedeck: The main deck of the catamaran. The deck containing the helm. Underdeck refers to the bottom portion of the bridge deck exposed to the sea. Often used interchangeably.
Bridle: Cable wires running from the bow of each hull to the forestay, or lines to the anchor rode.
Broach: To swing or slew around toward the wind when running down wind and thus come broadside to the sea. Major causes of broaching are the inability of a vessel to accelerate down a wave or to react quickly enough. Common broaching is almost unheard of in cruising catamarans.
Broad Reach: Any point of sailing between a beam reach and the wind quartering aft. (from 90 to 165 degrees)
Broadside: The entire side of a boat, from stem to stern.
Bruce Numbers: Numbers devised by Edmund Bruce and introduced through the Amateur Yacht Research Society to examine sailboat performance. Bruce number = $\sqrt[2]{\text{sail area}}$ / $\sqrt[3]{\text{displacement}}$
Bulwark: a raised protective wall along the vessels gunwale.

Catenary: The arc created by a line between two fixed points. The arc or curve in an anchor rode.
Catamaran: A twin hulled vessel. According to Webster's unabridged dictionary: "A vessel noted for its safety."
Catboat/Cat-Rigged: A sailboat with only one sail, the mast stepped at the bow.
Center Of Effort: (CE) The pressure center of the combined area of all the sails being flown on a boat This changes when you change sails.
Center Of Lateral Resistance: (CLR) The center of the area of lateral resistance of a hull's underwater

profile. The resistance to leeway, the center of the area that prevents the hull from slipping sideways.
Centerboard: A hinged board that passes through a slot in the hull to provide lateral resistance for a sailboat. It is enclosed in a trunk and may be raised in shallow water and is adjusted during sailing.
Chainplate: The metal connectors attached to the hull of a boat to which the wire rigging is attached.
Chine: The transition point or area on a hull between the topsides and the bilge.
Chock: A lead for docklines or anchor lines from the cleats to the points of attachment of the vessel. Skene chocks are those that skew the line in order to gain entry to the chock, the purpose being to help keep the lines from jumping out.
Chord: A straight line intersecting a curve.
Cleat: A device used to belay or secure a line.
Clew: The corner of a sail at the juncture of the leech and foot. The attachment point for the sheets.
Closed-hauled: Sailing as close to the wind as possible, see also Beat.
Close Reach: A point of sailing between a beam reach and a beat.
Cockpit: The exterior protected deck space that accommodates the helmsman and crew. The operations center.
Come About: To tack. To put the bow through the wind to fill the sails on the opposite side.
Cutter: A single masted sailing vessel with the mast at the center of the waterline.

Daggerboard: A non-pivoting board passed down through a slot or trunk in a sailboat to provide lateral resistance.
Dinghy: A small boat usually used to tend a larger boat; any of several small sailing boats that have developed into one-design racing classes.
Displacement: Displacement is an often misunderstood term because people use it incorrectly. Technically, displacement equals the weight of the water displaced by a vessel as it sits afloat. That is where the term comes from. Following that through, the displacement of a vessel is its total weight. This is not difficult to understand. Where the problems come from are: designed displacement, laden displacement and unladen displacement. The generic term "weight" can be used interchangeably with the term "displacement" when referring to vessels. While the actual cubic volume of water displaced will vary with the salinity of the water, the weight of the water displaced will be the same.

1. Designed displacement, which is the weight the designer estimates that the vessel will weight at launching. Most designers and builders use this number in their advertising because it is the most viable number they have, since it is often impossible to obtain an actual finished weight. The true weight of a vessel is most often never determined unless it is actually hung from a scale as is done when weighing a boat for a racing handicap certificate. What you can actually place on board in addition to the designed vessel weight is your payload or load carrying ability. You should always, and I emphasize always, know that number and try not to exceed it, no matter what kind of boat you have. Total displacement is what your boat would weigh if you placed it on a scale with you, and everything else on board. (Note: the "net tons" referred to on US documented vessels is not a measurement of weight but an arcane formula for commercial internal volume.)
Displacement Hull: A hull designed to pass through the water rather than skim over the surface.

Down: Away from the wind as in down wind. Putting the helm *down* (archaic) refers pushing the tiller down because the vessel is heeled which turns the vessel into the wind.
Downhaul: (noun) A line for putting downward stress on a sail at the luff. (Verb) The act of tightening the downhaul.
Downwind: Sailing with the wind up to 15 degrees either side of exactly downwind. (165 through 195 degrees)
Draft: A boat's maximum depth of penetration into the water. Boats with centerboards, or daggerboards have a *Minimum Draft* (depth with boards up) and *Maximum Draft* (the depth with boards completely down); The fullness of a sail measured by the chord.

Eye: Eye of the wind, exactly into the wind direction. Eye splice, a spliced loop in the end of a line.

Face Dock: The straight docks that have no pilings to form boat slips.
Fat Head: A type of mainsail that is actually quadrilateral rather than triangular.
Fathom: A unit of depth measure equaling six feet.
Following Sea: Seas coming up from behind.
Foot: The bottom edge of a sail.
Footing: Sailing to windward slightly less than close hauled on the theory that the added speed more than offsets loss in pointing. The faster the catamaran or trimaran, the more this is true.
Fore-And-Aft: Running, acting, or lying along the general length wise line of a boat.
Foresail: The first working sail immediately forward of the main; the staysail in a cutter rig.
Forestay: The forward mast-supporting cable on which the jib luff is usually attached.
Forward: At, near, or belonging to the fore part of a boat; also, more toward the bow than the stern.
Fouled: Tangled or caught, e.g., a fouled line or sheet.
Founder: To fill with water and sink.
Fractional Sloop: A sail plan in which the jib is a fraction of the mast height. Usually 7/8 or 3/4.
Freeboard: The area on the sides of the hull, from the water line to the gunwale.
Frugalphile: A term coined by me. A person who believes in frugality as a lifestyle. To live in a frugal manor as a passion.
Furl: To roll up a sail and secure it to a spar.

Gelcoat: The final resin gloss coating or a fiberglass hull. It contains the color pigment.
Genoa Jib: A large, overlapping jib with the foot parallel to the deck.
Ghoster: A sail capable of providing comparatively good speed in very light air.
Ghosting: Sailing in very light air.
Gooseneck: A metal two way swivel device used for securing the boom to the mast.
Gudgeons: The female half of the rudder hinge.
Gunkhole: A cruising term defining exploring little known, out of the way or normally inaccessible places.
Gunwale: The junction of the side of the boat and the deck.

Halyard: A line for hoisting or lowering a sail or a flag.
Hard Chine: An abrupt angle at the intersection of the hull topside and bilges.
Haul: To pull with force; to remove a boat from the water; to sail closer to the wind.
Head: The upper, top section of a sail.
Head: Heads, a marine toilet.
Headboard: A stiffening board inserted in the head of the sail.
Header: A wind shift that makes you steer down from your normal course to avoid luffing or losing speed.
Heading: Direction of travel of a boat; a course, normally expressed in compass degrees.
Headsails: Any sails forward of the mainmast.
Headstay: The forward most stay supporting the mast.
Head-To-Wind: With the bow headed into the wind.
Heave To: To keep a boat nearly stationary and headed into the wind, as when riding out a storm by the arrangement of sails.
Heel: (verb) To heel. (adverb) The lateral rotation of a sailboat under pressure of the wind. (tipping, leaning,) (2) The base of the mast section.
Helm: (noun) The steering device of the vessel. The helm can be a wheel or tiller. Helm: (verb) To helm, to steer .
Helm down: Turning to windward. *Archaic*. Refers to pushing the tiller down, on a heeled vessel.
Hermaphrodite: Showing the characteristics of both. In this book referring to foredeck structures.
Hull: The basic external structure of a vessel.
Hull Speed: The maximum speed to which a displacement-type hull can be driven. It is limited by the length of the hull's water-line. The formula most commonly seen is, $1.4 \times \sqrt{wl}$.. However, this formula is disputed as the 1.4 is considered a variable depending upon the L/B ratio. (See pg. 74)

Irons/In Irons: The point at which a sailboat is dead in the water, facing directly into the wind, sails flogging, making no way.

Jib: A triangular sail usually attached to the forestay and forward of the mainmast.
Jibe: Passing from one tack to another by swinging the stern of the boat through the eye of the wind.
Jury-Rig: Rig for temporary use.

Keel: The lower external edge or backbone of a vessel extending along the center of the bottom from bow to stern. (2) A lateral resistance device.
Keelson: The internal backbone of a vessel so constructed.
Ketch: A two-masted vessel with the mizzen mast (shorter mast) mounted forward of the rudder post.(See yawl, schooner.)
Knot: The speed of one nautical mile per hour.

Lateral Resistance: The ability of a hull, by means of the area of submerged surfaces, to resist being driven sideways by the wind or the pressure of wind on the sails.

Glossary

Leach: The back edge of a sail. The edge usually supported with battens is called the roach.
Lee: Shaded, protected, as "in the lee of..." An area shielded from the wind. The side of a boat away from the direction of the wind (opposite of weather side).
Lee Helm: A tendency to bear off the wind due to poor sail balance. It is caused by the center of effort being too far forward of the center of lateral resistance.
Lee Rail: The rail away from the wind. The junction of the deck and the topsides. The lee gunwale.
Lee Shore: The shore towards which the wind in blowing. Usually referred to in the context of danger.
Leeward: The side away from the wind, the opposite of windward.
Leeway: The amount a boat is pushed sideways by the wind, measured in degrees.
LOA: Length over all. The total length of a boat, including appendages such as davits, bowsprits, etc.
Luff: (noun) The leading edge of a sail. The fixed portion.
Luff: (verb) To luff: to flog the sails; alt. to deliberately take someone's wind.
Luff: To take the wind pressure off a sail by easing the sheet or by heading into the wind, causing the sails to flog.
LWL: length of the water line.

Masthead: The top of a mast. Often referred to as the truck.
Masthead Sloop: A sloop rig in which all sails reach to the masthead.
Multihull: A vessel with two or more hulls, typically referring to a catamaran, trimaran or proa.

Nautical Mile: A distance equal to one minute of arc of a great circle of the earth (actually 6,080.20 feet, but usually given as 6,080 feet). When reading a chart, it's handy to remember that a nautical mile equals one minute of latitude. One degree of latitude equals sixty nautical miles.

Pintle: An upright pivot pin that forms a hinge on a rudder. Pintle (male) fits into gudgeon (female).
Pitchpole: To capsize a boat forward, stern over bow.
Pointing: A boat's level of efficiency in sailing to windward; also, sailing as close to the wind as the boat's design will allow.
Point of Sail: Beat, reach, run are three major points. They denote the direction in relationship to the wind.
Polar Diagram: A diagram of relative potential speed on all points of sail as viewed from above. A diagram of concentric circles showing the boat at the center, denoting speed at the points of sail.
Port: The left side of a boat.
Port Tack: Sailing with the wind coming over the port side.
Pound: To strike the waves with a jarring force, as is characteristic of some hulls. See slamming,
Proa: A two-hulled vessel that has one hull smaller than the other, much like a trimaran with one ama omitted. There are basically two types, one that keeps the little hull to windward and one that keeps it to leeward. You do not tack these boats but "shunt" them end for end. Since they are not usually sold for cruising, a full discussion is beyond the scope of this book.

Rail: The top of the bulwarks or gunwales. The junction between the deck and the topsides
Rake: The inclination of a mast fore or aft, away from the perpendicular.
Reach: All points of sailing between a beat and a run. Close reach, beam reach, broad reach.
Reef: (adv) To shorten sails. Noun: shallow water.
Reef Points: Small grommets in the sail located in a row parallel to the foot and used for shortening sails.
Rhumb Line: A straight line between two points on a Mercator projection map or chart.
Rig: The type of sail arrangement; also, to step the mast and attach shrouds and stays while fitting out.
Rigging: Any rope, wires, or cable used to support (standing rigging) or control (running rigging) masts, spars, or sails.
Righting Moment: A disposition of weight in relationship to buoyancy, tending to prevent a boat from capsizing and to return her to her normal attitude of flotation.
Roach: That portion of the leech of a sail that extends aft of an imaginary line from the head to the clew.
Rocker: A hull characteristic, low in the middle and sweeping upward at the ends, such as a rocker sheer or rocker chine.
Roller-Reefing/Roller-Furling Gear: A device for rolling a sail up on its attachment point (forestay or spar) like a window shade for the purpose of making the sail smaller to match wind conditions; *reefed*. Not to be confused with roller furling which is similar, but only used as a sail storage device.
Round Up: To head into the wind.
Rub Rail: A protective or sacrificial molding around the hull to prevent the hull from being damaged during docking.
Rudder: The underwater mechanism that steers the boat.
Rudder Post: The shaft that connects the rudder to the quadrant or tiller arm or other mechanism that connects the rudder to the steering wheel.
Running Rigging: All sheets, halyards, guys, runners, and other lines used to control spars, sails, or the support of spars, not permanently fixed in place.

Saloon/Saloon Cabin: The main cabin area on a sailboat. Also called commonly called "salon".
Scantlings: The engineering data for a vessel; the sizes and weights of the various components.
Schooner: A two-masted vessel with the main mast aft of the foremast.
See-Saw Syndrome: The motion of a trimaran that has one hull out of the water at rest when it toggles to the opposite hull because of a weight shift or wave action.
Self-bailing: designating a boat capable of draining any water from its cockpit by gravity.
Shrouds: The wire rope stays supporting the mast laterally.
Skeg: A solid portion of the underbody that forms the forward portion of a rudder.
Skin Friction: Surface resistance of a hull as it passes through water.
Slamming, slam: Pounding. The action of the waves on bottom of the bridgedeck, referred to as the underdeck for clarity in this book.
Sloop: A single masted sailing vessel with the mast stepped forward of the center along the centerline length of the boat.(see masthead sloop, fractional sloop)

Snatch block: A block (pulley) that opens on one side in order to insert or remove a line without threading the line through from the end as with a fixed block. Most have snap shackles for convenient temporary use.
Spade rudder: Usually a balanced rudder under a vessel supported by a stock. The term spade comes from its shape.
Spaghetti: The random piles of line from sheets and other control lines that accumulate in the cockpit. Tangles of lines.
Spar: Any mast or boom that supports or extends the sail on a boat.
Speedophile: A term, coined by me, that describes a person who has speed under sail as his total rationale for sailing and has little, if any, patience for those that see other purposes for sailboats.
Spinnaker: A large light, symmetrical, balloon type headsail mostly used off the wind.
Squall: A sudden and violent burst of wind. A violent thunderstorm.
Stall: The slowing effect from sheeting the sails too tightly in relation to the wind direction or falling off without easing the sails.
Stanchion: The vertical support post for lifelines.
Standing Part: The part of a line made fast to something.
Starboard: The right side of a boat.
Starboard Tack: Sailing with the wind coming over the starboard side.
Stay: A wire rope used to support a mast or other spar in the fore and aft plane i.e. headstay, backstay, baby stay. Athwart ship supports are called shrouds, i.e. upper shrouds, lower shrouds.
Staysail: A sail with its luff attached to a stay.
Stem: The foremost timber (or steel bar) in a vessel.
Step: (verb)To set and affix a mast in position. (Noun) the base plate upon which a mast is affixed.
Stern: The after end of a boat. The back.
Sternway: Moving in reverse.
Swage: Swaging: A compression method of attaching end fittings for cable attachments to the ends of wires.

Tack: (noun) The corner of a sail at the juncture of luff and foot.
Tack: (verb) To come about; to change the course of the boat by bringing the bows through the wind so that the wind is now on the opposite side.
Tack: (pronoun) The relationship of a sailboat with respect to the wind. If the wind comes over the starboard rail, the boat is on starboard tack; if the wind comes over the port rail, the boat is on port tack.
Tackle: A purchase, consisting of one or more blocks (pulleys) and connecting falls, for multiplying the power of or changing the direction of a line.
Tang: A metal mast fitting to which a shroud or stay (wire) is attached.
Telltale: A short piece of wool, ribbon, plastic, or feather attached to sails and/or shrouds for the purpose of reading wind direction for monitoring sail trim.
Tiller: A steering handle attached to the rudder(s).
Topping Lift: A line by which the outer end of a boom is supported, usually from the masthead.

Topsides: The sides of a hull from the chine or the turn of the bilge to the sheer line or deck junction.
Traveler: A horizontal track with a movable car that is connected to the mainsheet for the purpose of controlling the boom and sail trim; also used for fore-aft and inboard-out-board jib lead locations.
Trim: To haul in on a sheet; (2) to set the sails at their optimum position for efficiency; (3) to balance the hull in a proper attitude in the water.
Trimaran: A three hulled vessel. The center hull being the main hull.
Trunk: A shaft in which daggerboards and centerboards are lowered and lifted.
Turnbuckle: A mechanical device, consisting of end screws of opposite threads, connected to a center barrel that can be rotated for adjusting the slack in a wire. *Rigging screws.*

Underdeck: (noun) The underside of a catamaran bridge deck exposed to the seas. The part that gets slammed.(pounded)

Wake: The temporary wave track left by a boat passing through the water. Also called "wash"
Washboards: Occasionally called drop boards. The conventional closure for the main entry hatch of both monohulls and trimarans.
Waterline: The line of intersection of a hull with the water when the boat is in its normal attitude.
Waterplane: The footprint that the boat has in the water exactly at the waterline. The actual area at the water surface at that intersection.
Weather: Indicating the side toward the wind, also known as windward; "to weather" is to windward.
Weather Helm: Having a tendency to head into the wind.
Weatherly: The ability of a sailboat to go to weather: A term usually used to describe the condition or ability of a vessel to go to windward: i e: "That is a very weatherly vessel." Weatherly: the acuteness of the angle between 000 and 045 is the arbiter of "weatherly".
Wetted Surface: The total area of the submerged portion of a hull and its appendages.
Whisker Pole: A spar used as a boom to hold the clew of a jib or spinnaker out and away from the boat.
Windward: The side from which the wind is blowing. The opposite of leeward.
Windward Ability: The ability of a vessel to sail into the wind. This ability varies considerably from vessel to vessel. Always of some importance, it is of paramount importance in certain circles.
Windward Leg: The portion of a sailed course that is towards the direction from which the wind is blowing.
Wing Mast: A rotating foil shaped mast with very large fore and aft axis and is a considerable portion of the total mainsail area.

Yacht: Any pleasure boat.
Yaw: To go off course or swing from side to side, normally associated with running before the wind in a sea.
Yawl: A two masted vessel with the mizzen mast mounted aft of the rudder post.

Zinc: Zinc sacrificial anode. The least noble of marine metals, used to protect other metals under the water from galvanic corrosion.

Index

Tel & Fax: 305-743-0626
Email:order@sailcopress.com

Mail order to: *SAILco Press*
P.O. Box 2099 Key Largo, FL 33037

ORDER FORM ORDER FORM ORDER FORM ORDER FORM ORDER FORM

Ordered by:

Name: ______________________________
Address: ______________________________
City: ____________________ State: ____________________ Zip: __________

Quantity	Title / Description	Price	Amount
New!	***Cruising in Catamarans***	**39.95**	
	CRUISING IS CONTAGIOUS	**14.95**	
	The CRUISING K.I.S.S. COOKBOOK	**24.95**	
	The GALLEY K.I.S.S. COOKBOOK	**13.95**	
Special!	**CRUISING K.I.S.S./Galley K.I.S.S. Combination**	**35.00**	
	Shipping and handling @ (see table)		
	Florida residents add 7% sales tax		
		TOTAL	

Please make check payable to:
SAILco Press, **P.O. Box 2099 Key Largo, FL 33037**

Shipping and Handling	
Up to 13.95	2.75
Up to 19.95	3.75
Up to 24.95	4.75
Up to 39.95	5.75
Above 39.95	6.75

Credit cards accepted: (check one) ☐ VISA ☐ MASTERCARD ☐ DISCOVER

Number: ____________________ **Exp. Date:** ____________________

Print Name On Card ______________________________

Signature ______________________________

Autograph to/Mail to:

Name:

Address:

City: **State:** **Zip:**

SAILco Press **Tel & Fax: (305) 743-0626** **Email: order@sailcopress.com**